EMISTRY OF THE ELEMENTS

r: **Earl Frieden**
Florida State University
Tallahassee, Florida

es in this series:

GENERAL PRINCIPLES OF
BIOCHEMISTRY OF THE ELEMENTS
Ei-Ichiro Ochiai

BIOCHEMISTRY OF THE LANTHANIDES
C. H. Evans

BIOCHEMISTRY OF THE ELEMENTAL HALOGENS
AND INORGANIC HALIDES
Kenneth L. Kirk

BIOCHEMISTRY OF HALOGENATED ORGANIC
COMPOUNDS
Kenneth L. Kirk

BIOCHEMISTRY OF COPPER
Maria C. Linder

BIOCHEMISTRY OF ZINC
Ananda S. Prasad

BIOCHEMISTRY OF NICKEL
Robert P. Hausinger

A BIOCHEMISTRY OF SCANDIUM AND YTTRIUM,
PART 1: PHYSICAL AND CHEMICAL
FUNDAMENTALS
Chaim T. Horovitz

B BIOCHEMISTRY OF SCANDIUM AND YTTRIUM,
PART 2: BIOCHEMISTRY AND APPLICATIONS
Chaim T. Horovitz

Order Plan is available for this series. A continuation order will bring delivery of each new volume on publication. Volumes are billed only upon actual shipment. For further information please contact

Biochemistry
Scandium an
Part 2: Bioche
and Applicatio

BIOCH

Series Edit

Recent volu

Volume 7

Volume 8

Volume 9A

Volume 9B

Volume 1

Volume 1

Volume 1

Volume 1

Volume 1

A Continuat
immediately
the publishe

Biochemistry of Scandium and Yttrium, Part 2: Biochemistry and Applications

Chaim T. Horovitz

Rehovot, Israel

SPRINGER-SCIENCE+BUSINESS MEDIA, LLC

Library of Contress Cataloging-in-Publication Data

Horovitz, Chaim T.
Biochemistry of scandium and yttrium/Chaim T. Horovitz
p. ; cm. — (Biochemistry of the elements; v. 13A–13B)
Includes bibliographical references and index.
Contents: pt. 1. Physical and chemical fundamentals—pt. 2. Biochmistry and applications.
ISBN 978-1-4613-6935-6 ISBN 978-1-4615-4311-4 (eBook)
DOI 10.1007/978-1-4615-4311-4
1. Scandium—Physiological effect. 2. Yttrium—Physiological effect.
I. Title. II. Series.
[DNLM: 1. Scandium—metabolism. 2. Yttrium—metabolism. 3. Scandium—chemistry.
4. Yttrium—chemistry. QU 130 B6144 1980 v. 13A–B]
QP535.S4 H67 2000
572′.52401—dc21

99-086353

Chaim T. Horovitz
Emeritus Professor
Hershensohn Str. 64
Rehovot 76484 Israel

c-horov@zahav.net.il.

ISBN 978-1-4613-6935-6

Originally published by Kluwer Academic / Plenum Publishers in 2000
Softcover reprint of the hardcover 1st edition 2000

To Elena, Simona, Sergej, Rita, and Hardy

In Memoriam

Professor Earl Frieden
1921–1996

Earl Frieden, Professor Emeritus of Chemistry, Florida State University, was born in Norfolk, Virginia, on December 31, 1921. He earned a Bachelor of Arts degree from UCLA in 1943 and a Ph.D. in biochemistry from the University of Southern California in 1949. He joined the chemistry faculty of Florida State University that same year and contributed to the transformation of the university into a major research institution. Frieden was elected chairperson of the chemistry department in 1962, and in the six years of his chairmanship three chemistry buildings were planned, funded, and built. The department also received a National Science Foundation of Excellence grant. In 1969, Frieden was designated Distinguished Professor by the university in recognition of his teaching, his research, his publications, and his service to the university. With his more than 80 graduate students and postdoctoral students, he developed extensive research activity in biochemistry.

He published four books and more than 150 articles and research papers. Frieden was a member of the editorial boards of several scientific journals. In 1971, he conceived the idea for the series *Biochemistry of the Elements*. In 1976, Plenum Press of New York published the first volume, and Frieden served as series editor of the series' first thirteen volumes. This book and those to be published in the future in the series *Biochemistry of the Elements* are a continuing testament to the vision of Professor Earl Frieden on the importance of inorganic biochemistry in medicine, agriculture, and environmental problems.

Chaim T. Horovitz and Esther Frieden

Preface

I became interested in scandium because it is considered rare in nature, it has unusual physical and chemical properties, it is less investigated, and I doubted that it is a "ballast element" of Life. Later, my preoccupation extended to its close brother-element yttrium, because scandium and yttrium are "neither here nor there" in the series of the lanthanides and transition elements. This pursuit continued during my years of work in Romania, Russia, Germany, Israel, and the United States. It is a continuation of my earlier book *Scandium* (1975).

It has taken more than six years of work to complete writing "The Book," to which it was referred when people close to me asked: "Chaim, how is the book coming?" My answer was invariably the same during the last five years: "It is close to the end." Nevertheless, the end was endlessly postponed because I tried to update and constantly improve earlier versions.

I owe my perseverance during the thorough investigation to my workaholic and energetic mother Marietta ז"ל, and to the passion for the study of my father Todres ז"ל. He was a simple man who only finished primary school, but he dedicated his whole life to studying the Holy Book and its commentaries. He fulfilled his dream to move and live in the Holy Land and spent with my mother the happiest, last years of their life together in Jerusalem.

The book *Biochemistry of Scandium and Yttrium* is intended to present existing information about these elements gathered from the whole arrays of disciplines, but with emphasis on biochemistry. It is divided in two parts. Part 1 dealt with their 200 years of history, their birth in the cosmos, and their occurrence in nature, and, in particular, in living organisms. Their inroads in inorganic chemistry and in physics, the difficulties that they present in analytical chemistry, and their interactions with molecules of biological interest were also considered.

This volume begins by considering the interactions of scandium and yttrium within cells, cellular organelles, and tissues. The interactions of these elements are thought to be limited to the outer surface of the plasmalemma, for which they show a high affinity and binding capacity. However, in several microorganisms they may penetrate through the membrane and become involved in cellular functions. The binding capacity of yttrium to bone is explained by its binding to phosphorus-containing compounds, mucopolysaccharides, and fluoride complexes.

Highly positive or negative correlations between scandium and yttrium and other elements are known. Several investigations have found specific stimulatory effects on some fungi and plant species, but inhibitory effects of equivocal data have been reported for other species. Despite the similarities between scandium and yttrium in size, bonding coordination geometry and donor atom preference, there are several differences in their body distribution. Scandium tends to accumulate in the liver and kidney, whereas yttrium prefers bone, the poor intestinal absorption of both elements in mammals is evident. A significant percentage of radioactive yttrium may be transferred into the fetus, and both elements may be transferred from the chicken body to the egg yolk. Scandium and yttrium show an ability to replace calcium in many biochemcial events. Increased or decreased concentrations of scandium in various tissues have been linked with various disease stakes and pathological conditions. Inhibitory effects of scandium and yttrium may be reversible with changes in pH and at low concentrations. Only a few studies have examined the effects of scandium and ytrrium on genetic and immunological processes.

Scandium and yttrium chlorides are less toxic, acutely or chronically, than several lanthanides. Most of the existing data on the toxic or carcinogenic effects pertain to lower organisms or animal species, and the extrapolation of this information to humans has to be done with reservation. The potentially harmful effects of several scandium and especially yttrium radioisotopes are of increasing concern because of their long half-lives.

The dominant opinion about the biological significance of scandium and yttrium is that these elements, like most elements of the periodic table, are not essential. However, a given element may also be essential, toxic, carcinogenic, or beneficial in various ways (growth stimulation, highly positive or negative correlations and interactions with other elements, therapeutic effects, diagnostic agents, etc.). These properties may depend upon its concentration, chemical form, tissue specificity, species properties, and several other factors. Essentiality of an element seems to manifest as a dynamic property that may change, under the influence of known and, at present still unknown, factors. Experimental data exists that scandium manifests an essential role in some species. Scandium and yttrium may figure as "supporting" (mimic or interfere) in relation to the central biochemical roles played by well-established essential macro- and trace elements.

The book ends with a survey of the uses of scandium and yttrium in medicine, environmental problems, biotechnology, and many other fields. Several long-lived and short-lived radioisotopes have found extensive use as inert markers in nutrition research, in investigations of blood flow and metabolic processes, and in radiodiagnostics and radioimmunotherapy. Further studies are needed to investigate their stimulatory or beneficial effects, and possibly unexpected biochemical properties.

The bibliography included in this book represents only a part of the documentation consulted: some 2,300 books and unpublished documents, and more than 7,000 journal articles.

Without the constant help and support of my sister Rita, and my brother-in-law Hardy, this book would not have come to life. Meinhard E. Mayer was always beside me when the computer or the printer broke down, and when I needed midnight advice on technical and scientific problems. His being a *mensch* deserves my endless recognition. I am especially indebted to Geraldine, who gave me so much support and help during the six years I spent in California working on this book. I also would like to express my gratitude to my brother Jehoshua and to Eva for their constant support.

Several chapters or chapter sections were improved immensely through critical reviews, sometimes harsh, by Kalman Burger, C. H. Evans, J. J. R. Frausto da Silva, Karl A. Gschneidner, Jr., Patrick J. Haley, Donald J. Hnatowich, Michael LaBarbera, Harri Lônnberg, Walter Mertz, Forest H. Nielsen, G. N. Schrauzer, Mette Strand, Steven Vogel, Robert J. P. Williams, and Helmward Zollner. To all of them I express my deep recognition for their professional advice and personal support. The subjective and personal presentation and interpretations of the data and ideas, the selection from the huge literature, and any mistakes are mine alone. I owe special gratitude to Stephen Bondy and Daniel Menzel for allowing me to do laboratory work during my sabbatical stay at the University of California at Irvine.

I am greatly indebted to Herman Boodman, Forrest H. Nielsen, Jacqueline Sedman and Geraldine Solnick, who contributed much to the improvement of the language and helped with the editing.

Finally, I would like to express my gratitude to Kenneth Howell for his continual support and help, to Amelia McNamara for her supervision, to Arne Olsen for his editorial assistance, to Richard Johnson for his editorial work, to Jacqueline Sedman for her diligent copy editing, and to the entire staff of Kluwer Academic/Plenum Publishers. The publication of this book completes my life long endeavor in the form of a much needed book on bioinorganic chemistry. Hopefully this book will be useful for further research and applications.

Chaim T. Horovitz
Corona del Mar, California

Contents

Interactions of Scandium and Yttrium within Cells, Cellular Organelles, and Tissues

7

There is no effect without a cause, modestly replied Candide, everything is necessarily linked, and arranged for the best.

Voltaire [François-Marie Arouet], 1694–1778

7.1 Introduction

A living organism is considered an extraordinarily complex, self-replicating, self-assembling, dynamic, open, physical and chemical system that is maintained in or near the steady state by the continual entry and exit of matter and energy. In such a system, inorganic ions and organic metabolites are transported through membranes in different locations by interacting in a myriad of chemical regulatory possibilities (Tosteson, 1989). Virtually all the biological macromolecules—proteins, nucleic acids, lipid components of membranes, and polysaccharides—are polyelectrolytes. These interact with each other and with small ions while performing their biological functions (Williams, 1991). Ionic groups of natural and synthetic polymers play an important role in controlling the interactions of these macromolecules in a biological milieu (Bloomfield and Carpenter, 1993).

The concept of chemical interactions has different approaches, which may show equivocal connotations of interaction and interdependence of effects in biological systems. The basic classes of chemical interactions in biological systems are additivity, synergy, and antagonism. However, these do not provide information on mechanisms, nor are they useful for regulatory applications. The interactions of inorganic elements in biological systems are usually emphasized with respect to environmental contaminants, such as aluminum and phosphate or fluoride. Few, if any, multiple chemical interactions have been investigated in detail (Dingle and Gordon, 1981; Calabrese, 1991). The biochemical interactions between metal ions and the broad classes of ion-containing natural and synthetic polymers find numerous applications, such as imaging and antimicrobial agents and substrates for bioseparation and blood-compatible materials (Voycheck and Tan, 1993).

Many of the experimental approaches employed in the investigation of trace element (TE) interactions are fragmentary and static, because they consider only partial aspects of the full scale of interrelationships between two or more elements (Kirchgessner *et al.*, 1981). As exemplified with microorganisms, four types of interactions can be identified. These are interactions at an interface (boundary between any two phases in a heterogeneous system), adhesion (energy involved in the formation of the adhesive junction), flocculation (involves cell–cell interactions, rather than cell–substratum interactions), and flotation (interaction between microorganisms and air–water interfaces) (Marshall, 1991). The cell-to-cell pathway concept considers the organism as a self-contained system of cell interconnections, which transmits information for the control of cellular growth and differentiation. In this interconnected community, ions play an essential role; for example, the interactions of calcium with hydrogen are important in binding to cell membranes, as well as in several biological mechanisms (Loewenstein, 1989). The competition of the lanthanides (Ln) with calcium is evident in many biological processes. A “whole-body” approach for modeling the gross conformation of assemblies was proposed by Harding (1989). This approach is used in the study of the cell surfaces of multicellular organisms, which are equipped with receptors and ligands for recognition of free molecules, specific substrates, and the surfaces of other cells. It involves carbohydrate ligands, which play a role in many cell–substrate and cell–cell interactions, as well as in tissue and epithelial–epithelial interactions (Harding, 1989).

Except in relation to the relatively well studied phenomena of transport, and the interactions between some essential elements, limited data are available for most chemical elements. It was asserted that trivalent lanthanides (Ln^{3+}) and Y^{3+} have binding characteristics much like those of Ca^{2+}; the strength of the interactions between these metal ions and ligands increases as the ionic radius decreases (Moeller, 1973). The available information about the interactions between Ln^{3+} or other trivalent metal ions, such as aluminum, gallium, and indium, with cells, cell organelles, tissues, and organs may be useful in this context, because several physical and chemical properties of these ions are similar to those of scandium and yttrium.

7.2 Membrane Interactions

The eukaryotic cells have a variety of membrane-bound compartments, called organelles. These have many functions, including the transport of solutes and control of cell–cell interactions. A variety of ion currents are created through at least three classes of membrane transport, i.e., passive, facilitated, and active mechanisms (Yeagle, 1993). Several reports indicated that toxic metals can be transported across the plasma membranes of bacteria into the cytoplasm

(Beveridge, 1989). The interaction between inorganic species within membranes can apparently play a significant role in their biological effects. Membranes may prevent the entry of various species into cells or limit the damaging effects of toxic species. Many interactions of inorganic species with membranes can result in direct damage to the membranes or can lead to an amplification of toxicity. Direct damage can include structural alterations, such as disruption of vital lipid–protein associations or introduction of new surface charges. More substantial damage can occur when inorganic species impair membrane function catalytically, e.g., providing ion conductance pathways or by promoting free-radical reactions which may cause massive oxidation of membrane lipids (Mehlhorn, 1986).

Numerous investigations have concluded that the primary site of interaction of the Ln with living cells is at the external surface. The Ln induce mitochondrial membrane aggregation by bonding on the outer surface and neutralizing the surface potential (Stoner *et al.*, 1978). The binding of Ln^{3+} ions on cell surfaces increases the membrane potential and the specific membrane resistance. It is asserted that cell surfaces possess more than one class of Ln^{3+} binding sites, differing in affinity and binding capacity. However, the specific receptor molecules have so far not been well characterized (Evans, 1990). The major components of the plasmalemma are lipids, proteins, and polysaccharides. It is assumed that whereas Ca^{2+} binds only to proteins in the membranes, Ln^{3+} ions bind to both membrane proteins and the phospholipid bilayer (Uyesaka *et al.*, 1976). Ligands may change the capacity of brush-border membrane vesicles to bind the Ln, and even only trace amounts of these metals seem to be internalized within the vesicles (Bingham and Dobrota, 1995).

The cell envelope of gram-negative bacteria consists of three different layers, i.e., the outer membrane (OM), the murein, and the inner membrane. The OM contains transmembrane channels, called porins. Most of the porins sort solutes mainly according to molecular mass and have little solute specificity; however, some of them have specific binding sites for specific classes of solutes (Benz, 1994). Variations in ultraviolet (UV) absorption frequencies were found for di- and trivalent metal systems with basement membranes, which represented specialized forms of extracellular matrices. Basement membranes contain a complete network of insoluble glycoproteins and are thought to allow an interaction between metallic ions and specific inside ligands (Fan *et al.*, 1992). The role of phospholipids and other membrane lipids in the functioning of membrane structures was recently reviewed (Harris *et al.*, 1991), but the biochemistry of their interactions with TE has scarcely been investigated.

Prokaryotes, and, in particular, bacterial walls, show a high affinity for heavy metals (Hughes and Poole, 1989; Collins and Stotzky, 1989). Ln^{3+} and some other ions have a high affinity for and binding capacity to the native OM (i.e., OM not treated with metal). Suspension of purified cell walls of *Bacillus subtilis* in solutions of 40 different metals showed that the reaction of some metals can be specific in terms of both uptake and staining response. A comparative investigation of the uptake, electron-scattering power, and staining reactions of scandium, yttrium, and some other metals in *Escherichia coli* showed that several transition elements displayed a high affinity for the wall fabric. Scandium and cerium were suitable as contrasting agents for electron microscopy (Table 7-1). Large amounts of hafnium and osmium [up to 0.4 μmol/mg dry weight (d.w.)], intermediate amounts of magnesium and manganese, and small amounts of scandium and other metals (< 0.1 μmol/mg d.w.) were detected. The gram-negative envelopes do not bind as much metal on a dry-weight basis as do gram-positive envelopes. The gram-negative envelope of *E. coli* was able to interact with and immobilize many metal ion species (Beveridge, 1978).

Murein (peptidoglycan) is a cross-linked polysaccharide–peptide complex of the inner cell layer of gram-negative bacteria. Murein and the OM show distinct chemical localization of the reactive groups. This may suggest that metal deposition occurs at the polar head-group regions of the constituent membranes or along the peptidoglycan layer. The membrane integrity did not seem to be disturbed under these ionic conditions. It was also suspected that the carboxylate groups play a role in the process of metal accumulation (Beveridge and Koval, 1981). Analysis of the native OM showed that several of the tested metals were indigenous to the native situation (Hoyle and Beveridge, 1983). Like the

Table 7-1. Comparison of Metal Uptake, Electron-Scattering Power, and Wall Properties of *Bacillus subtilis* in Aqueous Solutions of Various Metals[a]

Metal	Possible aqua ion	Metal uptake[b] (μmol/mg cell wall)	Electron-scattering power[c]	Wall thickness (μm)
Ca^{2+}	$[Ca(H_2O)_6]^{2+}$	0.399	+	ND[e]
Sc^{3+}	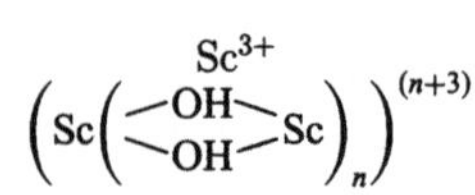	10.990	+++	19.8
Y^{3+}	$[Y(H_2O)_6]^{3+}$	1.380	++	16.3
Fe^{3+}	$[Fe(H_2O)_6]^{3+}$	3.581	+++	27.6

[a] Data reprinted from Beveridge (1978), with permission of the author and the National Research Council of Canada.
[b] The reaction consisted of suspending 1.0 mg dry weight of cell walls in a 5 mM metal solution for 10 min at 22°C.
[c] Compared with the embedding medium.
[d] Electron scattering confined to the central region of the wall.
[e] ND, not determined.

peptidoglycan of *B. subtilis*, that of *E. coli* was able to bind substantial amounts (> 1 μmol/mg d.w.) of scandium (Table 7-2) and some other TE (Ce^{4+}, Hf^{3+}, La^{3+}, Zr^{3+}) from solution. Seemingly, the bound metal may act as the nucleation site for deposition of more metal (Hoyle and Beveridge, 1984).

Various metal ions bind to cation-free bacteriorhodopsin of the purple membrane of *Halobacterium halobium* cells ("blue membrane," BR) with a specific geometry (Fig. 7-1). Four distinct metal-binding sites in BR were described: a pair of high-affinity sites, and another pair of low-affinity sites. Scandium, yttrium, and the Ln, as well as all alkali and alkaline-earth metals, bring about a shift in the absorption maximum of BR, whereas Hg^{2+} and Pt^{4+} do not. Sigmoidal spectroscopic titration curves and nonsigmoidal binding curves suggest that there are two high-affinity sites for cations in BR. Ln ions markedly inhibited the photocycle M_{412} decay at pHs above 6.3. This effect resulted from the metal binding to lipid phosphoryl groups (Ariki and Lanyi, 1986). Trivalent cations binding to bacteriorhodopsin release larger amounts of protons, as compared to divalent cations (Table 7-3). This conclusion is based on the assumption that the shift of the absorption maximum involves proton release from carboxyl groups. So far, no biological significance has been attached to such differences.

Metal interactions with algae occur through strong coordination bonding for certain transition metals with oxygen, nitrogen, and sulfur ligands. Experiments with marine algae showed that adsorption of aluminum and other polyvalent metal ions causes release of hydrogen, calcium, and magnesium metal ions from the cells (Crist *et al.*, 1992).

Mitochondria of eukaryotic species contain channels in their outer membrane. Metal ions interfere with these finely tuned systems. Scandium—as

Table 7-2. Concentrations of Metals in Native Outer Membrane (OM) and of Metals Bound by the OM, Envelope, and Peptidoglycan of *Escherichia coli*[a]

	Concentration (μmol/mg dry weight)			
Metal	Native OM[b]	OM	Envelope	Peptidoglycan
Mg^{2+}	0.009	0.019	0.256	0.035
Ca^{2+}	0.010	0.020	0.035	0.038
Fe^{3+}	0.096	0.233	0.200	0.100
Sc^{3+}	0.002	0.044	0.096	1.403
Mo^{3+}	0.017	0.043	0.226	5.545

[a] Data reprinted from Hoyle and Beveridge (1983 and 1984), with permission of the authors and the National Research Council of Canada.
[b] Native OM was defined as the OM that was extracted under metal-free conditions.

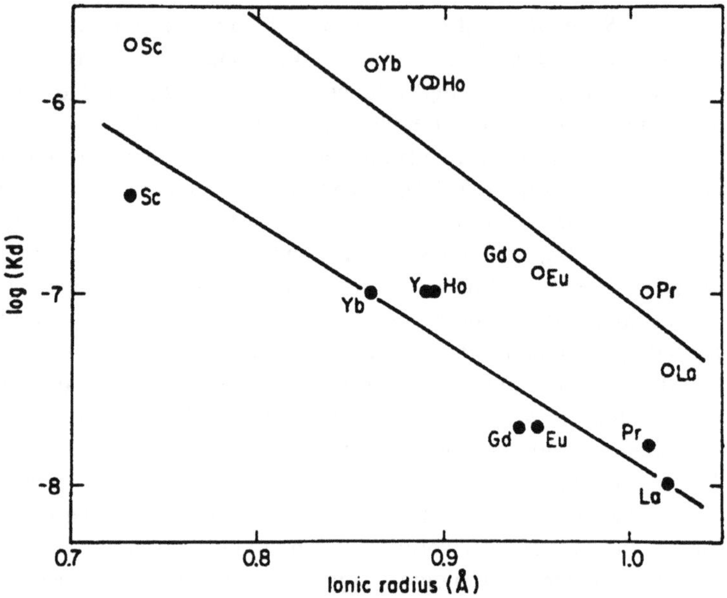

Figure 7-1. Relationship between dissociation constants (K_D) of scandium, yttrium, and other lanthanides for bacteriorhodopsin ("blue membrane") of *Halobacterium halobium* and the crystal ionic radii of the corresponding metal ions. ●, First high-affinity site; ○, second high-affinity site. Reprinted from Ariki and Lanyi (1986), with permission of the authors and the American Society for Biochemistry and Molecular Biology.

Table 7-3. Dissociation Constants for Metal Ions Binding to High-Affinity Sites of Bacteriorhodopsin from *Halobacterium halobium*[a]

	Dissociation constant (μM)	
Metal	First site	Second site
Ba^{2+}	0.33	10.4
Ca^{2+}	0.23	7.5
Sc^{3+}	0.29	2.1
Y^{3+}	0.090	1.3
La^{3+}	0.0081	0.043

[a] Data reprinted from Ariki and Lanyi (1986), with permission of the authors and the American Society for Biochemistry and Molecular Biology.

well as aluminum, chromium, iron, gallium, and indium, but not lanthanum and divalent metal ions—is capable of forming trihydroxide complexes with hydroxyl groups at the binding site of the voltage-dependent anion-selective channels (VDAC) (Table 7-4). It is presumed that the M^{3+} trihydroxide is the active species that interacts with polar molecules of the membrane and interferes with the opening and closing of the VDAC (Colombini, 1991).

Spectral studies of the complexation of Ln^{3+} ions with thylakoid membrane surfaces of spinach chloroplasts showed that these metal ions adsorb, on the membrane surface, to neutrally exposed segments of membrane-embedded protein complexes (Karuckstis *et al.*, 1995). It is anticipated that scandium and yttrium metal ions would exhibit similar effects (Karuckstis, 1996), on chlorophyll absorption and fluorescence spectra to those caused by Ln^{3+} (Karuckstis and Gruber, 1986).

Further aspects of the involvement of scandium and yttrium in membrane metabolism and transport processes are presented in Section 8.11.

7.3 Interactions with Cells and Cellular Organelles

The term "biochemical messengers" refers to various extracellular signal molecules (hormones, neurotransmitters, and growth factors). These, as well as

Table 7-4. Comparison of the Characteristics of Metals that are Effective and Ineffective in Inhibiting the Voltage-Dependent Closure of Voltage-Dependent Anion-Selective Channels[a]

Metal	Surface charge density ($e/Å^2$)	Ionic radius (Å)	Lowest effective or highest tested concentration (M): $M(OH)_z$[b]	Total
Effective				
Al^{3+}	0.92	0.51	8×10^{-7}	1×10^{-6}
Sc^{3+}	0.45	0.73	7×10^{-7}	1×10^{-5}
Fe^{3+}	0.58	0.64	$<4 \times 10^{-4}$	5×10^{-4}
Ineffective				
Mg^{2+}	0.37	0.66	—	5×10^{-3}
Ca^{2+}	0.16	0.99	—	5×10^{-3}
La^{3+}	0.23	1.02	0[c]	5×10^{-4}

[a] Data reprinted from Colombini (1991), with permission of Libreria Cortina Editrice SRL.
[b] Calculated from lowest effective concentration for effective metals and highest concentration tested for ineffective metals at pH 7, ionic strength = 1 M.
[c] Essentially not formed at pH 7.

the intracellular signal molecules, play a central role in cell–cell signaling. In multicellular organisms, the differentiation of cells ensures coordination of movement, metabolism, and growth. The permeability of gap junctions allows the movement of small molecules, such as ions or metabolites, directly from one cell to the next (Hardie, 1991).

Analysis of sarcoplasmic reticulum vesicles isolated from white muscle of rabbit showed a scandium concentration of less than 0.1 nmol/mg protein. A possible role of several TE in the modification of SH groups and in lipid peroxidation was suggested (Papp *et al.*, 1985).

Intravenous administration of ^{46}Sc citrate to tumor-bearing rats resulted in tumor uptake of ^{46}Sc and binding of ^{46}Sc mainly to the lysosome fraction of the mitochondria of the liver and tumors (Table 7-5). The subcellular binding of ^{46}Sc was similar to that of other radioisotopes. However, whereas ^{46}Sc citrate was mainly bound to acid polysaccharides with molecular weights exceeding 4000, $^{51}Cr^{3+}$ chloride was bound to acid mucopolysaccharides with molecular weights of about 10,000, and large amounts of ^{51}Cr were excreted (Ando *et al.*, 1987).

Sc^{3+} binds *in vivo* specifically to rabbit plasma transferrin at two iron-binding sites (see Part 1, Fig. 6-11). When quantities of scandium below the free iron-binding capacity of serum were added to whole serum *in vitro*, the only protein complex observed was the scandium–transferrin complex (Ford-Hutchinson and Perkins, 1971). The binding of ^{46}Sc to erythrocytes initially

Table 7-5. Evolution with Time of the Subcellular Biodistribution of ^{46}Sc Citrate in Experimental Tumors and Host Livers of Rats[a]

Time after administration	Tumor				Host liver			
	Nuc	Mit	Mic	Sup	Nuc	Mit	Mic	Sup
Ehrlich tumor								
10 min	7.7	4.2	8.7	79.4	7.3	5.2	12.1	75.4
60 min	8.5	6.4	7.3	77.8	14.8	15.0	13.7	56.5
3 h	9.5	14.1	11.2	65.2	18.0	36.1	13.1	32.8
24 h	18.5	36.7	28.0	16.8	20.3	53.1	22.2	4.4
48 h	17.4	35.7	29.4	17.5	17.9	54.4	23.7	4.0
Hepatoma AH109A								
10 min	5.5	7.2	16.4	70.9	8.1	17.1	22.1	52.7
60 min	10.0	22.1	23.0	44.9	10.3	31.2	19.4	39.1
3 h	23.1	34.4	23.4	19.1	26.9	44.7	16.3	12.1
24 h	23.6	36.9	27.8	11.7	23.8	58.7	10.9	6.6
48 h	25.6	39.0	27.1	8.3	27.0	55.4	13.0	4.6

[a] Values are percentages in each of four cellular fractions: Nuc, nuclear fraction; Mit, mitochondrial fraction; Mic, microsomal fraction; Sup, supernatant fraction. Data reprinted from Ando *et al.* (1987), with permission of the authors and Elsevier Science.

accounted for less than 5% of the total radioactivity in the blood (Fig. 7-2). No specific binding of scandium to plasma and erythrocytes of rabbit was observed. The radioactivity of ^{46}Sc was associated with either the bile pigments or transferrin naturally excreted in the bile pigments (Ford-Hutchinson and Perkins, 1971).

Sc^{3+} and Fe^{3+} showed no effect on the precipitation of chromatin isolated from rat brain and liver, whereas Al^{3+}, In^{3+}, and Ga^{3+} were effective precipitants in the micromolar concentration range (Walker *et al.*, 1989; Fig. 7-3). On the basis of the effect of ionic strength on the sedimentation coefficient of chromatin, it was emphasized that metal ions may induce more subtle, physiologically relevant changes in chromatin structure at concentrations an order of magnitude lower than those required to produce precipitation (Walker *et al.*, 1989).

Numerous studies indicated that lanthanum and a large number of heavy metals accumulate in microorganisms. It is controversial whether this reflects their transport and intracellular accumulation or a passive attachment to the cell surface. Many experimental data indicate that Ln^{3+} ions do not enter the cytoplasm of living cells; however, a few other literature reports dispute this concept of impermeability (Evans, 1990).

Multivalent metal ions were tested for their ability to replace the Ca^{2+} requirements of aggregation factor complex from the marine sponge *Microciona* in activity, stability, and integrity assays. All the Ln, as well as yttrium, produced normal activity, but only at 10- to 40-fold lower metal ion concentrations than Ca^{2+}. Similarly, the Ca^{2+} requirement for the conversion of trypsinogen to

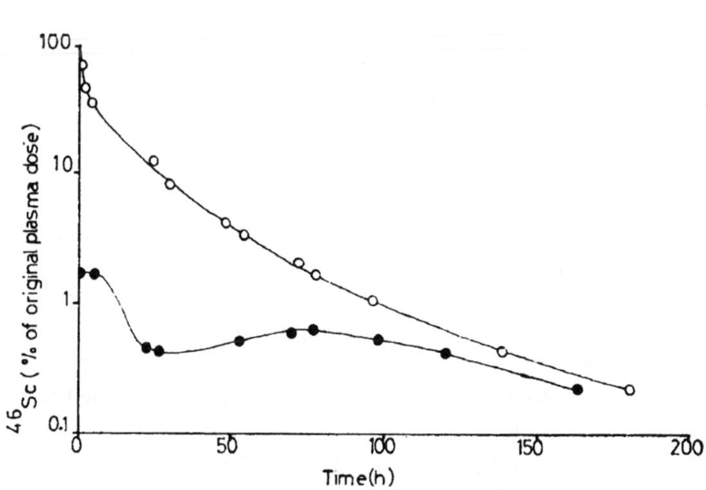

Figure 7-2. Binding of scandium *in vivo* to plasma (○) and erythrocytes (●) of rabbits following injection with 1 : 2 ^{46}Sc citrate complex (0.5 mg Sc^{3+}, 1 mCi). Reprinted from Ford-Hutchinson and Perkins (1972), with permission of Academic Press.

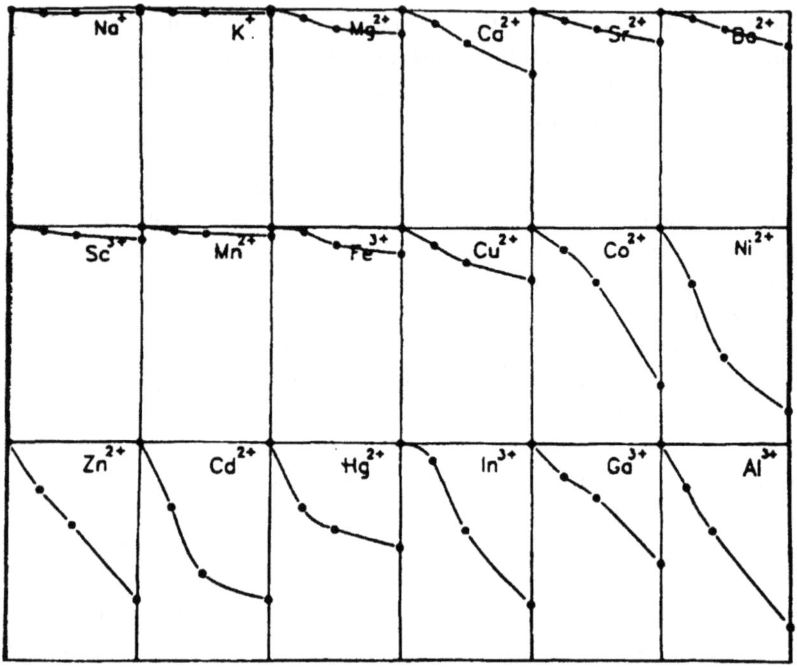

Figure 7-3. The ability of scandium and other metals at concentrations in the range of 0–500 μM to bind to chromatin isolated from livers and brain cortex of rats. In each panel within the chart, the percentage of chromatin remaining in the solution is plotted on the *y*-axis (0–100%) against aluminum concentration on the *x*-axis (0–500 μM). Reprinted from Walker *et al.* (1989), with permission of the American Chemical Society.

trypsin could be replaced by various Ln^{3+} ions and Y^{3+} at much lower concentrations (Rice and Humphreys, 1983).

A study of the interaction between yttrium and the soil bacterium *Pseudomonas fluorescens* found that, at the stationary phase of growth, 90% of yttrium was sequestered in an exocellular phosphorus-containing fraction (Appanna and Huang, 1992b). Intracellular binding of yttrium was assumed (Appanna and Huang, 1992a). A novel protein with an approximate molecular mass of 32 kDa was induced in yttrium-supplemented *P. fluorescens* culture. Three other proteins, with molecular masses of 160, 58, and 30 kDa, respectively, were absent in the electrophoregram from the yttrium-supplemented culture (Fig. 7-4). Various peak accumulations of yttrium in cell components were measured with X-ray fluorescence analysis during the growth stages (Fig. 7-5). Whether a differential-binding capacity for cell compounds may exist was not established. In a medium containing 0.5 mM yttrium nitrate, the metal-bound citrate (see Fig. 8-7) was rapidly consumed by the bacterium. Thus, the bacterium easily adapted

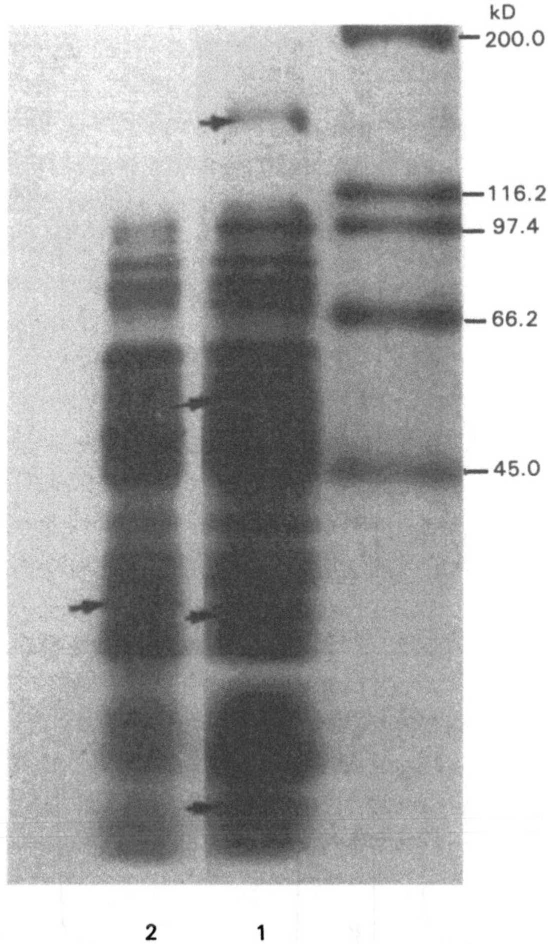

Figure 7-4. Sodium dodecyl sulfate–polyacrylamide gel electrophoresis (SDS-PAGE) analysis of cytoplasmic proteins from *Pseudomonas fluorescens* grown in medium supplemented with 0.5 mM yttrium (lane 2) and control, without yttrium (lane 1). Reprinted from Appanna and Huang (1922b), with permission of the authors and Faculty Press.

to yttrium complexed to its sole source of carbon by insolubilizing the trivalent metal ion (Appanna and Huang, 1992b).

The calcium-binding proteins, which play important roles in many biological systems, exhibit calcium-site specificity and bind other metal ions present in the medium. The D-galactose and D-glucose receptor (also termed the galactose-binding protein) of *E. coli* showed a high specificity of binding metal ions from groups IA, IIA, and IIIA and the Ln. Divalent group IIA metal ions and trivalent

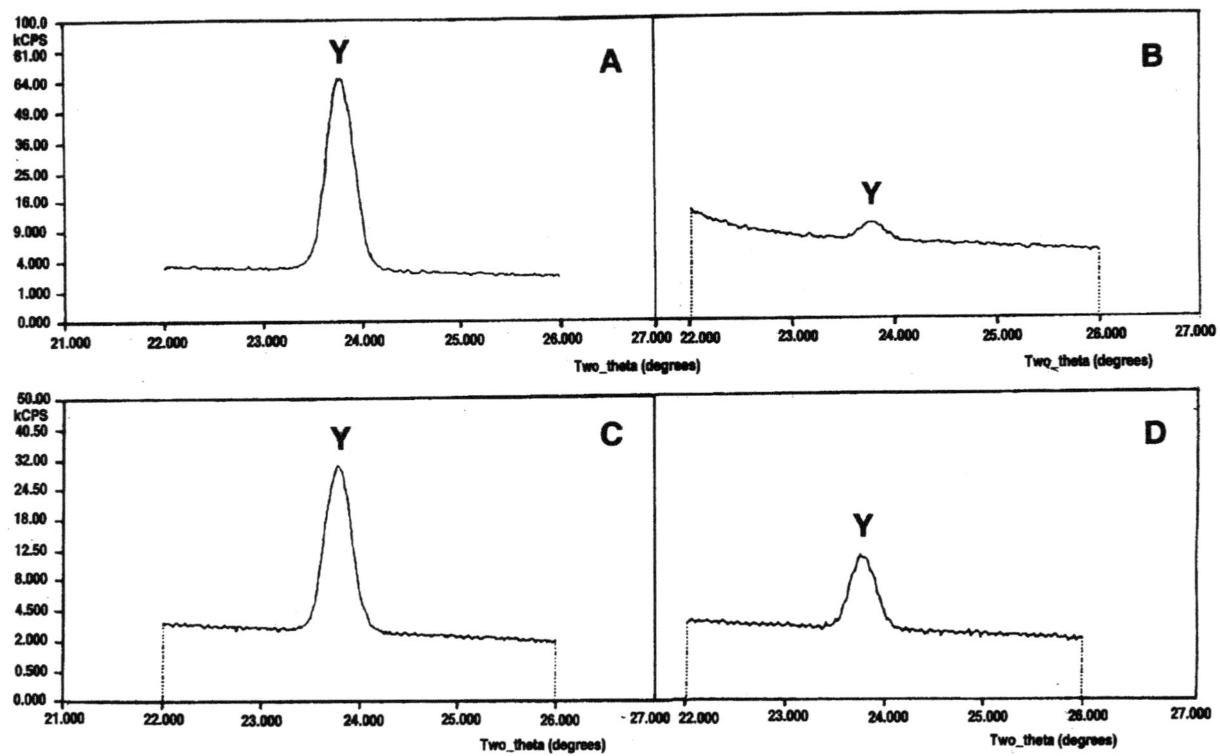

Figure 7-5. X-ray fluorescence analysis of bacterial cells of *Pseudomonas fluorescens* grown in medium supplemented with yttrium Y = A, Yttrium peak. (A) At incubation time; (B) from supernatant at stationary phase of growth; (C) in bacterial cells at cessation of cellular multiplication; (D) in pellet isolated at stationary phase of growth. Reprinted from Appanna and Huang (1992a), with permission of the authors and Springer-Verlag.

metal ions exhibited size-dependent affinities, with an optimal effective ionic radius between those of Sc^{3+} (0.81 Å) and Yb^{3+} (0.925 Å) (Snyder *et al.*, 1990). Sites containing a smaller side chain bound the largest cations up to 50-fold more tightly than the native site. Sites containing an engineered negative side chain exhibited a preference for Sc^{3+}, Y^{3+}, or La^{3+} over divalent cations. These sites were up to 1900-fold higher in number than the native sites. Apparently, the ion specificity of the prokaryotic electrostatic field EF-hand-like site is controlled by the interaction between the metal ion and the site, rather than the energetic contribution of metal dehydration (Falke *et al.*, 1991; Fig. 7.6). The carboxylate cluster Mg^{2+}-binding site of the *E. coli* chemotaxis pathway binds divalent and trivalent cations, but it effectively excludes monovalent cations. Comparing of the ion specificity of the Mg^{2+} site with the specificity of the EF-hand class of Ca^{2+} sites shows that both sites exhibit similar charge selectivity. The Ca^{2+} site is highly size-selective and prefers divalent and trivalent ions with radii similar to that of Ca^{2+}. The different size specificities stem from fundamentally different coordination schemes (Needham *et al.*, 1993). The dissociation constants of metal binding to the EF-hand-like binding site of Ca^{2+} in *E. coli* are presented in Table 7-6.

Competitive inhibitory effects of the Ln on calcium influx in rat brain synaptosomes were found. These effects were due to binding of the Ln to the outer side surface of the channels. Two classes of Ln^{3+}-binding sites on the surface of intact cells were distinguished, i.e., high- and low-affinity sites (see Table 8-27). The rapid inhibitory effect of Ln^{3+} was quickly reversible, with K_D values of 0.2 μM for Y^{3+} and 0.7 μM for La^{3+} (Nachsen, 1984).

7.4 Interactions with Antibodies

Antibodies, which are adapter molecules linking antigen and effector, are able to localize tumor sites. MAbs are likely therapeutic and or diagnostic functions for selectively carrying cytotoxic or diagnostic agents. Methods of coupling cytocidal or image-producing radioisotopes to antibodies were worked out for the treatment or visualization of malignancies (Quadri *et al.*, 1993). Experiments with laboratory animals in which bifunctional chelating agent (BCA) complexes which were employed as therapeutic agents suggested further applications of these complexes in human medicine. Improved techniques of radiolabeling tumor-associated MAbs with metal chelators as potential therapeutic or diagnostic agents have been intensively investigated in numerous institutions (Beyer *et al.*, 1992; Buchsbaum *et al.*, 1993; Mardirossian *et al.*, 1993; Camera *et al.*, 1994; Takenouchi *et al.*, 1993; DeNardo *et al.*, 1995; Kosmas *et al.*, 1995; Schomaker *et al.*, 1995; Antoniw *et al.*, 1996).

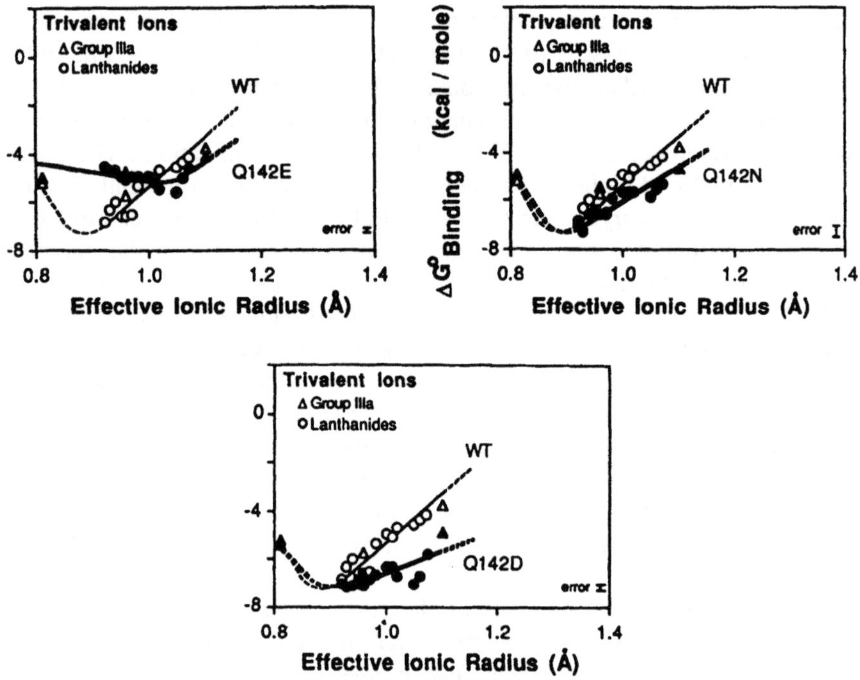

Figure 7-6. Dependence of binding free energy on cation charge and size of trivalent ions at the Q142E, Q142N, and Q142D engineered sites of the *Escherichia coli* D-galactose and D-glucose receptor. Binding free energies for the engineered site (filled symbols, bold curve) and the native site (open symbols, fine curve) are plotted against effective ionic radius for 7-fold coordination. Trivalent ions are Sc^{3+} and Y^{3+} (Δ) and the Ln^{3+} (O). All measurements were at 25°C and 2.5 μM protein, 100 mM KCl, and 10 mM PIPES, pH 6.0. Reprinted from Falke *et al.* (1991), with permission of the authors and the American Chemical Society.

Table 7-6. Metal Ion Dissociation Constants (K_D) of the Ion-Specific EF-Hand-Like Ca^{2+} Binding Sites WT, Q142N, Q142E, and Q142D in *E. coli*[a]

Metal	Effective ionic radius (Å)	K_D (M)			
		WT	Q142N	Q142E	Q142D
K^+	1.46	≥3.9	≥1.8	≥4.3	≥13
Mg^{2+}	0.81	570×10^{-3}	≥7	≥1.1	≥1.6
Ca^{2+}	1.06	13×10^{-6}	15×10^{-6}	0.18	2.8×10^{-3}
Sc^{3+}	0.81	170×10^{-6}	170×10^{-6}	250×10^{-6}	150×10^{-6}
Y^{3+}	0.96	74×10^{-6}	75×10^{-6}	390×10^{-6}	150×10^{-6}

[a] Measurements were made at 25°C with 2.5 μM protein, 100 mM KCl, and 10 mM PIPES, at pH 6.0. Data reprinted from Falke *et al.* (1991), with permission of the authors and the American Chemical Society.

These substances may be used for immunoassay *in vitro* by employing radioactive metals or for *in vivo* imaging of patients with the use of radioisotopes (Inoue and Gokel, 1990). Comparison of eight radioisotopes conjugated to IgG, $F(ab')_2$, and Fab antibody forms indicated that ^{90}Y, ^{153}Sm, and ^{186}Re are the most promising for radioimmunotherapy (Yorke *et al.*, 1991). ^{87}Y and ^{88}Y have also been used for imaging in addition to ^{90}Y, and therapy (Moi *et al.*, 1990). The stable trivalent metal ions In^{3+}, Bi^{3+}, and Y^{3+} form antibody complexes with good thermodynamic and kinetic stability, which prevents dissociation of these complexes in aqueous media. The ligands which are expected to be most useful for M^{3+} are the chelating agents ethylenediaminetetraacetic acid (EDTA) and diethylenetriaminepentaacetic acid (DTPA) and the cryptands (Fig. 7-7). Since the early 1980s, attachment of scandium (Scheinberg *et al.*, 1982; Reardan *et al.*, 1985) and yttrium (Hnatowich *et al.*, 1985) radioisotopes to antibodies by means of BCAs has been explored. Both ^{47}Sc and ^{90}Y have a high yield of energetic β-particles, and their half-lives are 81.6 and 64.1 h, respectively. Plots of the stability constants of the complexes of Y^{3+} and Sc^{3+} with DTPA in aqueous solutions at various pHs are depicted in Fig. 6-25 of Part 1.

Sufficient radioactivity must be deposited in the target tissue to kill the pathological cells, while the uptake in nontarget tissue is kept to a minimum. Selection of the chelating agent is critical for each metal, in order to prevent inappropriate release of the radiometal *in vivo* (Kozak *et al.*, 1989). BCAs and macrocyclic bifunctional chelating agents (MBCAs), such as the 12-membered macrocycle, can hold radiometals with high stability in human serum under

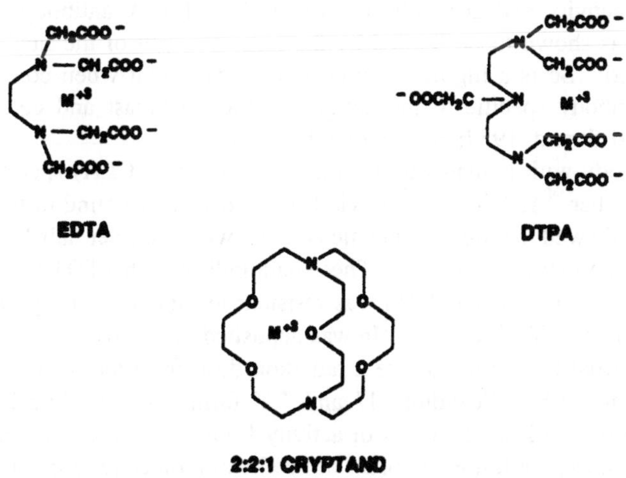

Figure 7-7. Three types of metel chelates used for radiolabeling of monoclonal antibodies with ^{90}Y and other radioisotopes. Reprinted from Gansow (1991), with permission of the author and Elsevier Science.

physiological conditions (Moi *et al.*, 1990; Fig. 7-8). Derivatives of polyazamacrocycles, bearing a carbon-substituted functional group for antibody attachment, showed remarkable kinetic inertness (Gansow, 1991). Pendant-type macrocyclic BCAs were developed, and the rate of complex formation with ^{88}Y and the stability of the resulting complexes in human serum were determined (Takenouchi *et al.*, 1993). Novel methods of radiolabeling antibodies have been developed, among which are those relying upon chelators to stably bind ^{90}Y or ^{111}In to proteins. Nitrogen-linked amines have been used to prepare such chelators. The 9N3 and 12N4 macrocyclic chelators provided superior results in term of labeling efficiency (Wu *et al.*, 1992). Vallera *et al.* radiolabeled the MAb which recognizes Ly1, the homolog of CD5, by binding it to ^{90}Y; the ^{90}Y-anti-Ly1 was selectively localized in lymphoid tissues, but not in blood, bowel, lung, muscle, or skin (Fig. 7-9). Binding constants of di- and trivalent metal ions were compared with those of indium bound to different chelators (Reardan *et al.*, 1985). In^{3+}, Fe^{3+}, and Cd^{2+} best labeled the L-benzyl-EDTA chelate, but Sc^{3+} and Mn^{2+} also were chelated (Table 7-7). *In vivo* chemical stability and clearance of radiolabeled antibodies were assessed in serum from mice injected with labeled antibody or free radioisotope. One hundred percent of the radioactivity present in serum could be accounted for by antibody- or transferrin-associated activity of ^{46}Sc, ^{90}Y, ^{111}In, ^{125}I, and ^{153}Gd. Only ^{46}Sc and ^{111}In showed a significant transfer to transferrin, with 30% of scandium and 10% of indium being lost within 24 h. A study of the stability of ^{46}Sc-DTPA–antibody conjugate *in vivo* found that 25% of the radioisotope was bound to transferrin and the remaining 75% was attached to antibody conjugate (Anderson and Strand, 1985). The specific activity and concentration of the ^{47}Sc-DTPA–antibody conjugate preparation was shown to be limited by the ionic strength of the final solution. The short-lived ^{47}Sc is a highly specific tumoricidal agent when complexed to B72.3, an antibody specific to the antigen of human breast and colon cancer (Anderson and Strand, 1987; St. Raymond, 1990).

An antibody with human carcinoembryonic antigen (CEA) specificity was labeled with either ^{90}Y, ^{47}Sc, or ^{111}In via DTPA covalently bound to the protein. Scandium-47 showed the highest specific activity, with a shorter half-life whereas ^{111}In showed the lowest (Table 7-8). These data indicated that DTPA was able to extract ^{90}Y and ^{47}Sc, whereas ^{111}In was resistant to ligand exchange *in vivo*. A high proportion of ^{90}Y, ^{47}Sc, and ^{111}In was deposited in the liver, which is one of the critical normal tissues to receive a radiation dose from the material injected (Vaughan *et al.*, 1987). Scandium-47 and ^{90}Y form a more labile bond with DTPA, which *in vivo* leads to a loss of activity from the DTPA. Associated with the blood clearance, the following order of uptake of radioactivity in the liver was found: $^{111}In < ^{90}Y < ^{47}Sc$ (Fig. 7-10).

Yttrium-90 and ^{111}In were attached to an anti-adenocarcinoma chimeric L6 MAb after prechelation to a tetraazocyclododecanetetra acetic acid (DOTA)-peptide isothiocyanate, and the desired neutral radiochelate was obtained by

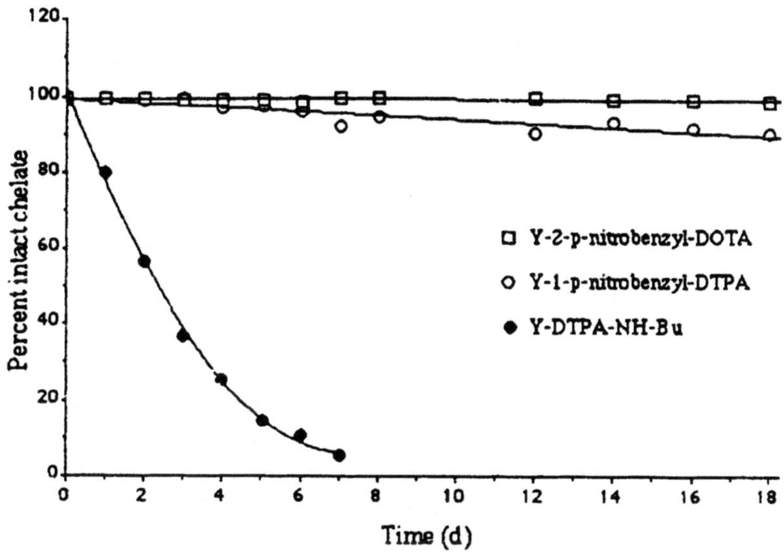

Figure 7-8. Rate of decomposition of yttrium chelates in human serum. The total ligand concentration was 10^{-7} M, while the total yttrium concentration was $\sim 10^{-11}$ M. Reprinted from Moi *et al.* (1990), with permission of the American Association for Cancer Research, Inc.

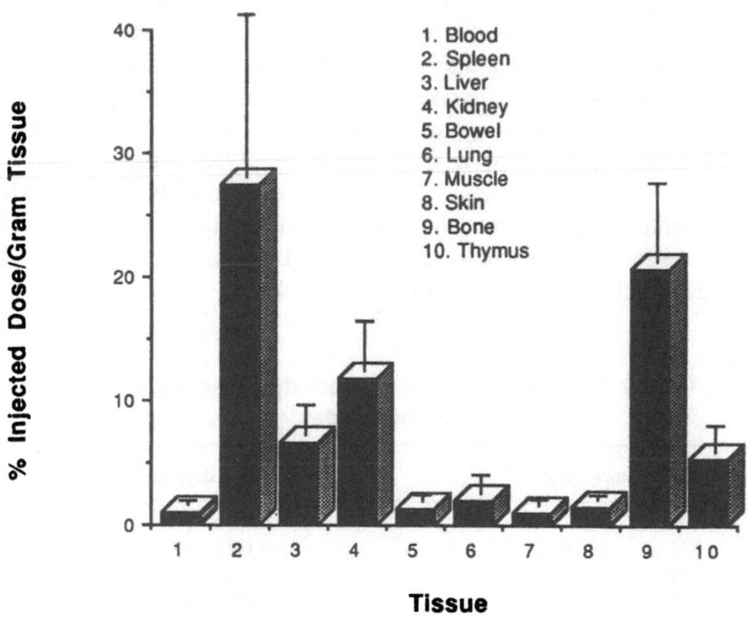

Figure 7-9. Biodistribution of ^{90}Y-anti-Ly1 in mice. Animals were injected with 20 μCi of ^{90}Y-anti-Ly1, and after 5 days the radioactivity in tissues of sacrificed animals was measured and expressed as percent of the injected dose per gram of tissue. Reprinted from Vallera *et al.* (1991), with permission of the American Association for Cancer Research, Inc.

Table 7-7. Relative Binding Capacities of Metal L-Benzyl-EDTA Chelates to Monoclonal Antibodies[a]

Metal	r_{ion} (Å)	log K_b (EDTA)	K_b (CHA255)	K_b (CHB235)
In^{3+}	0.81	25.0	4.0×10^9	1.1×10^8
Fe^{3+}	0.64	25.1	1.8×10^8	2.4×10^6
Sc^{3+}	0.81	23.1	3.0×10^6	5×10^4
Ga^{3+}	0.62	20.3	5×10^5	5.2×10^4

[a] K_b values are the antibody-binding constants of each metal L-benzyl-EDTA chelate for the two MAbs studied, CHA255 and CHB235. Data reprinted from Reardan *et al.* (1985), with permission of the authors and *Nature* (London).

purification. The observed specific activity of ^{90}Y in the tumor and its minimal bone accumulation (Fig. 7-11) indicate that the therapy dose (KCi) correlated with the amount and duration of response. A favorable radioimmunoconjugate therapeutic index may be achieved by using this agent (DeNardo *et al.*, 1995).

In addition to accumulating in femur, pelvis, and vertebrate, ^{88}Y acetate also accumulated, although in much smaller amounts, in kidneys, liver, and spleen of mice bearing human colon carcinoma (Table 7-9). The stability of ^{88}Y-bifunctional macrocyclic chelating agent attached with a pendant-type carboxymethylamino ligand and nitrobenzyl group in serum demonstrated its usefulness (Roselli *et al.*, 1989). Studies of the biodistribution of ^{88}Y-labeled MAb B72.3 IgG in animal models showed that the more highly substituted bifunctional DTPA ligands resulted in a higher uptake of ^{88}Y in the tumor (Fig. 7-12).

A revised method of labeling mouse IgG with ^{90}Y used isothiocyanatobenzyl-EDTA acid as a chelating agent (Arbab *et al.*, 1995). Instant thin-layer-chromatographic separation showed the labeling efficiency before and after transchelation with Na_2EDTA; the labeled IgG remained at the origin while the unlabeled acetate form or transchelated form moved to the front (Fig. 7-13). An alternative way to achieve high selective tumor uptake of

Table 7-8. Blood Clearance of Antibody Labeled with Various Radioisotopes[a]

Radioisotope label	Half-life in days ± SD
^{47}Sc	1.37 ± 0.07
^{90}Y	1.63 ± 0.09
^{111}In	1.92 ± 0.11
^{125}I	2.20 ± 0.13

[a] A single exponential for clearance was assumed. Data reprinted from Vaughan *et al.* (1987), with permission of the National Cancer Institute.

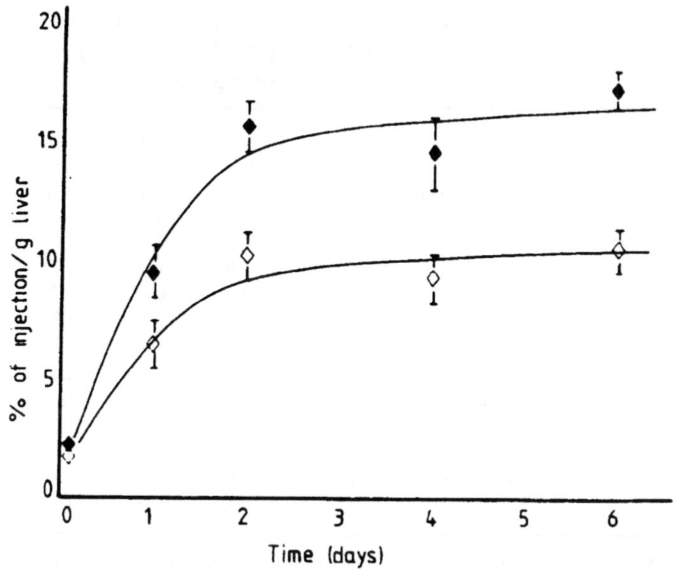

Figure 7-10. Uptake of activity into liver of BALB/c mice after injection of antibodies labeled with either ^{47}Sc (◆) or ^{111}In (◇). Reprinted from Vaughan *et al.* (1987), with permission of the author.

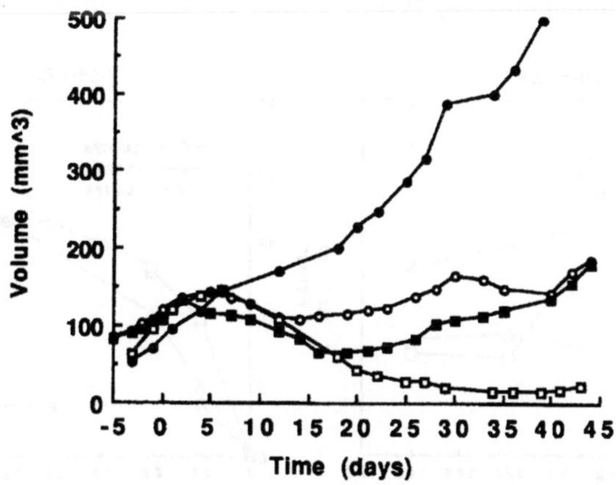

Figure 7-11. Tumor growth of HBT 3477 human breast cancer xenograft in four groups of nude mice: untreated (●) or treated with 100 (○), 150 (■), or 250 μCi of ^{91}Y-ChL6 (□). Tumors were taken for studies from the 150-μCi therapy group at 3, 6, and 24 h after injection. Reprinted from DeNardo *et al.* (1995), with permission of the authors and the American Association for Cancer Research, Inc.

Table 7-9. Biodistribution of ^{88}Y Acetate and ^{111}In Citrate Administered Intravenously to Athymic Mice Bearing Human Colon Carcinoma Xenografts[a]

	^{88}Y			^{111}In		
	24 h	48 h	168 h	24 h	48 h	168 h
Tumor	2.7	2.4	1.6	6.5	5.2	2.5
Blood	0.1	0.06	0.05	3.6	1.0	0.2
Liver	9.2	7.8	3.2	9.3	10.8	12.2
Spleen	1.6	1.7	1.0	4.8	5.1	6.8
Kidney	13.6	10.5	5.5	36.3	25.4	13.7
Lung	1.7	1.5	1.0	4.6	3.3	2.6
Heart	1.1	0.9	0.5	2.6	1.8	2.0
Intestine	0.9	0.6	0.2	8.1	5.4	1.1
Vertebrae	18.6	21.8	16.0	5.7	4.9	3.7
Pelvis	17.6	21.9	14.5	6.0	5.6	3.6
Femur	22.8	28.0	16.3	7.0	5.8	4.5

[a] Athymic mice bearing the LS-174T human colon carcinoma xenograft were injected intravenously with 0.5 mCi of ^{88}Y acetate or 5 mCi of ^{111}In citrate and sacrificed at the indicated time. The results are expressed as percent of the initial dose per gram. Data adapted from Gansow (1991), with permission of the author and Elsevier Science.

^{90}Y-labeled DOTA with simultaneous reduction of non-target-tissue background involved a three-step pretargeting technique, which was developed for radioimmunoscintigraphy (Goodwin *et al.*, 1994). The pretargeting of ^{90}Y hapten-specific MAb Janus-DOTA that was obtained with this technique showed good promise for radioimmunotherapy (Fig. 7-14). A modified carbon backbone of the

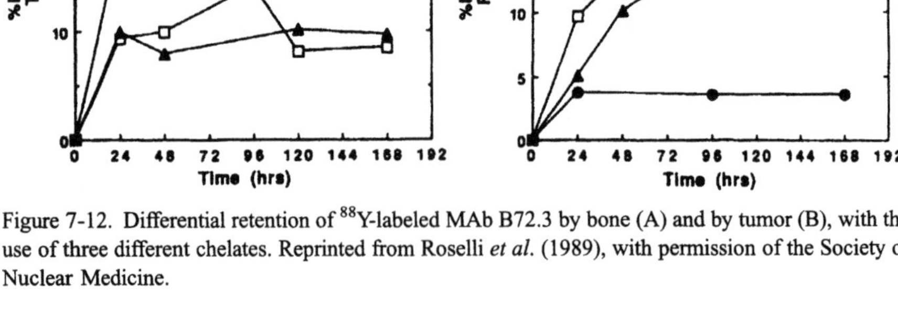

Figure 7-12. Differential retention of ^{88}Y-labeled MAb B72.3 by bone (A) and by tumor (B), with the use of three different chelates. Reprinted from Roselli *et al.* (1989), with permission of the Society of Nuclear Medicine.

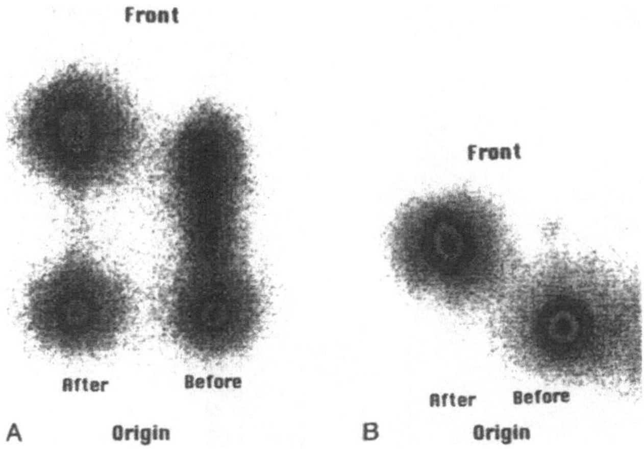

Figure 7-13. Instant thin-layer-chromatographic (TLC) tracing showing the labeling efficiency of mouse IgG with ^{91}Y acetate (A) and ^{111}In acetate (B) using isothiocyanatobenzyl-EDTA, before and after NaEDTA challenge. Labeled radio metals remained at the origin of the Sephadex-25 gel column, while unlabeled and tranchelsted forms moved toward the front. Reprinted from Arbab *et al.* (1995), with permission of the authors.

acyclic ligand DTPA that permitted involvement of all five carboxylic acid groups in complexation of yttrium was synthesized; the labeling efficiency of the complex was demonstrated *in vivo* in experiments with mice (Harrison *et al.*, 1991). Although the ^{90}Y-DOTA complex may not be completely inert *in vivo*, compared to carcinoma tumor-DTPA, and carboxylic acid-DTPA, it is taken up less by the bone, which allows a greater amount of ^{90}Y to be available for tumor uptake (Fig. 7-15).

Recently, a third generation of acyclic chelators was introduced; these demonstrated increased stability *in vivo* compared to DTPA. They include nitro-1B4M-DTPA, nitro-MX-DTPA, and nitro-CHX-A-DTPA, which chelate yttrium instantaneously with a chelator : metal molar ratio of 1 : 1. Another class of macrocyclic chelators, nitro-DOTA, nitro-PADOTA, and nitro-TRITA, chelated ^{90}Y at a slower rate and, except for nitro-TRITA, exhibited exceptional stability at pH 2.0 (Stimmel *et al.*, 1995). The MAb CO17-1A was found to possess specificity for colorectal and pancreatic carcinomas. Radiation-induced myelosupression was reduced by employing granulocyte colony-stimulating factors (G-CSF). Nude mice bearing xenografted human colorectal carcinomas and receiving G-CSF clearly demonstrated tumor reduction (Crook *et al.*, 1990; Fig. 7-16). The radiation-induced myelosupression could be ameliorated by employing G-CSF. *In vivo* binding of ^{90}Y-labeled DOTA^0,D-Phe^1.Tyr^3-octreotide (DOTATOC) in rat cerebral cortex microsomes caused a 2–6 fold uptake of radioactivity in the octreotide receptor-expressing tissues (pancreas, pituitary,

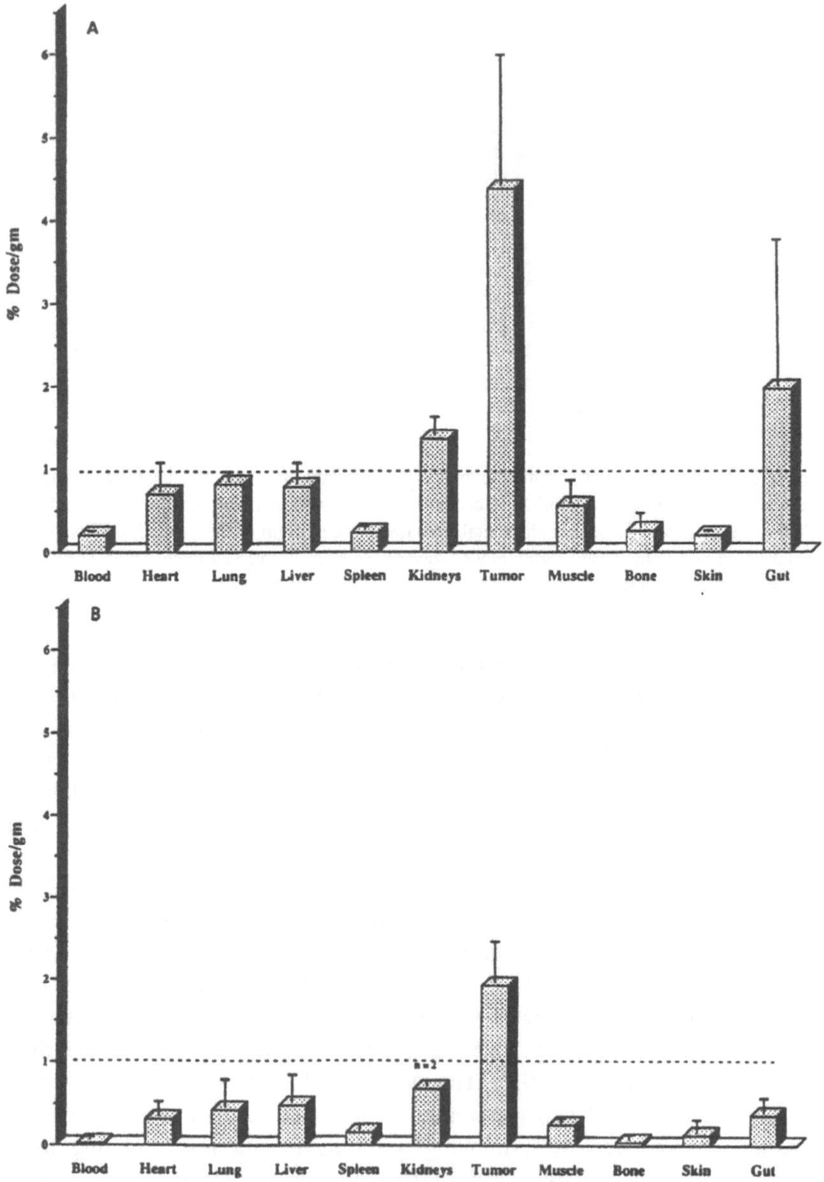

Figure 7-14. Organ distribution of pretargeted ^{89}Y-Janus-DOTA in athymic mice bearing human colon carcinoma xenografts after 3 h (A) and 24 h (B). All organs except kidneys and guts contain less than 1% of the dose after 1 h, and the percentages in the latter organs decrease to less than 1% at 24 h. Note the exceptionally low concentrations in blood and bone at both times. Reprinted from Goodwin *et al.* (1994), with permission of the American Association for Cancer Research, Inc.

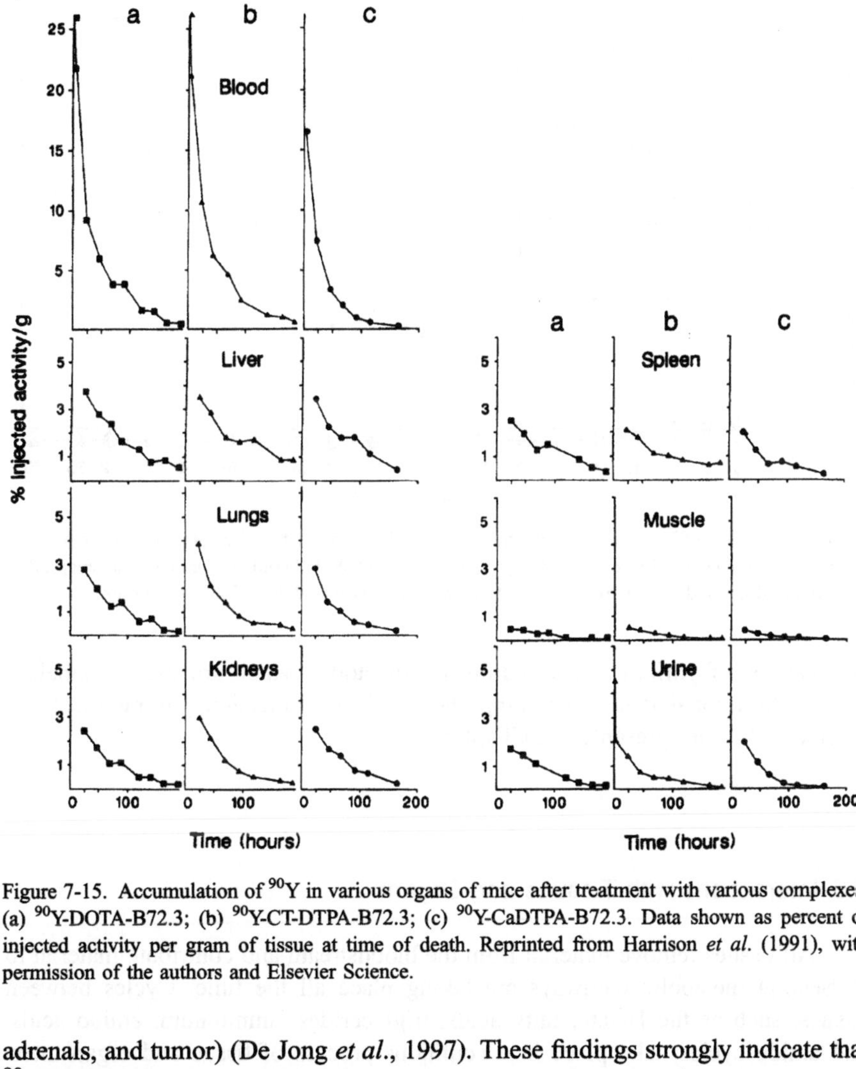

Figure 7-15. Accumulation of ^{90}Y in various organs of mice after treatment with various complexes: (a) ^{90}Y-DOTA-B72.3; (b) ^{90}Y-CT-DTPA-B72.3; (c) ^{90}Y-CaDTPA-B72.3. Data shown as percent of injected activity per gram of tissue at time of death. Reprinted from Harrison *et al.* (1991), with permission of the authors and Elsevier Science.

adrenals, and tumor) (De Jong *et al.*, 1997). These findings strongly indicate that ^{90}Y-DOTATOC is a promising radiopharmaceutical in the treatment of neuroendocrine tumors (De Jong *et al.*, 1997).

Evidence was presented for the existence of two different breast epithelial antigens. The two antigens, which are major components of the human milk fat globulin membrane, are breast mucin and M46000 glycoprotein. When labeled with ^{90}Y, some MAbs raised against these antigens in mice showed promise as radioimmunotherapeutic agents (Peterson *et al.*, 1995).

Metabolic aspects of the ^{90}Y labeled metabolites of MAbs (Rosenblum *et al.*, 1991) are presented in Section 8.8. Table 10-3 summarizes medical

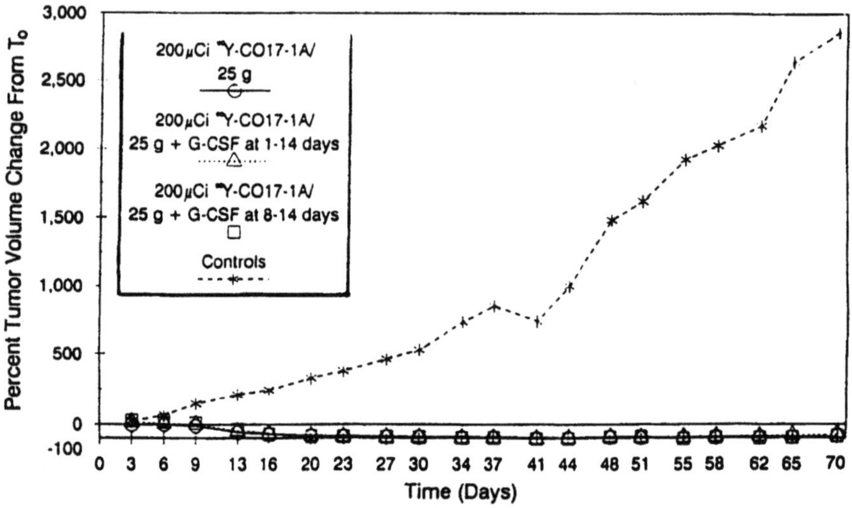

Figure 7-16. Effect of ^{90}Y-CO17-1A and granulocyte colony-stimulating factors (3 μg/mouse per day) administered at days 1–14 or 8–14 on the growth of human SW 948 colorectal carcinoma xenograft in nude mice. Reprinted from Crook *et al.* (1990) with permission of John Libbey, Eurotex.

applications of yttrium and scandium radioisotopes which are used in labeling MAbs. Additional data concerning the medical applications of radioisotope-labeled MAbs are presented in Chapter 10.

7.5 Interactions with Tissues and Organs

All tissues remove material from the bloodstream and contribute material to it. Several metabolic pathways are taking place all the time. Cycles between tissues, such as the lactate, fatty acids, triglycerides, ammonium, amino acids, and amide cycles, take place in a continous process of tissue and organ interrelationships (Banks *et al.*, 1976).

Following injection of ^{91}Y citrate into *Drosophila melanogaster*, the radioisotope was completely retained in the body (Oftedal, 1961). The retention site after 48 h was located in a narrow band of cells in the endodermal mid-gut. Injection of ^{91}Y into Musca sp. did not bring any radioactivity in the pericardial cells or the Malpighi tubules, but most of the activity was found in the mid-gut. The amount of radioactivity that was concentrated in the pericardial cells and the thoracic nephrocytes, or in the haemocytes, depended upon the type of aggregate formed by ^{91}Y citrate (Oftedal, 1961).

A soluble acid phosphate of the complex ScF^{2+}, administered to rats, was deposited in various skeletal tissues (cranial bone, femural epiphysis and diaphysis, and incisors) (Anbar and Ernst, 1962). These tissues may compete effectively for fluoride under physiological conditions. The rate of fluoride clearance wass enhanced in the presence of scandium, and the extent of fluoride accumulation in bone tissues was markedly reduced both relatively and absolutely. These results are in accordance with the fact that only 20% of the scandium in circulation is deposited in bone. The effect of yttrium metal ions on fluoride distribution *in vivo* indicated the existence of a rather stable complex, which may bind fluoride under physiological conditions (Anbar and Ernst, 1962). A high affinity of yttrium for bone in areas of high osteoblastic activity and a decrease of the relative concentration of fluoride in bone with increasing dose of yttrium was found (Dudley, 1955; Dudley and Greenberg, 1958).

Injection of mixture of carrier-free radioisotopes into rats bearing left tibial fractures in a state of repair showed that the avidity of healing bone for ^{91}Y increased sharply during a 3–10-day interval; the ^{35}S deposition became most effective, whereas the retention values for ^{91}Y increased only slightly, if at all (MacDonald *et al.*, 1957; Fig. 7-17). Yttrium fixation in the healing bone was more closely related to events leading to organic matrix formation than to mineralization of the new tissue. The data suggested that formation of the sulfated mucopolysaccharides of osteoid matrix and fibro-cartilaginous callus began by at least the 3rd-day, if not earlier. Another investigation showed that after i.p. injection of ^{91}Y to puppies, the radioisotope taken up in the skeleton appeared on nongrowing bone surfaces (Jowsey *et al.*, 1958). The radioisotope remained in the inorganic fraction of the bone. Jowsey *et al.* (1958) concluded that bone mineral does take up REE from plasma *in vivo* and that they are localized on the available resorbing and inactive surfaces of bone tissue. The assertion that yttrium binds mainly to inactive and mineral phases of the bone structures disputes the result of *in vitro* experiments in which ^{91}Y deposited in the skeleton was found to be present in the bone collagen fractions: the radioisotope bonded with the osteoid base was combined with mucopolysaccharides (Razumovsky *et al.*, 1960). The yttrium deposition in bone was related to its ability to enter the reticuloendothelial system and the inorganic bone structure (Ramsden, 1961; Herring *et al.*, 1962; Triffitt, 1985). Intravenous administration of ^{88}Y to rats as citrate, diphosphonate, or chloride complexes demonstrated a high affinity to the epiphyses of the femur (Kutzner *et al.*, 1981; Fig. 7-18). Adding inactive yttrium to the Y^{3+} citrate complex that was i.v. administered to rats caused a reduction in bone uptake and an increased accumulation in the liver, where it was transformed into a colloidal form. The uptake of yttrium in the lung indicated the formation of macroparticles of this metal (Kutzner *et al.*, 1986).

A considerable amount of yttrium radioisotopes administered intravenously to tumor-bearing mice was eliminated via urine, whereas much smaller amounts

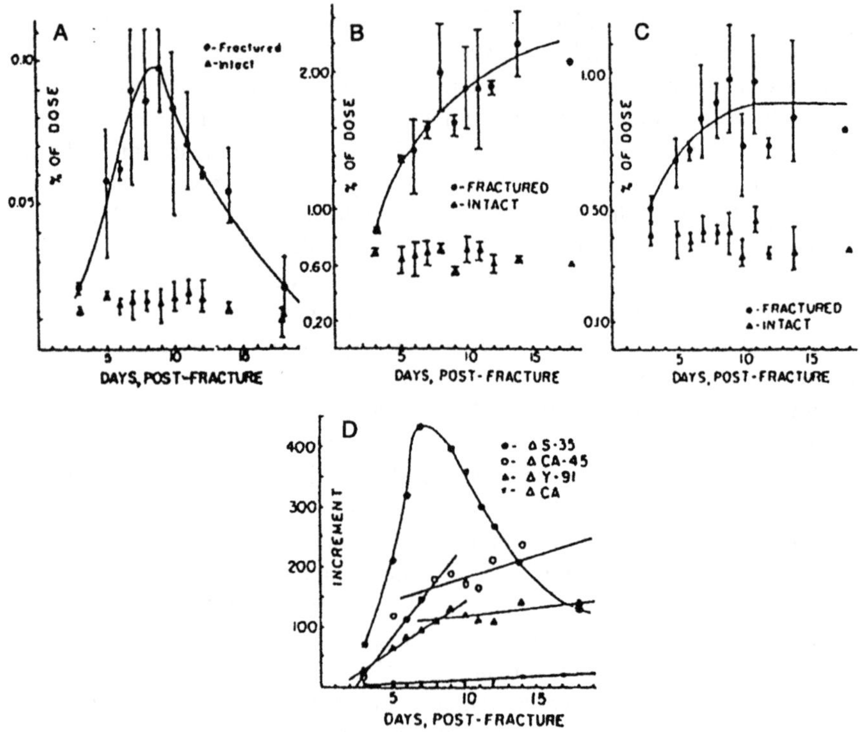

Figure 7-17. Percentage of injected doses of ^{35}S (A), ^{45}Ca (B), and ^{91}Y (C) recovered in rat tibias. Rats were injected 14 days after fracture. Each point represents the mean value and each vertical line represents the range of values for that group of rats. Panel (D) shows the incremental changes in the amounts of calcium and of radioisotopes in healing tibial fractures; e.g., ΔCa = (mg calcium in fractured tibia − mg calcium in intact tibia)(100)/(mg calcium in intact tibia). The mean value of ΔCa was 10, while the mean value of Δ^{45}Ca was 238. Reprinted from MacDonald *et al.* (1957), with permission of the American Physiological Society.

were found, in decreasing order, in femur, kidney, and liver (Beyer *et al.*, 1992). Of interest are the relatively large differences in the biodistribution of ^{87}Y and ^{88}Y, evidently because of their physical and chemical properties (Table 7-10). The lower renal excretion of ^{88}Y was due to pathological states causing a high retention in the kidneys. The liver accumulation of radioactive yttrium was clearly lower with ethylenediaminetetramethylenephosphonic acid (EDTMP) as ligand than with citrate as ligand. The ability of this ligand to protect against M^{3+} accumulation may have practical importance with respect to diagnostic imaging and endo-radionuclide therapy (Beyer *et al.*, 1992).

Administration of yttrium in various forms to human subjects showed, following i.v. injection of the ionized form, the occurrence of high and sustained

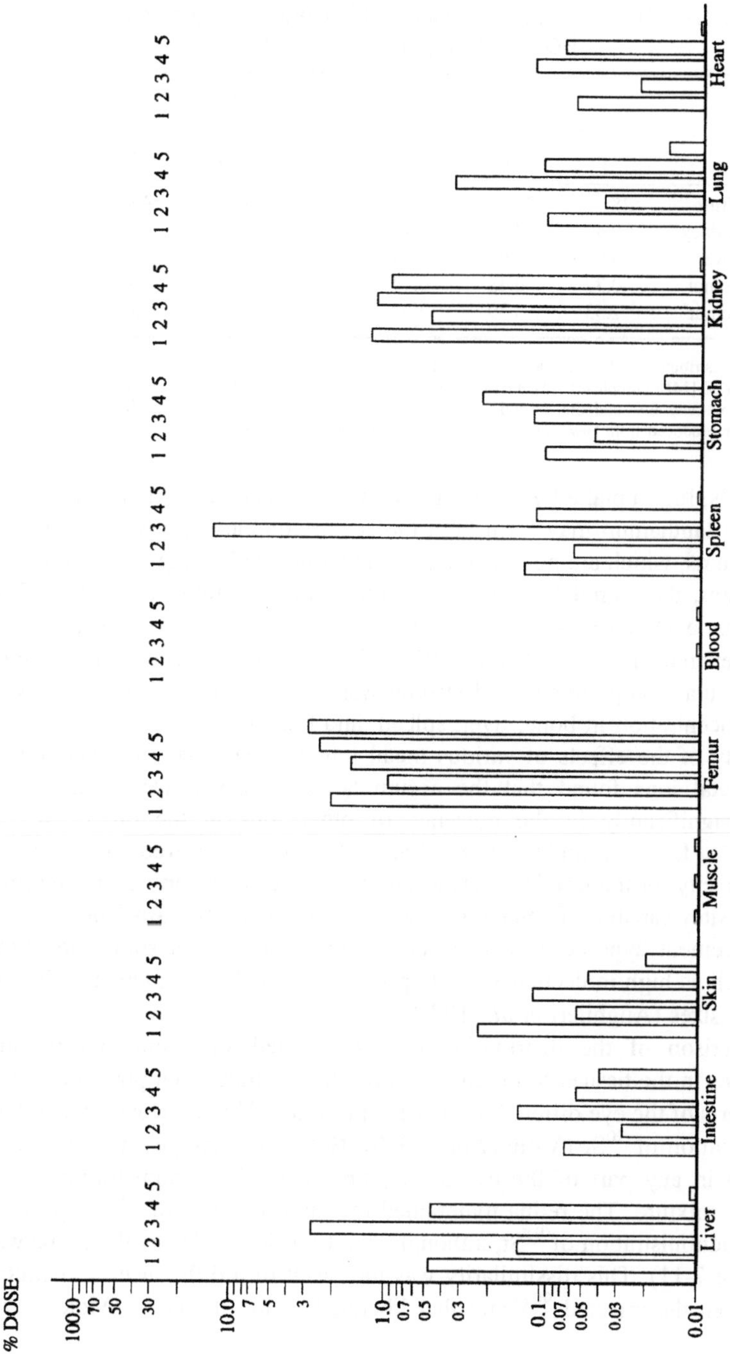

Figure 7-18. Percent accumulation of yttrium complexes in various organs of rats, 48 hours after i.v. injection of ^{88}Y-labeled compounds. Reprinted from Kutzner *et al.* (1981), with permission of the authors and F. K. Schattauer Verlagsgesellschaft mbH.

Table 7-10. Biodistribution (%/g) of ^{87}Y and ^{88}Y Injected as EDTMP Complexes in Rats[a]

	^{88}Y			^{87}Y		
	1 h	2 h	4 h	1 h	2 h	4 h
Urine	28	24	27	24	45	45
Kidney	0.4	2.6	2.1	0.3	0.3	0.4
Liver	0.03	0.05	0.54	0.93	0.03	0.03
Tumor	0.37	0.48	0.44	0.39	0.29	0.34
Femur	2.53	3.39	3.92	2.24	2.63	3.64

[a] Distribution of ^{87}Y- and ^{88}Y-ethylenediaminetetramethylenephosphonic acid (EDTMP) complexes was measured 5 h after second injection. The time intervals between first and the second injection are indicated above each column. Data reproduced from Beyer *et al.* (1992), with permission of Elsevier Science.

plasma levels, high uptake by the liver, and a low urinary excretion (Fig. 7-19). Yttrium and lanthanum may form high molecular complexes, possibly with the plasma proteins, which are not excreted by the kidney. Following i.v. injection of metal chelates, the cumulative urinary excretion ranged from $<5\%$ to $>85\%$; ^{90}Y-NTA was excreted least, while ^{90}Y-EDTA in the presence of excess Ca^{2+}-EDTA was excreted most (Hart *et al.*, 1955). There was no evidence of net reentry into the vascular compartment, and yttrium was localized in various tissues with areas of concentration in bone, liver, spleen, and kidney.

A study of the effects of various metal ions on the *in vivo* distribution of trivalent metal ions found that the accumulation of carrier-free ^{67}Ga citrate decreased significantly in the presence of other ions in the order: $In^{3+} > Sc^{3+} > Fe^{3+} > La^{3+}$ (Anghileri *et al.*, 1982). Most tissues showed a competitive binding capacity for indium, iron, and scandium citrate. This implies the presence of binding sites capable of interacting with ions of their size. An isomorphous ionic replacement hypothesis was presented, which stated that some biological ligands, such as high-molecular-weight species, are capable to interact with ions of different sizes (Anghileri *et al.*, 1982).

Comparison of the distribution of accumulated strontium and yttrium radioisotopes in the tissues of the rabbit's eye, demonstrated that physiologically different parts of the eye differed in their retention of ^{91}Y to a greater extent than in their retention of ^{90}Sr (Mole *et al.*, 1959). Both radioisotopes were rarely in equilibrium in any part of the eye at any time after the administration of an equilibrium mixture. The retina maintained the largest excess of ^{90}Y, especially after daily administration of ^{91}Sr, although direct uptake of ^{91}Y by the retina was small (Table 7-11). This dissimilarity was the result of a difference in chemical form between the injected ^{90}Y and that liberated within the body (Mole *et al.*, 1959).

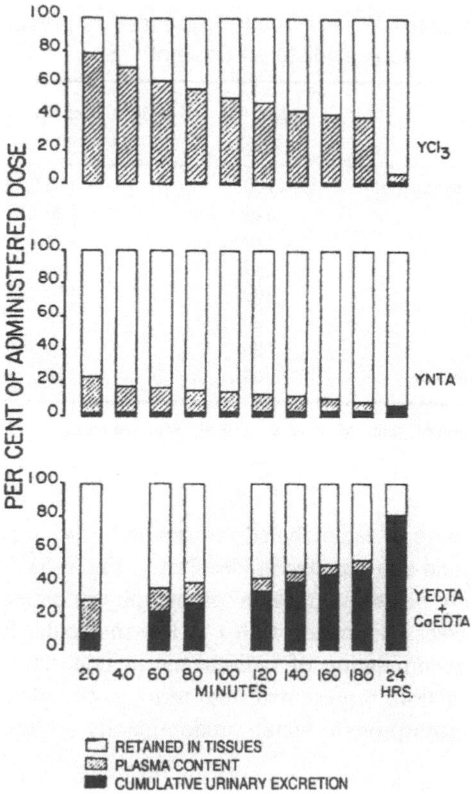

Figure 7-19. Changes in ^{90}Y distribution, as a function of time, in liver, bone, lung, and other tissues of patients to whom yttrium chelates were administered. Plasma content and cumulative urinary excretion are also plotted. Reprinted from Hart *et al.* (1955), with permission of Mosby-Year Book, Inc.

TE may be incorporated into skin, hair, and nails from several sources, both exogenous and endogenous. The mechanism of TE attachment to chemical groups of the hair structure is poorly understood. It is believed that the most of the metal content in human hair exists as an integral part of fiber structure, i.e., in salt bridges or coordination complexes with the side chains of pigments or proteins (Dawber, 1991).

Estimates of the bremstrahlung radiation dose of ^{90}Y indicated that the contribution of the absorbed dose from an organ to itself is very small, compared to that from the beta dose; nonetheless, the contribution to other organs is not always negligible (Table 7-12), especially when large amounts of ^{90}Y may be involved (Stabin *et al.*, 1994).

Table 7-11. Amount of ^{90}Y in Various Eye Tissues of Rabbits, at the Time of Death, Expressed as Percent of the Administered Dose of ^{90}Sr[a]

Tissue	Mean for 30 days	Mean after continuous oral administration
Iris with ciliary body	100	110
Retina	310	800
Choroid	110	140
Cornea	9	—
Sclera	100	150
Vitreous	20	20
Lens	80	90
Striated muscle	60	120

[a] Data reprinted from Mole *et al.* (1959), with permission of *Nature* (London).

The *in vitro* binding of radiolabeled yttrium and other metal ions to human blood serum protein and apo-transferrin (see Part 1, Fig. 6-5) was further studied *in vivo* in mice and by positron-emission tomography in patients suffering from multiple bone metastases (Neumaier, 1996). A low-molecular fraction was found responsible for the accumulation of radiometals in bone and bone metastases; a high-molecular fraction represented the ternary complex between $^{90}Y^{3+}$, PO_4^{3-} and human serum protein, which predominantly accumulated in the liver (Fig. 7-20).

7.6 Summary

The predominant view is that most biochemical responses of isolated cells, tissues, and organs to lanthanides, and apparently scandium as well as yttrium, can be explained in terms of their interactions with the outer surface of the plasmalemma. It is largely believed that, despite their avid binding to the cell surface, Ln^{3+} ions are unable to penetrate natural and artificial bilayers and, as a result, cannot enter healthy cells. Unlike the accepted view of the biochemistry of essential trace elements, this concept applies also to the majority of all other nonessential trace elements. However, recent findings show that in several microorganisms scandium and yttrium may penetrate through the membrane and become involved in the functions of the cell.

Scandium and yttrium show similar high affinities for and binding capacities to the native outer membrane of many microorganisms. At the same time, these two metals show different capacities to bind to blue algae, with diverse binding

Table 7-12. "S", Absorbed Dose per Unit Cumulated Activity for Bremsstrahlung Dose to Various Target Organs from a Uniform Source of ^{90}Y in the Liver and of a Standard Adult Phantom Spleen[a]

Target organ	S-value in the liver (mGy/MBg-h)	S-value in the spleen (mGy/Mbq-h)
Adrenals	1.01×10^{-4}	1.04×10^{-4}
Brain	3.70×10^{-7}	3.70×10^{-7}
Breast	1.64×10^{-5}	1.09×10^{-5}
Gallbladder wall	2.07×10^{-4}	2.69×10^{-5}
Lower large intestine wall	3.65×10^{-6}	1.00×10^{-5}
Small intestine	2.55×10^{-5}	2.20×10^{-5}
Stomach wall	3.27×10^{-5}	1.74×10^{-4}
Upper large intestine wall	4.16×10^{-5}	2.19×10^{-5}
Heart wall	5.24×10^{-5}	3.57×10^{-5}
Kidneys	6.65×10^{-5}	1.61×10^{-4}
Liver	6.30×10^{-4}	7.27×10^{-5}
Lungs	4.81×10^{-5}	3.86×10^{-5}
Muscle	1.81×10^{-5}	2.56×10^{-5}
Ovaries	8.60×10^{-6}	7.68×10^{-6}
Pancreas	8.38×10^{-5}	3.24×10^{-4}
Red marrow	1.95×10^{-5}	2.01×10^{-5}
Bone surfaces	2.38×10^{-5}	2.46×10^{-5}
Skin	9.14×10^{-6}	9.22×10^{-6}
Spleen	1.59×10^{-5}	3.89×10^{-3}
Testes	6.97×10^{-7}	8.62×10^{-7}
Thymus	1.34×10^{-5}	7.73×10^{-6}
Thyroid	2.37×10^{-6}	2.16×10^{-6}
Urinary bladder wall	3.41×10^{-6}	2.67×10^{-6}
Uterus	7.41×10^{-6}	6.49×10^{-6}

[a] Data reproduced from Stabin *et al.* (1994), with permission of the Society of Nuclear Medicine.

sites in chromatin and actin. Also, differences exist in their affinities for inhibiting bacteriorhodopsin and their binding ratios to transferrin. Few studies have been made to compare scandium with the lanthanides, whereas the properties of yttrium have more often been paralleled to those of the more explored lanthanide series. When the bonding of Y^{3+} to organic substrates is compared with that of the Ln^{3+} ions, the binding constants of Y^{3+} are similar to those of ions of the first half of the lanthanide series.

The well-known phenomenon of "bone-seeking elements," such as yttrium and plutonium, is explained by their ability to bind to the bone inorganic fraction and bind phosphorus-containing compounds and mucopolysaccharides. Also, fluoride complexes of scandium and yttrium may be involved in the binding of

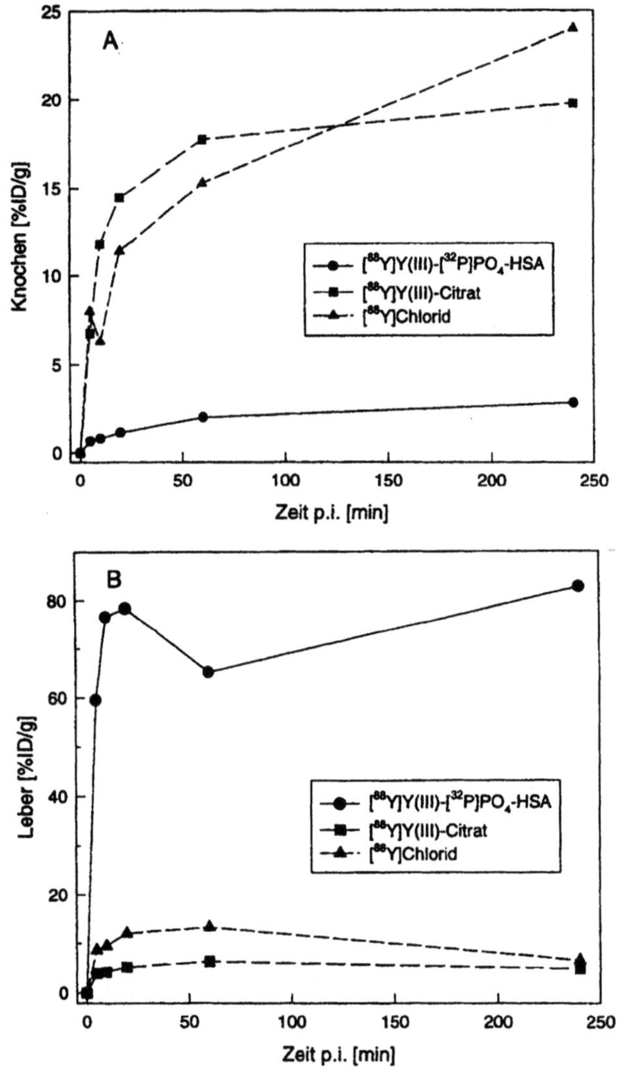

Figure 7-20. Pharmacokinetics of ternary complex between $^{90}Y^{3+}$, PO_4^{3-}, and human serum protein (●), ^{90}Y citrate (□), and $^{90}YCl_3$ (▲) in bones (A) and liver (B) of NMRI mice. Reprinted from Neumaier (1996), with permission of the author and Institut für Nuklearchemie, Jülich.

yttrium with bone tissues. The liver, kidney, and spleen are also known for their capacity to accumulate yttrium, as well as lanthanides and other trivalent metal ions. Detailed experiments revealed the localization of the scandium citrate complex in mitochondrial and nuclear fractions.

The high tumor-specific incorporation and much lower toxicity of ^{90}Y- and ^{47}Sc-labeled antibodies compared to other radioisotopes demonstrated in numerous investigations warrant further evaluation of the therapeutic efficacy of these compounds. The formation of artificially produced cationic complexes of siderophores with scandium, yttrium, and other lanthanides offers new possibilities in the study of weaker bonding interactions of various complexes with metal ions in solution. Such types of interactions may be also of significance in biosorption and migration processes in microorganisms under natural conditions. Extensive applications of scandium and especially of yttrium compounds and isotopes for labeling and as agents for immunoassay or for therapy require further studies of the biochemical and physiological implications of these biomedical applications. Novel approaches have been developed using inorganic biochemistry, genetic engineering, immunology, and molecular biology, with the objectives of regulating the concentrations of metal ions in various tissues and modifying the biological response to metal ions.

Several metals may interfere with specific interactions that exist between important molecules. Such interactions include those between proteins or nucleic acids and small molecules or polymeric systems, such as enzyme–substrate, hormone–receptor, neurotransmitter–receptor, and drug–receptor interactions, as well as interactions in membranes and supramolecular systems, and hydrophobic interactions. The essential characteristics of living systems are dominated by numerous regulatory and coordination processes, in which long-range interactions and long-range correlations are required. The finding of multiple interactions between multivalent ions and biomembranes shows the need for further exploration of the mechanisms in membrane transport systems. These systems apparently exist with directional, energetic, specificity, and other properties tailored to cellular needs. So far, only limited data have been reported for scandium and yttrium.

References

Anbar, M., and Ernst, N., 1962. A distribution study of ^{18}F-labeled Ca ions fluorocomplexes in rats, *Int. J. Appl. Radiat. Isot.* 13:47–51.

Anderson, W. T., and Strand, M., 1985. Stability, targeting and biodistribution of ^{46}Sc- and ^{67}Ga-labeled MAb in erythroleukemic mice, *Cancer Res.* 45:2154–2158.

Anderson, W. T., and Strand, M., 1987. Radiolabeled antibody. Iodine versus radiometal chelates, *Natl. Cancer Inst. Monogr.* 3:149–151.

Ando, A., Ando, I., Yamada, N., Hiraki, T., and Hisada, K., 1987. Distribution of ^{46}Sc and ^{51}Cr in tumor-bearing animals and the mechanism of accumulation in tumor and liver, *Nucl. Med. Biol.* 14:143–151.

Anghileri, L. J., Thouvenot, P., Brunotte, F., Marchal, C., and Robert, J., 1982. Effects of various cations on the *in vivo* distribution of ^{67}Ga citrate, *Int. J. Nucl. Med. Biol.* 9:195–196.

Antoniw, P., Farnsworth, A. P. H., Turner, A., Haines, A. M. R., Mountain, A., Mackintosh, J., Shochat, D., Humm, J., Welt, S., Old, L. J., Yarranton, G. T., and King, D. J., 1996. Radioimmunotherapy of colorectal carcinoma xenografts in nude mice with ^{90}Y A33IgG and Tris-Fab, *Br. J. Cancer* 714:513–524.

Appanna, V. D. and Huang, J., 1992a. Bio-accumulation of yttrium, in *Pseudomonas fluorescens*. *Bull. Environ. Contam. Toxicol.* 49:620–625.

Appana, V. D., and Huang, J., 1992b. Microbial precipitation of yttrium, *Microbios* 72:129–136.

Arbab, A. S., Kaizumi, K., and Araki, T., 1995. A revised method of labeling mouse IgG with ^{90}Y, *Yamanashi Med. J.* 10:121–124.

Ariki, M. and Lanyi, J. K., 1986. Characterization of metal ion-binding site in bacteriorhodopsin, *J. Biol. Chem.* 261:8167–8174.

Banks, P., Bartley, W., and Birt, L. M., 1976. *The Biochemistry of the Tissues*, John Wiley and Sons, London.

Benz, R., 1994. Uptake of solutes by rough bacterial outer membranes, in: *Bacterial Cell Wall* (J. M. Ghuysen and R. Hakenbeck, eds.), Elsevier, Amsterdam, pp. 397–423.

Beveridge, T. J., 1978. The response of cell walls of *Bacillus subtilis* to metals and to electron microscopic stains, *Can. J. Microbiol.* 24:89–104.

Beveridge, T. J., 1989. Metal ions and bacteria, in: *Metal Ions and Bacteria* (T. J. Beveridge and R. J. Doyle, eds.), John Wiley and Sons, New York, pp. 1–39.

Beveridge, T. J., and Koval, S. F., 1981. Binding of metals to cell envelopes of *Escherichia coli* K-12, *Appl. Environ. Microbiol.* 42:325–335.

Beyer, G. J., Bergmann, R., Kampf, G., Mäding, P., and Rösch, F., 1992. Simultaneous study of the biodistribution of radio-yttrium complexed with EDTMP and citrate ligands in tumor-bearing rats, *Nucl. Med. Biol.* 19:201–203.

Bingham, D., and Dobrota, M., 1995. Binding of lanthanides to cell membranes in the presence of ligands, *J. Inorg. Biochem.* 59:39–52.

Bloomfield, V., and Carpenter, I. L., 1993. Biological polyelectrolytes, in: *Polyelectrolytes* (M. Hara, ed.), Marcel Dekker, New York, pp. 77–125.

Buchsbaum, D. J., Lawrence, T. S., Roberson, P. L., Heidorn, D. B., Ten, H., Randall, K., and Steplewski, Z., 1993. Comparison of ^{131}In and ^{90}Y-labeled MAb 17-1A for treatment of human colon cancer xenografts, *Int. J. Radiat. Oncol. Biol. Phys.* 25:629–638.

Calabrese, E. J., 1991. *Multiple Chemical Interactions*, Lewis Publishers, Chelsea, pp. 9–18.

Camera, L., Kinuya, S., Gamestani, K., Wu, C., Brechbiel, M. W., Pai, L. H., McMurry, T. J., Gansow, O. A., Pastan, I., Paik, C. H., and Carrasquillo, J. A., 1994. Evaluation of the serum stability and in vivo biodistribution of CHX-DTPA and other ligands for yttrium labeling of MAb, *J. Nucl. Med.* 35:882–889.

Collins, Y. E., and Stotzky, G., 1989. Interaction of microbial cell walls with metals, in: *Metal Ions and Bacteria* (T. J. Beveridge and R. J. Doyle, eds.), John Wiley and Sons, New York, pp. 38–52.

Colombini, M., 1991. Aluminum and membrane channels, in: *Aluminum in Chemistry, Biology and Medicine* (M. Nicolini *et al.*, eds.), Cortina International, Verona, pp. 33–43.

Crist, R. H., Oberholser, K., McGarrity, J., Crist, D. R., Johnson, J. K., and Brittsan, J. M., 1992. Interaction of metals and protons with algae with emphasis on lead and aluminum, *Environ. Sci. Technol.* 26:496–502.

Crook, J. E., Washburn, L. C., Lee, Y. C. C., Sun, T. T. H., Byrd, B. L., Holloway, E. C., Ju, H. S. J., and Steplevski, Z., 1990. Review of ^{90}Y radiolabeled antibodies, in: *Metal Ions and Medicine* (P. Collery *et al.*, eds.), Libbey Eurotext, Paris, pp. 538–542.

Dawber, R., 1991. Hair follicle structure, keratinization and the physical properties of hair, in: *Diseases of the Hair and Scalp* (A. Rook and R. Dawber, eds.), Blackwell Scientific, London, pp. 18–41.

De Jong, M., Bakker, W. H., Krenning, E. P., Breeman, W. A., van der Pluijm, M. E., Bert, B., Visser, T. J., Eduard, J., and Martini, B., 1997. ^{90}Y and ^{111}In labeling, receptor binding and biodistribution of DOTA^{0},D-Phe^{1},Tyr^{3}-octreotide, a promising somatostatin analog for radionuclide therapy, *Eur. J. Nucl. Med.* 24:368–371.

DeNardo, S. J., *et al.*, 1995. Pharmacokinetics of chimeric L6 conjugated to ^{111}In- and ^{90}Y-DOTA-peptide in tumor-bearing mice, *J. Nucl. Med.* 36:829–836.

Dingle, J. T., and Gordon, J. L., 1981. *Cellular Interactions*, Elsevier, Amsterdam.

Dudley, H. C., 1955. Influence of chelates on the metabolism of ^{90}Y, *J. Lab. Clin. Med.* 45:792–799.

Dudley, H. C., and Greenberg, J., 1958. Influence of chelates on the metabolism of radioyttrium, *J. Lab. Clin. Med.* 52:533–540.

Evans, C. H., 1990. Interactions of lanthanides with tissues, cells and cellular organelles, in: *Biochemistry of the Lanthanides*, Plenum Press, New York, pp. 211–283.

Falke, J. J., Snyder, E. E., Thatcher, K. C., and Voertler, C. S., 1991. Quantitating and engineering the ion specificity of an EF-hand-like Ca^{2+} binding site, *Biochemistry* 30:8690–8697.

Fan, B. T., Xiao, L., and Lapluye, G., 1992. Interaction of metallic ions with basement membranes, *J. Chim. Phys.* 89:1911–1930.

Ford-Hutchinson, A. W., and Perkins, D. J., 1971. The binding of scandium ions to transferrin *in vivo* and *in vitro*, *Eur. J. Biochem.* 21:55–59.

Ford-Hutchinson, A. W., and Perkins, D. J., 1972. ^{46}Sc metabolism. Binding to metalloproteins *in vivo* and *in vitro*, *Radiat. Res.* 51:244–248.

Gansow, O. A., 1991. Newer approaches to the radiolabeling of MAb by use of metal chelates, *Nucl. Med. Biol.* 18:369–381.

Goodwin, D. A., Meares, C. F., Watanabe, N., McTigue, M., Chaovapong, W., McK. Ransone, C., Renn, O., Greiner, D. P., Kukis, D. L., and Kronenberger, S. I., 1994. Pharmacokinetics of pretargeted MAb 2D12.5 and ^{88}Y-Janus 2-DOTA in BALB/c mice with KHJJ mouse adenocarcinoma. A model for ^{90}Y radioimmunotherapy, *Cancer Res.* 54:5937–5946.

Hardie, D. G., 1991. *Biochemical Messengers. Hormones, Neurotransmitters and Growth Factors*, Chapman and Hall, London.

Harding, S. E., 1989. Modelling the gross conformation of assemblies using hydrodynamics. The whole body approach, in: *Dynamic Properties of Macromolecular Assemblies* (S. E. Harding and A. J. Rowe, eds.), Royal Society of Chemistry, London, pp. 32–73.

Harris, N., Exner, T., Hughes, G. R. V., and Asherson, R. A., 1991. *Phospholipid-Binding Antibodies*, CRC Press, Boca Raton, Florida.

Harrison, A., Walker, C., Parker, D., Jankovski, K., Cox, J., Craig, A., Sansom, J., Beeley, N., Boyce, R., Chaplin, L., Eaton, M., Farnsworth, A., Millar, K., Millican, A., Randall, A., Rhind, S., Secher, D., and Turner, A., 1991. The *in vivo* release of ^{90}Y from cyclic and acyclic ligand–antibody conjugates, *Nucl. Med. Biol.* 18:469–476.

Hart, H. E., Greenberg, J., Lewin, R., Spencer, H., Stern, K. G., and Laszlo, D., 1955. Metabolism of lanthanum and yttrium chelates, *J. Lab. Clin. Med.* 46:182–192.

Herring, G. M., Vaughan, J., and Williamson, H., 1962. Preliminary report on the localization and possible binding agent for yttrium, americium and plutonium in cortical bones, *Health Phys.* 8:717–724.

Hnatowich, D. J., Virzi, F., and Doherty, P. W., 1985. DTPA-coupled antibodies labeled with ^{90}Y, *J. Nucl. Med.* 26:503–509.

Hoyle, B. D., and Beveridge, T. J., 1983. Binding of metallic ions to the outer membrane of *Escherichia coli*, *Appl. Environ. Microbiol.* 46:749–752.

Hoyle, B. D., and Beveridge, T. J., 1984. Metal binding by the peptidoglycan sacculus of *Escherichia coli* K-12, *Can. J. Microbiol.* 30:204–211.
Hughes, M. N., and Poole, R. K., 1989. *Metals and Microorganisms*, Chapman and Hall, London.
Inoue, Y., and Gokel, 1990. *Cation Binding by Macrocycles*, Marcel Dekker, New York.
Jowsey, J., Rowland, R. E., and Marshall, J. H., 1958. The deposition of the rare earths in bone, *Radiat. Res.* 8:490–501.
Karuckstis, K. K., 1996, personal communication, with permission.
Karuckstis, K. K., and Gruber, S. M., 1986. Effect of trivalent lanthanide cations on chlorophyll fluorescence and thylakoid membrane stacking, *Biochim. Biophys. Acta* 851:322–326.
Karuckstis, K. K., Kao, M. Y., Savin, D. A., Bittker, R. A., Klaphengst, K. J., Emetarom, C. M., Naito, N. R., and Takamoto, D. Y., 1995. Spectral studies of lanthanide interaction with membrane surfaces, *J. Phys. Chem.* 99:4339–4346.
Kirchgessner, M., Reichlmayr-Lais, A. M., and Schwarz, F. J., 1981. Interactions of trace elements in human metabolism, in: *Nutrition in Health and Disease and International Development* (A. E. Harper and G. K. Davis, eds.), Alan R. Liss, New York, pp. 180–183.
Kosmas, C., Maraveyas, A., Gooden, C. S., Snook, D., and Epenetos, A. A., 1995. Anti-chelate antibodies after i.p. injection of ^{90}Y-labeled MAb immunoconjugates for ovarian cancer therapy, *J. Nucl. Med.* 36:746–753.
Kozak, R. W., Raubischek, A., Mirzadeh, S., Brechbiel, M. W., Junghaus, R., Gansow, O. A., and Waldmann, T. A., 1989. Nature of the bifunctional chelating agent used for radioimmunotherapy with ^{90}Y MAb, *Cancer Res.* 49:2639–2644.
Kutzner, J., Mittas, M., Grimm, W., 1981. Distribution in rats after i.v. injection of ^{88}Y compounds, *Nucl. Med.* 22:35–39 (German).
Kutzner, J., Alfter, G., and Grimm, W., 1986. Effect of carrier on the organ distribution of radioactive yttrium in rat, *Nucl. Med.* 25:78–80 (German).
Loewenstein, W. R., 1989. From cell theory to cell connectivity. Experiments in cell-to-cell communication, in: *Membrane Transport. People and Ideas* (D. C. Tosteson, ed.), American Physiological Society, Bethesda, Maryland, pp. 291–302.
MacDonald, N. S., Lorick, P. C., and Petriello, L. I., 1957. Healing bone fractions and simultaneous administration of radioisotopes of sulfur, calcium and yttrium., *Am. J. Physiol.* 191:185–188.
Mardirossian, G., Wu, C., and Hnatowich, D. J., 1993. The stability in liver homogenates of ^{111}In and ^{90}Y attached to antibody via two popular chelators, *Nucl. Med. Biol.* 20:65–74.
Marshall, K. C., 1991. The importance of studying microbial cell surfaces, in: *Microbial Cell Surface Analysis* (N. Mozes, ed.), VCH, Weinheim, pp. 3–20.
Mehlhorn, R. J., 1986. The interaction of inorganic species with biomembranes, in: *The Importance of Chemical Speciation in Environmental Processes* (M. Bernhard *et al.*, eds.), Springer-Verlag, Berlin, pp. 85–97.
Moeller, T., 1973. The lanthanides, in: *Comprehensive Inorganic Chemistry* (J. C. Bailar *et al.*, eds.), Pergamon Press, New York, Vol. 4, pp. 28–39.
Moi, M. K., DeNardo, S. J., and Meares, C. F., 1990. Stable bifunctional chelates of metals used in radiotherapy, *Cancer Res.* 50(Suppl.):789s–793s.
Mole, R. H., Pirie, A., and Vaughan, J. M., 1959. Differential distribution of radioactive strontium and yttrium in the tissues of the rabbit's eye, *Nature* 183:802–807.
Nachsen, D. A., 1984. Selectivity of the calcium binding site in synaptosome calcium channels. Inhibition of calcium influx by multivalent metal cations, *J. Gen. Physiol.* 83:941–967.
Needham, J. V., Chen, T. Y., and Falke, J. J., 1993. Novel ion specificity of a carboxylate cluster Mg^{2+} binding site. Charge selectivity and weak size selectivity, *Biochemistry* 32:3363–3367.
Neumaier, B., 1996. Radiochemical investigation of the binding of proteins and affinity of bone metastases of Y^{3+} complexes, Research Center Jülich, Report 3236, pp. 1–112.

Oftedal, P., 1961. Studies with radioactive yttrium in flies. 1. Retention and distribution in *Drosophila* after injection. 2. Retention and distribution in *Drosophila* and *Musca*, *Int. J. Radiat. Biol.* 3:211–221, 222–230.

Papp, S., Rutzke, M., and Martonosi, A., 1985. The effect of chelating agents on the elemental composition of sarcoplasmic reticulum, *Arch. Biochem. Biophys.* 243:254–263.

Peterson, J. A., Couto, J. R., Taylor, M. R., and Ceriani, R. L., 1995. Selection of tumor-specific epitopes on target antigens for radioimmunotherapy of breast cancer, *Cancer Res.* 55(Suppl. 23):5847s–5851s.

Quadri, S. M., Shao, Y., Blum, J. E., Leichner, P. K., Williams, J. R., and Vriesendorp, H. M., 1993. Preclinical evaluation of i.v. administered ^{111}In and ^{90}Y-labeled B72.3 immnoconjugate (GYK-DTPA) in beagle dogs, *Nucl. Med. Biol.* 20:559–570.

Ramsden, E. N., 1961. A review of experimental work on radioyttrium comparing: the tissue distribution, the mechanism of deposition in bone, and the state in the blood, *Int. J. Radiat. Biol.* 3:399–410.

Razumovsky, N. O., Torchinskaya, O. L., and Balabukha, V. S., 1960. Type and stability of ^{91}Y bonds in bone tissues, in: *Khimicheskaya Zashchita Organizma ot Ioniziruyushchikh Izluchenii*, Moscow, pp. 130–136 (Russian). CA 55:27492.

Reardan, D. T., Meares, C. F., Goodwin, D. A., McTigue, M., David, G. S., Stone, M. R., Leung, J. P., Bartholomew, R. M., and Frincke, J. M., 1985. Antibodies against metal chelates, *Nature* 316:265–268.

Rice, D. J., and Humphreys, T., 1983. Two Ca^{2+} functions are demonstrated by the substitution of specific divalent and lanthanide cations for the Ca^{2+} required by the aggregation factor complex from the marine sponge *Microciona prolifera*. *J. Biol. Chem.* 258:6394–6399.

Roselli, M., Schlom, J., Gansow, O. A., Raubitschek, A., Mirzadeh, S., Brechbiel, M. W., and Colcher, D., 1989. Comparative biodistribution of yttrium- and indium-labeled MAb B72.3 in athymic mice bearing human colon carcinoma xenografts, *J. Nucl. Med.* 30:672–682.

Rosenblum, M. G., Kavanagh, J. J., Burke, T. W., Warton, J. T., Cunningham, J. E., Shanken, L. J., Silva, E. G., Thompson, L., Cheung, L., Lamki, L., and Murray, J. L., 1991. Clinical pharmacology, metabolism and tissue distribution of ^{90}Y labeled MAb B72.3 after intraperitoneal administration, *J. Natl. Cancer Inst.* 83:1629–1636.

Scheinberg, D. A., Strand, M., and Gansow, O. A., 1982. Tumor imaging with radioactive metal chelates conjugated to MAb, *Science* 215:1511–1513.

Schomaker, K., Hientzsch, B., Franke, W., and Scheidhauer, K., 1995. Studies on the biokinetics of tumor-affine yttrium preparations, *Nuklearmedizin* 34:32–39 (German).

Snyder, E. E., Buoscio, B. W., and Falke, J. J., 1990. Ca^{2+} site specificity: Effect of size and charge on metal ion binding to EF-hand-like site, *Biochemistry* 29:3937–3943.

Stabin, M. G., Eckerman, K. F., Ryman, J. C., and Williams, L. E., 1994. Bremsstrahlung radiation dose in ^{90}Y therapy application, *J. Nucl. Med.* 35:1377–1380.

Stimmel, J. B., Stockstill, M. E., and Kull, F. C., Jr., 1995. ^{90}Y chelation properties of tetraazatetraacetic acid macrocycles, diethylenetriaminepentaacetic acid analogues, and a novel terpyridine acyclic chelator, *Bioconjug. Chem.* 6:219–225.

Stoner, C. D., Sirak, H. D., and Richardson, M., 1978. Magnesium induced inner membrane aggregation in heart mitochondria. Prevention and reversal by carboxyatractyloside and bongkrekic acid, *J. Cell Biol.* 77:417–426.

St. Raymond, P. A., 1990. ^{47}Sc chelated MAb, PhD thesis, University of Maryland. Diss. Abstr. 52:873b.1991.

Takenouchi, K., Watanabe, K., Kato, Y., Koike, T., and Kimura, E., 1993. Novel bifunctional macrocyclic chelating agents appended with a pendant-type carboxymethylamino ligand and nitrobenzyl group and stability of the ^{88}Y complexes, *J. Org. Chem.* 58:1955–1958.

Tosteson, D. C., 1989. Introduction. Membrane transport in physiology, in: *Membrane Transport* (D. C. Tosteson, ed.), American Physiological Society, Bethesda, Maryland, pp. 1–14.

Triffitt, J. T., 1985. Receptor molecules, coprecipitation and ion exchange processes in the deposition of metal ion in bone, in: *Metals in Bone* (N. D. Priest, ed.), MTP Press, Lancaster, England, pp. 3–20.

Uyesaka, N., Kamino, K., Ogawa, M., Ionouye, A., and Machida, K., 1976. Lanthanum and some other cation-induced changes in fluidity of synaptosomal membrane studied with nitroxide stearate spin labels, *J. Membr. Biol.* 27:283–295.

Vallera, D. A., Schmidberger, H., Buchsbaum, D. J., Everson, P., Snover, D. C., and Blazar, B. R., 1991. Radiotherapy in mice with ^{90}Y-labeled anti-Ly1 MAb. Therapy of established graft-versus-host disease induced across the major histocompatibility barrier, *Cancer Res.* 51:1891–1897.

Vaughan, A. T. M., Yankuba, S. C. S., and Anderson, P., 1987. Antibodies labeled with metallic radionuclides, *Natl. Cancer Inst. Monogr.* 3:141–144.

Voycheck, C. L., and Tan, J. S., 1993. Ion-containing polymers and their biological complexes, in: *Polyelectrolytes* (M. Hara, ed.), Marcel Dekker, New York, pp. 299–388.

Walker, P. R., LeBlanc, J., and Sikorska, M., 1989. Effects of aluminum and other cations on the structure of brain and liver chromatin, *Biochemistry* 28:3911–3915.

Williams, R. J. P., 1991. Complexation and catalysis in biology, in: *A Study of Enzymes* (S. A. Kuby, ed.), CRC Press, Boca Raton, Florida, Vol. 2, pp. 83–103.

Wu, C., Virzi, F., and Hnatowich, D. J., 1992. Investigations of N-linked macrocycles for ^{111}In and ^{90}Y labeling of proteins, *Nucl. Med. Biol.* 19:239–244.

Yeagle, P., 1993. *The Membranes of Cells*, Academic Press, Orlando, pp. 1–17.

Yorke, E. D., Beaumier, P. L., Wessels, W. B., Fritzberg, A. R., and Morgan, Jr., A. C., 1991. Optimal antibody-radionuclide combinations for clinical radioimmunotherapy. A predictive model based on mouse pharmacokinetics, *Nucl. Med. Biol.* 18:827–835.

Biochemistry and Physiology of Scandium and Yttrium*

8

> Quite simply, in central biochemistry, everything depends on everything. Many things have to be before a machine, built along the lines of modern organisms, can start to work at all.
>
> A.G. Cairns-Smith, 1982

8.1 Introduction

The concentrations of metal ions in biological systems are selectively controlled by their charge, size, and hard or soft ligand preference in cell metabolism. The uptake pathways of calcium, iron, and a few other metals have been investigated in much detail, but little is known about most other metal ions (Lippard and Berg, 1994). Because of their physical and chemical properties, the biochemical effects of scandium and, in particular, yttrium have been compared with those of the Ln. Attention has been focused on their ability to mimic and interfere with the metabolism of calcium, magnesium, and other essential elements.

Brown *et al.* (1990) reviewed the role of rare earth elements (REE) in biological systems. These elements display a range of physiological and biochemical effects on plant and animal growth and development. The review concludes that the REE bind in biological systems at superficially located Ca^{2+} sites and that their binding at these sites is less reversible than that of Ca^{2+}; therefore, the REE may influence biochemical reactions that involve Ca^{2+} (Brown *et al.*, 1990).

Earlier investigations of the biochemistry and physiology of scandium and yttrium were summarized in books (Evans, 1990), book chapters (Horovitz, 1975; Schroeder and Darrow, 1973; Xu and Xiao, 1985; Nielson, 1986), dissertations (Reece, 1965; Duncan, 1966; Coomes, 1968; Farrar, 1988) and review articles (Spencer, 1960; Ramsden, 1961; Kyker, 1962; Ellis, 1977; Brown *et al.*, 1990; Deuber and Heim, 1991; Horovitz, 1995; Burgess, 1996).

*I am grateful to Professor Forrest H. Neilsen for his suggestions and improvements to this chapter.

8.2 Cytological and Morphological Patterns

No visible morphological changes were noted in several bacterial or yeast cells exposed to low or even relatively high concentrations of yttrium and other REE. However, the hyphae of all filamentous fungi exhibited structural changes that included morphological aberrations, multiple terminal branching, swelling, lateral branching, and the rupture of hyphae strands by a process similar to plasmoptysis (Talburt and Johnson, 1967).

Intravenous (i.v.) administration of 10–50 mg YCl_3/kg to rats did not cause obvious changes in the liver, whereas deposition of a brown substance in the spleen was observed (Fig. 8-1). Correspondingly, higher doses of yttrium caused the formation of xenobiotic granuloma in the spleen (Nakamura *et al.*, 1991a).

Addition of yttrium nitrate to the drinking water of rats changed incisor pigmentation and molar morphology; intraperitoneal (i.p.) injection of this salt caused wider fissures and thicker dentine in molars. These effects were similar to those caused by strontium. Also, an i.v. injection of yttrium before eruption of the

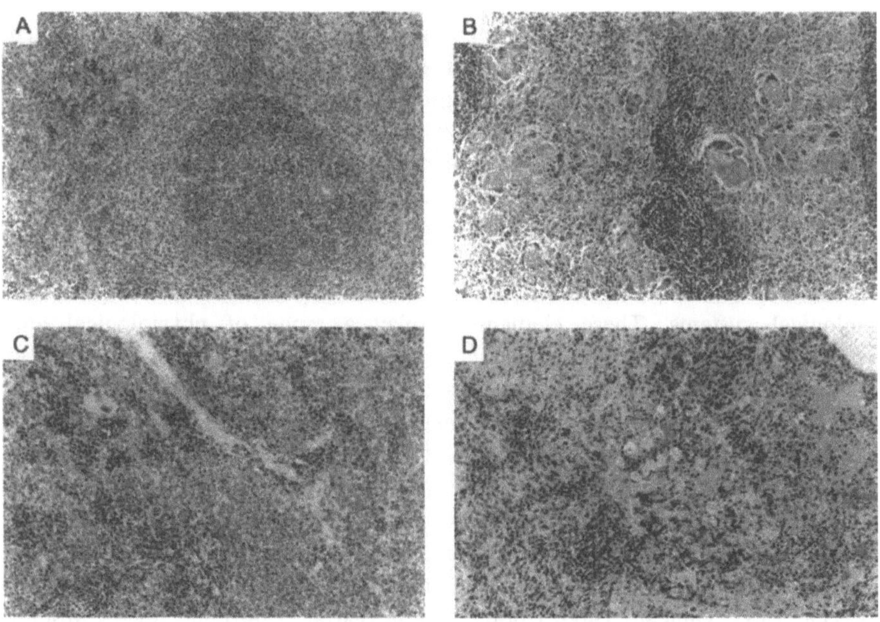

Figure 8-1. Liver and spleen of rats 8 days after i.v. injection of 10 mg YCl_3/kg. A photomicrograph was obtained with hematoxylin and eosin stain, after fixation with 10% formalin. (A) Liver of control; (B) liver of an animal administered yttrium; (C) spleen of control; (D) spleen of an animal administered yttrium. Reprinted from Nakamura *et al.* (1991b), with permission of the authors and the Pharmaceutical Society of Japan.

molars caused an alteration in their morphology, which could modify carries susceptibility. During tooth development, yttrium and strontium may modify the cells responsible for the formation of enamel and dental matrices. Similarly to fluorine, yttrium changed the character of the apatite crystals (Castillo Mercado and Bibby, 1973). Whether a reduction in caries and changes in tooth morphology are caused by the administration of yttrium is unknown (Castillo Mercado and Ludwig, 1973).

Throughout a 12-week period of feeding $ScCl_3$ in the diet of rats, no significant hematological response was observed. At autopsy, the histopathological examination of the internal organs of all groups revealed normal morphological structures, without outward signs of damage attributable to the ingestion of scandium (Haley *et al.*, 1962).

No detrimental effects on the developing ova nor any reduction in fertilizing capacity of rabbit sperm were observed to result from treatment with a $^{90}Sr/^{90}Y$ equilibrium mixture. Yttrium-90 accounted for 99.6% of the activity of the $^{90}Sr/^{90}Y$ mixture and was specifically taken up by sperm tails, but not by the heads of rat sperm. A maximum yttrium uptake of 3×10^7 atoms/rat sperm was calculated (Coomes, 1968).

Following i.v. injection directly into the cavity of cystic craniopharyngioma of four patients, a remarkable resolution of the tumor size occurred (Szeifert *et al.*, 1990). Patho-morphological examination of biopsy specimens revealed that eight months after i.v. instillation the lining epithelial cell layer became destroyed and the cyst was destroyed (Fig. 8-2). A radioautographic investiga-

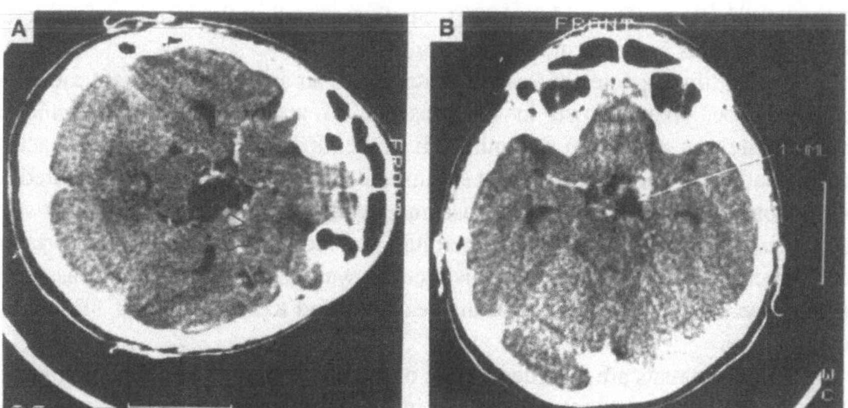

Figure 8-2. Pathomorphological changes in cystic craniopharyngiomas of patients following intracavitary ^{90}Y treatment. (A) Suprasellar cystic craniopharyngiomas before ^{90}Y treatment; (B) eight months after ^{90}Y treatment, the shrunken cyst can be seen. Reprinted from Szeifert *et al.* (1990), with permission of the authors and Springer-Verlag.

tion of the metabolism of ^{90}Y, administered to mice as a chloride salt, found structural differentiation of the inclusions in liver, kidney, and bones (Graul and Hundeshagen, 1959; Fig. 8-3). Supposedly, the accumulation of yttrium in the parenchymal cells of the liver increased with time; also, an increased diffuse activity of the Malpighi corpuscles in the spleen preparations occurred. A radioautographic wreath in the lumina of the alveoli and bronchi and a large accumulation of ^{90}Y in the nuclei of the hypophysis, as well as a special affinity for the Kupffer cells, were observed. The metal ion was presumed to be present in colloidal form (Graul and Hundeshagen, 1959). Intra-articular radiation synovectomy with ^{90}Y silicate applied to adult dogs did not cause morphological abnormalities in articular cartilage. The amounts of water and uronic acid, or the DNA concentration, did not differ from those in cartilage of control knees. However, synovium from irradiated knees displayed areas of necrosis and fibrosis, and less hyaluronate was synthesized *in vitro*, in comparison to contols (Myers *et al.*, 1989).

Other morphological modifications caused by higher concentrations of scandium and yttrium and under pathological conditions are given in Section 8.13, whereas the effects of toxic concentrations are presented in Chapter 9.

8.3 Biochemistry in Lower Organisms (Monera, Protoctista, Fungi)

Information regarding the biochemistry of TE in bacteria and in other lower forms of life is important because it relates, theoretically and practically, to all aspects of biology (Beveridge, 1989). Studies regarding the affinity of bacterial walls for metals are presented in Section 7.2.

Unidentified species of yeast, isolated from an amphibolite on a weathering rock surface, when grown in the presence of ground rock, were found to concentrate iron, scandium, and other TE (Rades-Rohkohl *et al.*, 1979). An active rock-weathering ability of the yeast strains and an uptake mechanism mediated by ionophores and organic acids were assumed. A differentiation was made between solubilized elements present in the medium and those associated with the washed cell fraction. Nevertheless, no distinction was made between an assumed adsorption of TE to the surface of the yeast cell and transportation of TE into the cell.

Microbial mats are regarded as one of the major sites of bacterial biotransfer. Microbial mat systems are widespread in marine intertidal zones, lake bottoms, exposed rock surfaces, and soils. Biotransfer (trapping, enriching, absorbing, precipitation) in microbial mats has been documented for at least 30 chemical elements, including the trivalent metals aluminum, germanium, and uranium, although scandium and yttrium were not investigated (Krumbein, 1986).

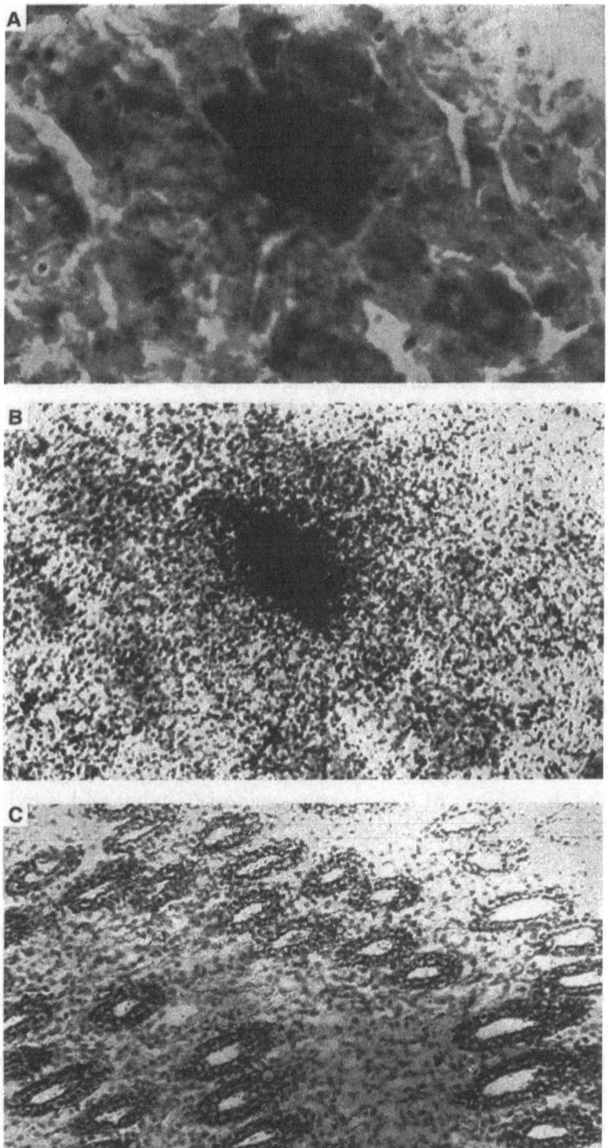

Figure 8-3. Radioautographs of liver and kidney of mice, after ^{90}Y administration. (A) Histology of liver section of Kupffer cells, ×1200; (B) radioautographs of liver cells, ×1200; (C) radioautographs of kidney, showing the zone of the straight collecting tubule, ×400. Reprinted from Graul and Hundeshagen (1959), with permission of Elsevier Science.

During the fermentation of *Saccharomyces cerevisiae*, yttrium and other REE inhibited the uptake of glucose and decreased the production of ethanol and carbon dioxide and the accumulation of dry cell weight. The range of effectiveness of the metals tested showed the following decreasing order of toxicity: $Y > Er > Ho \approx Lu > Eu \approx Ce \approx Dy \approx Nd$. The chelating agent EDTA greatly decreased the effect of these metals on fermentation. During aerobic metabolism, yttrium and other REE inhibited glucose uptake by *S. cerevisae*, but stimulated the utilization of internally stored carbohydrates. Oxygen uptake and carbon dioxide production were always increased beyond that in controls, while the final dry weight of the fermentation mixture was less than that of the original inoculum (Duncan, 1966).

A study in which the fungus *Aspergillus niger* was grown for 4 days in a culture medium containing glycerol as carbon source, and to which various TE were added, showed that copper, iron, gallium, manganese, molybdenum, and zinc were apparently necessary for growth, whatever the source of carbon; however, scandium exhibited biological activity only when the fungus was grown on glycerol (Steinberg, 1939). Addition of scandium caused a doubling of mycelial dry weight. Further, the growth of the pathogenic fungi *Rhizoctonia solani* in a calcium-deficient nutrient solution was stimulated only by scandium and strontium; none of the other 16 TE that were tested seemed capable of substituting for calcium, even partially. Addition of Sc^{3+} to the calcium-deficient nutrient solution at a concentration of 0.1 mg/l caused the largest dry weight production (Steinberg, 1950).

An optimal concentration of 0.01 mg Sc^{3+}/kg produced a 10-fold increase in the mycelial dry weight of *Cercospora granati* (Chahal and Rawla, 1977). The response to scandium was similar to that produced by iron, manganese, molybdenum, and zinc; none of the other chemical elements that were tested (boron, calcium, cobalt, copper, gallium, and tungston) were able to duplicate the specific action of scandium. Even a concentration of 1 ng Sc^{3+}/kg induced a detectable stimulation in growth after an optimum 16 days of incubation. This finding indicated that scandium is unconditionally essential for the growth of *C. granati* with glucose as carbon source (Chahal and Rawla, 1977).

Using the gradient-plate technique, Talburt and Johnson (1967) showed that Y^{3+}, and other REE at a concentration of 11 mM manifested different degrees of inhibition on the growth of five fungi and two bacteria (Fig. 8-4). Whereas *Aspergillus nigricans* and *Rhizopus nigricans* continued their growth across the gradient until a maximum was obtained, the other fungal species stopped growing within a few days. These fungi manifested a lower sensitivity to yttrium and REE than the bacteria (Talburt and Johnson, 1967). Proliferation of *Tetrahymena pyriformis* was stimulated by 10 mg YCl_3/l added to peptone–glucose media (Fig. 8-5). At a concentration of 300 mg/l, YCl_3 was inhibitory in the first 20 h of population growth; later, a partial recovery occurred. Growth was strongly

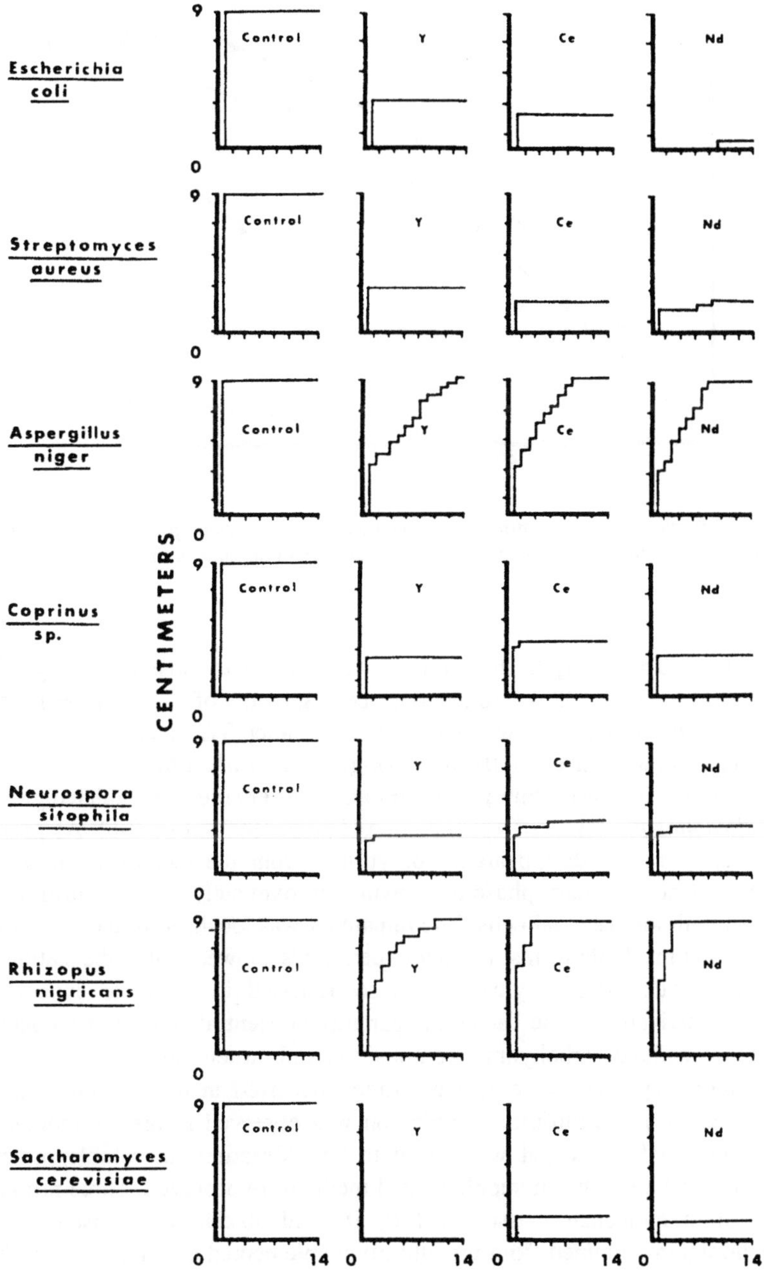

Figure 8-4. Daily growth responses of seven species of fungi and bacteria on yttrium, cerium, and neodymium gradient plates. Reprinted from Talburt and Johnson (1967), with permission of the New York Botanical Garden.

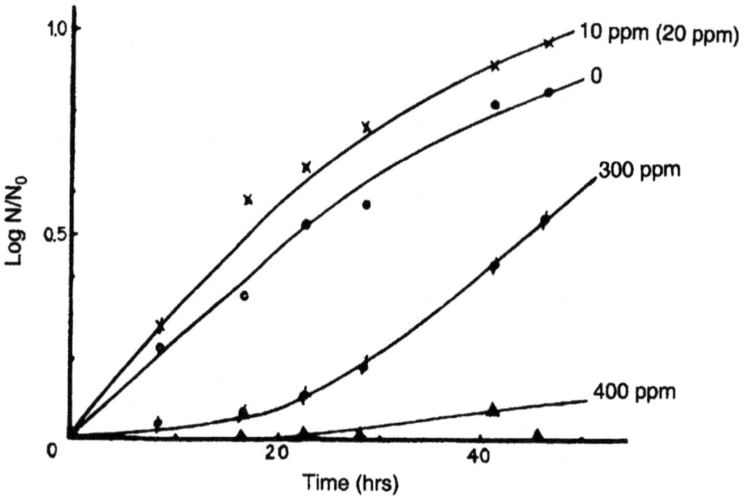

Figure 8-5. Effects of various yttrium concentrations on population growth of *Tetrahymena pyriformis*. Reprinted from Tang *et al.* (1985), with permission of the authors and Humana Press, Inc.

inhibited by 400 mg YCl_3/l. These effects were similar to those caused by other REE (Tang *et al.*, 1985). Analogous cell growth of *T. pyriformis* and *Chlamydomonas reinhardtii* was stimulated by other TE at concentrations in the culture medium between 0.01 and 20 mg/l (Liu *et al.*, 1986).

Addition of 0.5 mM yttrium nitrate into the nutrient media stimulated growth of *Pseudomonas fluorescens* (Appanna and Huang, 1992b; Fig. 8-6). As growth progressed, a gradual disappearance of yttrium from the culture medium was observed. At the stationary phase of growth, approximately 65% of yttrium was associated with the bacterial cells. The remainder was localized in the spent fluid in soluble and insoluble forms. Further experiments showed that at the stationary phase of growth, 90% of yttrium was sequestrated in insoluble exocellular phosphorus derivatives. The bacteria apparently concentrated yttrium intracellularly; the exocellular carbohydrate composition and protein decreased by almost 50%. A more rapid rate of citrate utilization occurred in the yttrium-enriched culture (Fig. 8-7). No yttrium precipitation was observed in the noninoculated medium, even when the pH was raised to 9.0. Variations in soluble proteins between control and yttrium-supplemented media were observed. Three products with apparent molecular masses of 160, 58, and 30 kDa were absent in the electrophoregram obtained from the yttrium-supplemented culture, and a normal protein with an approximate molecular mass of 32 kDa was induced in this medium. The potential use of this bacteria to eliminate soluble yttrium from the environment was emphasized (Appanna and Huang, 1992b).

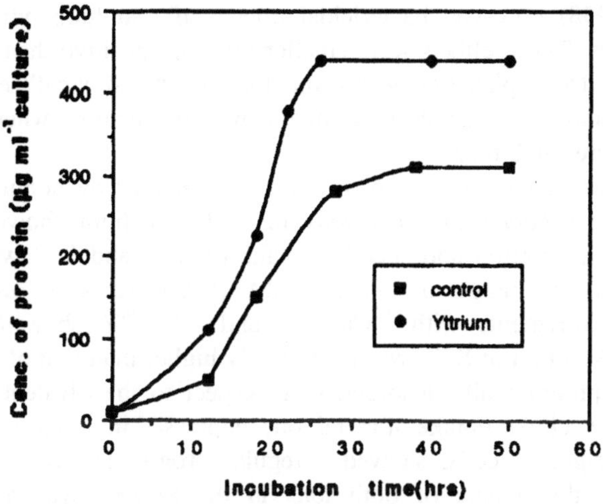

Figure 8-6. Effect of 0.5 mM yttrium on growth of *Pseudomonas fluorescens*. ■, Control medium; ●, yttrium-supplemented medium. Reprinted from Appanna and Huang (1992b), with permission of the authors and Faculty Press.

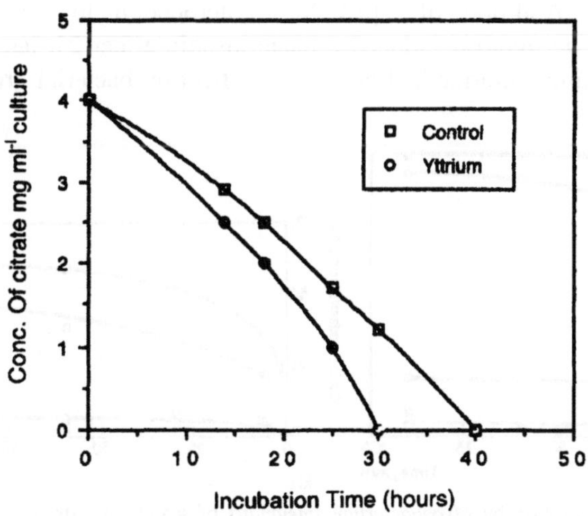

Figure 8-7. Citrate use by *Pseudomonas fluorescens*, in media without or with added yttrium. □, Control medium; ○, yttrium-supplemented medium. Reprinted from Appanna and Huang (1992a), with permission of the authors and Springer Verlag.

Muroma (1958) observed bactericidal effects of scandium, yttrium, and other Ln chlorides. These effects were smaller on gram-positive than on gram-negative bacteria. No explanation was given for the "paradoxical" effect; i.e., higher concentrations occasionally permitted growth of bacteria, whereas lower concentrations were inhibitory.

Measurements of the adsorption of several metal ions by nonproliferative suspensions of *Mycobacterium smegmatis* in acidic medium showed regular trends in percent adsorption and specific uptake of Sr^{2+} and Y^{3+} with carrier concentrations in the range of 50–2000 μM (MacCordick *et al.*, 1990). Comparative measurements with $^{47}Ca/^{47}Sc$ and $^{90}Sr/^{90}Y$ showed that the corresponding adsorption of Sc^{3+} was consistently higher than that of Y^{3+}. Both metal ions were preferentially adsorbed with respect to the divalent Sr^{2+} and Ca^{2+}. The specific cellular uptake of metal ions, expressed in molar units of ions per gram dry weight of cells, showed a regular progression with increasing concentration for the uptake of both ions in the Sr^2/Y^3 system. The ion separation efficiency in the Ca^{2+}/Sc^{3+} system was more pronounced at lower ionic concentrations (Fig. 8-8) and it was highest with the 100 μM solution. The observed trend in cellular uptake, Sc > Y > Eu, was thought to reflect differences in ion retention resulting from covalent bonding and the variation in charge/radius ratio for the trivalent metal ions concerned. The markedly higher and initially more rapid adsorption of the trivalent ions by the bacteria was considered useful for practical purposes of concentrating radioisotopes in effluents and other media (MacCordick *et al.*, 1990).

Pseudomonas aeruginosa represents an important cause of morbidity and mortality in man. Addition of 1 μM Sc^{3+}-enterochelin to serum before inoculation with *P. aeruginosa* induced a bacteriostatic effect 2 h later (Rogers *et al.*, 1984). Scandium chloride had no obvious effect on bacterial growth. The

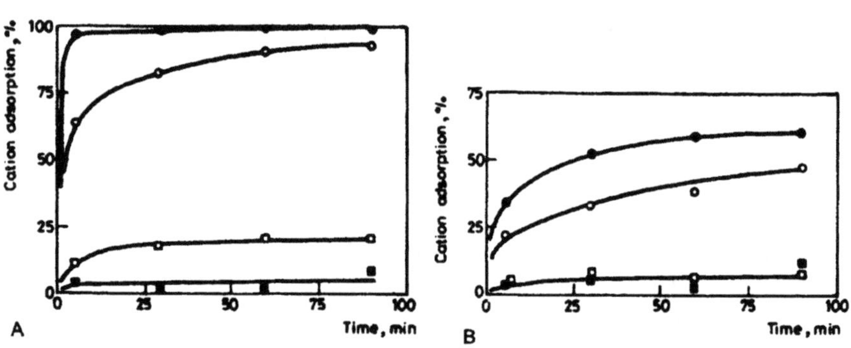

Figure 8-8. Cellular adsorption by *Mycobacterium smegmatis* of scandium, yttrium, calcium, and strontium ions at pH 1.0 in equimolar M^2/M^{3+} solution. (A) Concentration = 50 μM; (B) concentration = 2000 μM. ■, Calcium; ●, scandium; □, strontium; ○, yttrium. Reprinted from MacCordick *et al.* (1990), with permission of Akademiai Kiado.

Sc^{3+}-enterochelin complex showed an antibacterial effect on this bacterium (Fig. 8-9). The uptake of both Fe^{3+}- and Sc^{3+}-enterochelin complexes by *P. aeruginosa* was thought to be mediated by binding to a specific receptor. After translocation, the ligand is apparently rapidly degraded by an intracellular esterase, similar to those present in *E. coli* (Plaha and Rogers, 1983; Rogers *et al.*, 1984). Sc^{3+}, Y^{3+}, and other trivalent metal ions complexes of enterochelin protected mice from otherwise lethal doses of *P. aeruginosa* or *E. coli*. Animals injected with 0.2 ml of 1 mM suspensions of In^{3+}- or Sc^{3+}-enterochelin complexes remained healthy for 1 month (Table 8-1).

The growth and siderophore production of the fluorescent *Pseudomonas* species was influenced by the presence of trivalent metal ions (Fekete and Barton, 1991). The inhibition of growth by 1 mM Sc^{3+}, In^{3+}, and Ga^{3+} was elevated in Fe^{3+}-limited media by comparison with that in media containing 10 μM Fe^{3+}. Pseudobactin synthesis was inhibited in Fe^{3+}-limited culture containing also 1 mM Cr^{3+}, Sc^{3+}, or Y^{3+} during the first 36 h of bacterial growth. In the following period, up to 120 h, Sc^{3+} as well as Cr^{3+} stimulated pseudobactin synthesis and yield (Fig. 8-10). The data regarding the antibacterial effects of Sc^{3+} on the pseudomonad isolate contradicted the results of Rogers (1987), who observed that the free ion Sc^{3+} lacked antibacterial activity against pathogenic serotypes of *E. coli*. Nevertheless, when Sc^{3+} was supplied as the Sc^{3+}-enterobactin complex, the bacterial cells were immediately killed. No investigation has been undertaken on the possible formation of a Sc^{3+}-pseudobactin

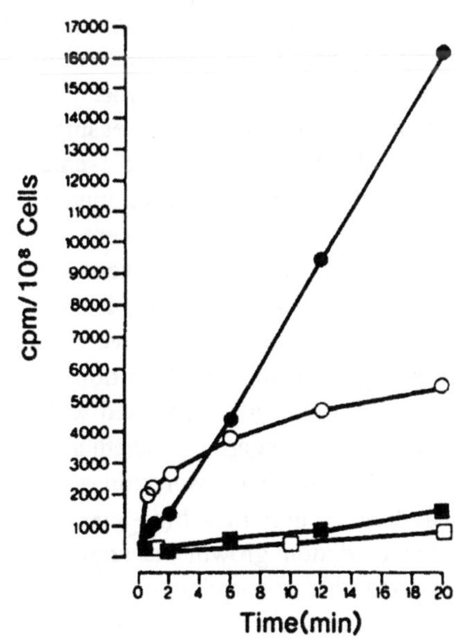

Figure 8-9. Uptake of label from 1 μM $^{59}Fe^{3+}$-enterochelin by *Pseudomonas aeruginosa* suspended in trypticase soy broth at pH 7.4 and 37°C. Uptake by iron-deficient (●, ○) or iron-replete (■, □) cells was measured in the absence (closed symbols) or the presence (open symbols) of 10 μM Sc^{3+}-enterochelin (added 10 min before the $^{59}Fe^{3+}$ complex). Reprinted from Rogers *et al.* (1984), with permission of the authors and Elsevier Science.

Table 8-1. Growth of *Escherichia coli* and Survival Time of Mice Infected with *E. coli* in the Presence of Various Enterochelin Complexes[a]

Enterochelin complex with metal chloride (10 mM)	Viable counts (no. of bacteria/ml)	Time to 50% death (6 h)
Control	$<10^3$	4.75
Sc^{3+}	$<10^3$	7.00
In^{3+}	$<10^3$	8.50
Al^{3+}	2.0×10^5	4.75
Y^{3+}	2.1×10^6	4.75
Fe^{3+}	2.3×10^7	—

[a] Data reprinted from Rogers *et al.* (1982), with permission of the authors and the Society for General Microbiology.

complex and the effects of the addition of this Sc^{3+}-siderophore complex to *Pseudomonas* sp. culture. Although the metals tested have the same oxidation state and relatively similar ionic radii, they markedly differed in their bacteriostasis and siderophore formation (Fekete and Barton, 1991). The scandium–enterochelin complex is assimilated and degraded to the ferric complex in both *Klebsiella pneumoniae* and *E. coli* (Rogers *et al.*, 1980, 1982). When treated with Sc^{3+}-enterochelin, cells of *K. pneumoniae* grew or divided more slowly than untreated cells (Plaha *et al.*, 1984). The effect of this complex could be detected only at the time of division (Fig. 8-11). Macromolecular synthesis was inhibited by the scandium complex, with the extent of inhibition decreasing in the order RNA > protein > DNA > phospholipid (Fig. 8-12). Sc^{3+}-enterochelin inhibited Fe^{3+}-enterochelin uptake by bacteria in a competitive manner. It was concluded that the Sc^{3+} complex enters the bacterial cell, where it is degraded by enterochelin esterase. The liberated Sc^{3+} ion possibly exerts some toxic effects on bacterial growth (Plaha *et al.*, 1984; Plaha and Rogers, 1983).

The removal of radioactive yttrium from seawater by algae was considered to be partly the result of adsorption and partly the result because of ion exchange with yttrium already associated with cells (Spooner, 1949). No definite evidence was provided regarding whether the assimilation of yttrium and other REE represents a passive absorption or an active uptake. A slight nonspecific effect of scandium on growth of blue–green algae species has been found; however, on the effect on diatom growth was apparently more likely (Koelling, 1971). Rice (1956) reported fast absorption of ^{90}Y by *Carteria* sp. cells, such that a balanced concentration was reached within 15 min. Also, *Anacystis nidulans* showed a very

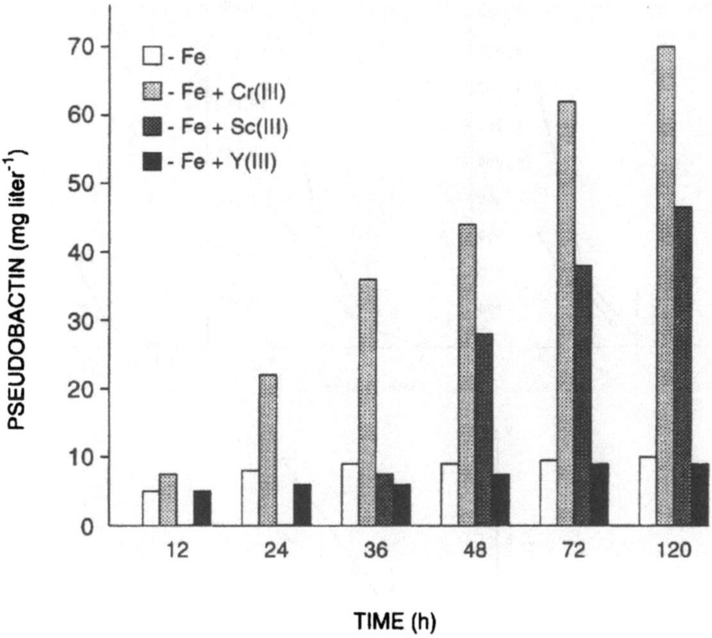

Figure 8-10. Production of ferric pseudobactin by *Pseudomonas* sp. in Fe^{3+}-limited control cultures and cultures supplemented with 1 mM Cr^{3+}, Sc^{3+}, or Y^{3+}. Reprinted from Fekete and Barton (1991), with permission of the authors and Springer-Verlag.

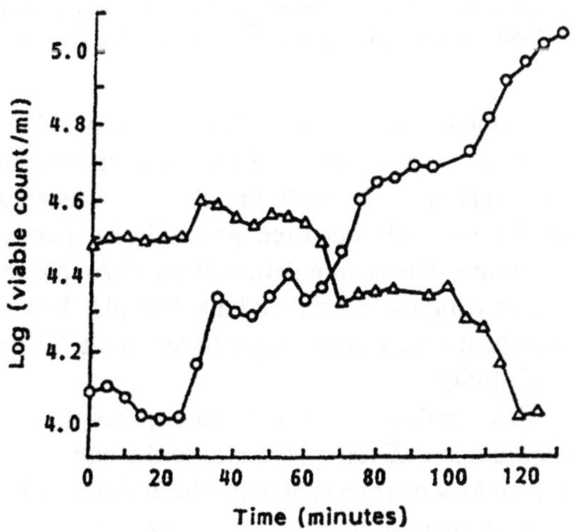

Figure 8-11. Effect of 0.4 μM Sc^{3+}-enterochelin on synchronously growing *Klebsiella pneumoniae* in heat-inactivated rabbit plasma. ○, Control; △, treated with Sc^{3+}-enterochelin from time zero. Reprinted from Plaha *et al.* (1984), with permission of the authors and the Pharmaceutical Society of Japan.

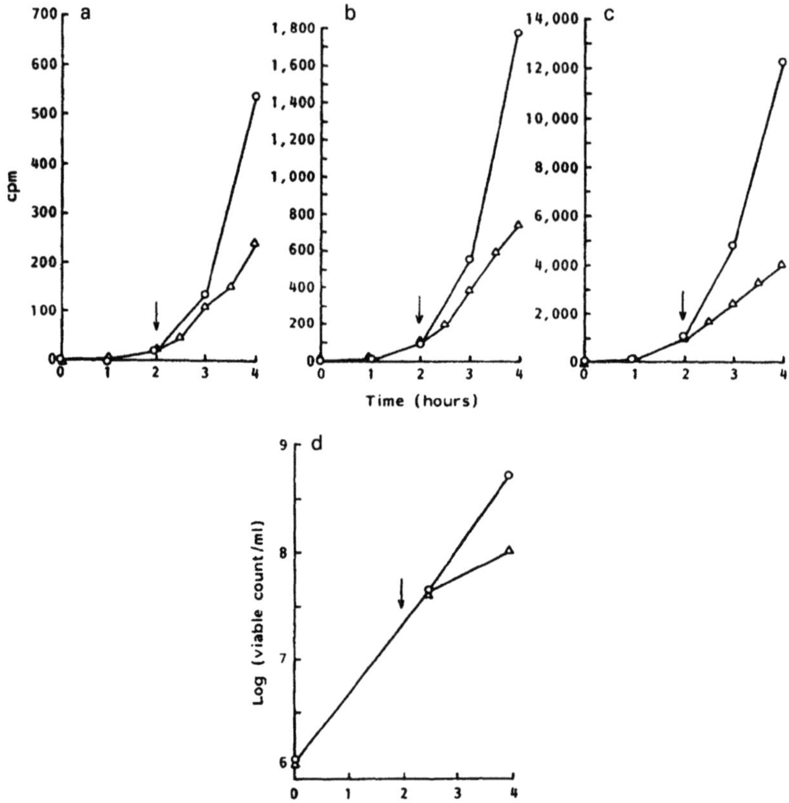

Figure 8-12. Effect of 10 μM Sc^{3+}-enterochelin on the incorporation of orthophosphate into phospholipids (a), DNA, (b), and RNA (c) and on growth (d) of *Escherichia coli* in heat-inactivated rabbit plasma at 37°C. ○, Control; △, Sc^{3+}-enterochelin added at 2 h (arrow). Reprinted from Plaha *et al.* (1984), with permission of the authors and the Pharmaceutical Society of Japan.

rapid uptake of ^{90}Y and ^{90}Sr; a linear relationship between the amount of radioisotopes accumulated in the cells and the amount present in the surrounding medium (Liu *et al.*, 1990; Fig. 8-13). It was assumed that the uptake of the Sr/Y pair by the cells occurred primarily by passive adsorption rather than active absorption. The accumulation of the Sr/Y pair by cells was low at acidic pH and relatively high at neutral and alkaline pH. This alga was suggested to be a useful biosorbent for concentrating and removing ^{90}Y and ^{90}Sr from polluted waters (Liu *et al.*, 1990).

The influence of moderate potassium deficiency and the addition of scandium at different pH values to the nutrient media of the green alga *Chlorella pyrenoidosa* has been studied. The addition of 0.1 mM $ScCl_3$ to complete nutrient media decreased the cell number, whereas a lower concentration (0.1 μM $ScCl_3$)

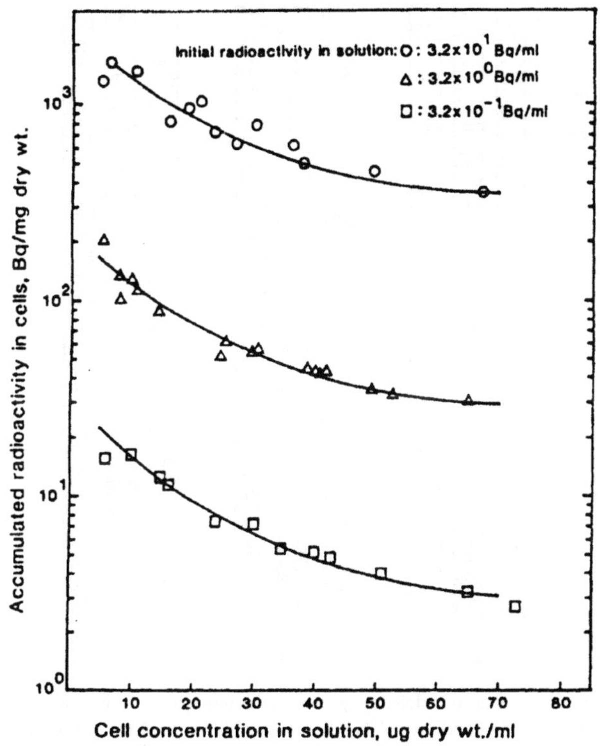

Figure 8-13. Effects of ^{90}Sr/^{90}Y concentration and cell concentration on the uptake of ^{90}Sr/^{90}Y by *Anacystis nidulans* cells. Reprinted from Liu *et al.* (1990), with permission of Elsevier Science.

resulted in an increase in the cell number. Analysis of the ^{14}C-labeled photosynthetic compounds showed that, at acid and neutral pH in complete media, scandium induced an increase in the relative content of sugar phosphates and a decrease in sucrose. The relative content of amino acids was increased by addition of scandium to complete media, whereas the opposite effect occurred with potassium deficiency (Fig. 8-14). A partial substitution of scandium for potassium with moderate potassium deficiency apparently occurred (Horovitz and Wegmann, 1975).

The marine algae *Ulva lactuca* and *Porphyra umbilicalis* absorbed 35% and 15%, respectively, more ^{46}Sc per unit weight in the light than in the dark. Algae transferred after 48 h from light to dark showed a definite retardation in the rate of ^{46}Sc uptake. High concentration factors (CF) were found for both species, 2600- and 1500-fold, respectively, relative to the external concentration. Adsorption accounted for 40–60% of the total uptake (Gutknecht, 1961). Similarly, *Euglena*

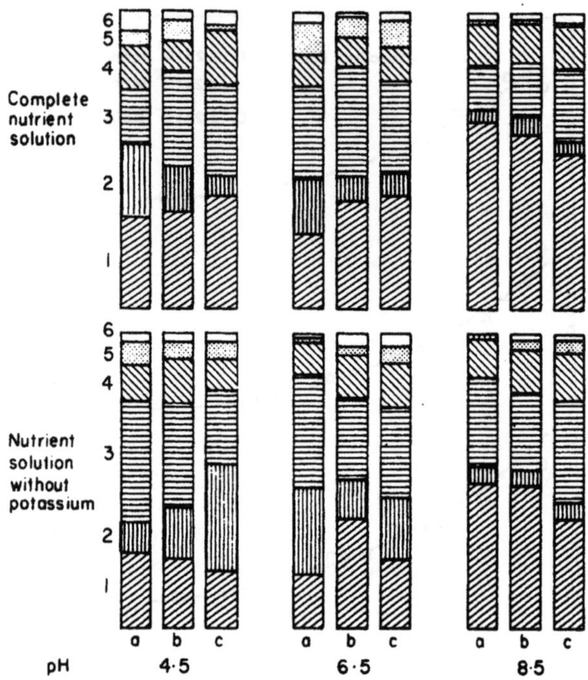

Figure 8-14. Influence of potassium deficiency and additional scandium in the nutrient solution at different pH values on the incorporation of ^{14}C into photosynthesis products of *Chlorella pyrenoidosa*. (A) Control medium without scandium; (B) medium with 0.5 mg Sc^{3+}/l; (C) medium with 5 mg Sc^{3+}/l. 1, Sugar phosphates; 2, saccharose; 3, main amino acids; 4, aspartic, fumaric, and succinic acids and phosphopyruvate; 5, other organic acids; 6, other compounds. Reprinted from Horovitz and Wegmann (1975), with permission of the editor-in-chief of *Physiologia Plantarum*.

cells showed a high accumulation capacity for stable scandium and some other REE from river water (Mann *et al.*, 1988).

8.4 Biochemistry in Plants

The ability of plants to build biological matter from CO_2 and incorporate inorganic salts into organic molecules such as proteins, phospholipids, and nucleic acids, is essential for cellular activity and growth. Bollard (1983) analyzed the involvement of chemical elements in plant growth and nutrition. He stressed that several "unusual" elements, such as aluminum, cobalt, chromium, nickel, selenium, and vanadium, may show stimulatory effects on different crops under some conditions. However, this was not considered proof of their essentiality. It is largely believed that, except for calcium, copper, iron, magnesium, molybdenum, zinc, and a few others, the TE are not needed for growth and development of plants (Anderson and Beardall, 1991).

Evans (1914) performed one of the first experiments on the influence of REE on plant growth. He added 1 μg La^{3+}/l, 0.7 μg Ce^{3+}/l and 1.7 μg Y^{3+}/l to tap water in which hyacinth (*Hyacynthus* sp.) rootlets were grown. Cerium and lanthanum produced a marked increase in cell division with regular arrangements of the cell, while yttrium produced a diminution in cell division with irregular cell arrangements. Young (1935) found that an application of 500 mg Y^{3+}/kg to sandy loam stimulated growth of grass timothy (*Phleum pratense*). A similar nonspecific effect on the rate of growth of grass timothy was obtained with 33 out of 35 TE when added at appropriate concentrations to the soil. A study of the accumulation of scandium, chromium, cesium, iodine, and thallium in corn plants found different rates of accumulation in the leaves (Yuan, 1955). Scandium was highly accumulated in the upper parts of the leaves and the sheath (Fig. 8-15). As expected, large amounts of scandium accumulated in the root system, and lesser amount accumulated in the reproductive organs.

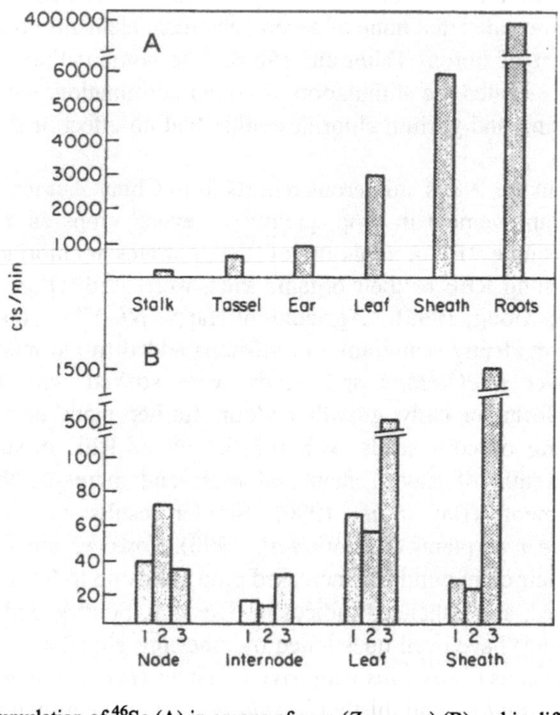

Figure 8-15. Accumulation of ^{46}Sc (A) in organs of corn (*Zea mays*) (B) and in different leaf parts. 1, Lower part of leaf; 2, middle part of leaf; 3, upper part of leaf. Redrawn from Yuan (1955). We made efforts, without results, to trace the copyright holders.

An optimal concentration of 100 μM Sc^{3+}, added to the nutrient solution as the sulfate salt, stimulated significantly the percent germination, number of embryonal roots, growth rate, and accumulation of dry matter of wheat (*Triticum vulgare*) and sunflower (*Hellianthus annuus*). The optimal stimulatory effects on pea (*Pisum sativum*) and sugar beet (*Beta vulgaris*) occurred with 0.1 mM and 1 mM Sc^{3+}, respectively. However, the apparent initial stimulatory effect decreased with further growth and development of the plants (Horovitz, 1965). An antagonistic relationship between potassium and scandium was found in corn (*Zea mays*) plants grown in moderately potassium-deficient nutrient solution. Addition of 10 μM Sc^{3+} to the nutrient solution induced a significant increase in growth of roots and first leaves, as well as a higher synthesis of dry matter (Horovitz, 1968). Accumulation of scandium in aerial organs of corn, cotton (*Gossypium* sp.), rape (*Brassica* sp.) and soybean (*Glycine soja*) plants during their growth periods until fruit formation was also noticed (Qi *et al.*, 1984).

Spraying of grape vine (*Vitis* sp.) leaves with trace concentrations of scandium sulfate during the flowering period accelerated ripening, increased the yield and sugar content, and lowered the acidity of the grapes (Dobroliubskyi, 1962). A study of the influence of various factors on pollen germination of *Petunia hybrida* revealed that none of several chemical elements could replace the stimulatory effect of boron (Fähnrich, 1964). The combination of $ScCl_3$ with boron markedly elevated the stimulation of pollen germination, while aluminum, indium, lanthanum, and yttrium chlorides either had no effect or showed a slight inhibitory effect.

Beginning in the 1980s, numerous reports from China claimed an increase in the yield and improvement in crop quality of several crops as a result of the application of a single REE or a mixture of REE. Nitrates or chlorides of complex fertilizers containing REE or their organic salts, were used (Qi *et al.*, 1984; Xu and Xiao, 1985; Xiong, 1995). "Agricultural Happy NL-1" (a mixture of REE, which included evidently scandium and yttrium) added to the nutrient solution, and to sunflower (*Helianthus* sp.) seeds were soaked with this mixture. "Beneficial" effects on early growth and on further plant development was observed. Sowing of corn seeds with 0.2–0.3 mg of REE mixture per gram increased dry weight of leaves, stem and roots and increased the content of chloroplast pigments (Bai *et al.*, 1990). Similar results were obtained with soybean (*Glycine max*) plants (Kastori *et al.*, 1990). However, the reported effects of the REE or their compounds of increased crop yields up to 50%, improvement of pasture grasses, and enhanced indices in livestock, poultry, and fishery up to 29% (Xiong, 1995) have been questioned by other investigations (Diatloff *et al.*, 1995). Seeds of beans (*Phaseolus vulgaris*), tomatoes (*Lycopersicon esculentum*), wheat (*Triticum vulgare*), and thistle (*Salsola pestifer*) were grown for 32 days in nutrient solutions that contained 1 mg/l $Y(NO_3)_3$, and ^{91}Y as a tracer. Within the range of pH 4–6, the low uptake efficiency of yttrium was proportional to the hydrogen-ion concentration. Tomatoes exhibited the greatest uptake efficiency for yttrium with a leaf: root ratio of 2×10^{-3}; bean was next with a ratio of 6×10^{-4}. Thistle and wheat followed with ratios of 6×10^{-5} and 3×10^{-5}, respectively (Fig. 8-16). Slight translocation to the aerial tissues, as well as little loss from the mas-sive accumulation on the roots, occurred, even when the plants were changed to solutions in which yttrium was absent. A CF (concentration in aerial tissues/concentration in soil) of 0.006 for barley grown on sandy loam was calculated (Rediske and Selders, 1954). The rate of ^{46}Sc uptake in the apical 3 mm of the root axis of onion (*Allium cepa*) was more rapid than elsewhere in the root (Clarkson and Sanderson, 1969). The uptake of scandium and of other metal ions proceeded in two distinct phases: phase I seemed to be a superficial adsorption, characterized by a rapid initial rate; phase II was slower, and the rate of uptake remained constant for 24 h. Autoradiography of roots treated for 30 min with ^{46}Sc showed that most of the radioisotope in the root tip was concentrated in a peripheral belt, which corresponded to the mucigel layer of the root cap. This was considered the site of phase I adsorption. The underlying root cap and epidermal cells retained very little scandium, but the amount retained increased steadily over 6 h to an estimated concentration of 30 mM, which was taken to represent phase II uptake (Fig. 8-17). Because scandium entered the dividing cells in advance of observable effects on cell division, it was concluded that this ion

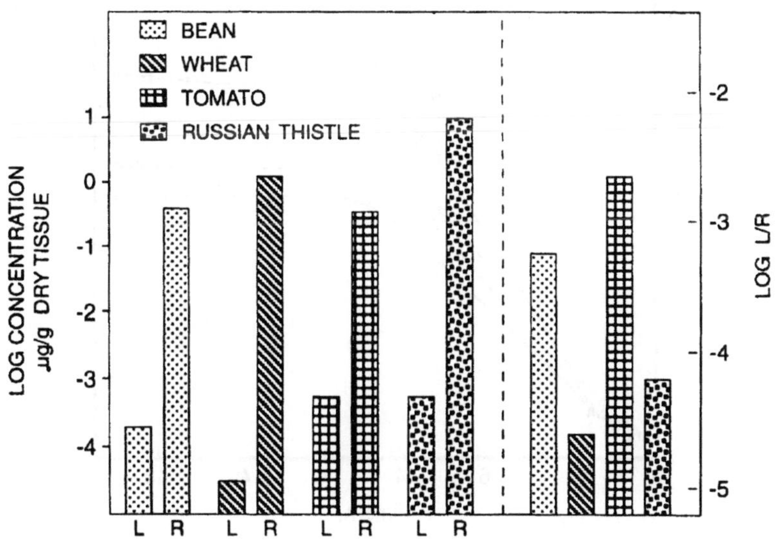

Figure 8-16. Uptake of ^{91}Y by the leaf (L) and roots (R) of four plant species grown in a nutrient solution containing 0.1 µg Y^{3+}/l as $Y(NO_3)_3$ at pH 6.0. The leaf: root uptake ratios are plotted at the right. Reprinted from Rediske and Selders (1954), with permission of the Botanical Society of America.

may inhibit the mitotic cycle (Clarkson and Sanderson, 1969). Further experiments revealed the effect of polyvalent metal ions on uptake and long-distance transport of calcium in barley (*Hordeum vulgare*) plants. Concentrations of 0.1 μM Fe^{3+} and Al^{3+} were not inhibitory on calcium uptake. Sc^{3+} caused a 25% inhibition, at a $Sc^{3+}:Ca^{2+}$ ratio of 1 : 1000. Elution experiments showed that Fe^{3+}, Al^{3+}, and Sc^{3+} reduced the amount of Ca^{2+} held in the water free space, and the Donnan free space, but increased both the exchangeable and the absorbed chloride content of the root. Autoradiography indicated that ^{46}Sc was restricted to the epidermis and outer rank of cortical cells, from where it controls calcium movement throughout the root (Fig. 8-18). This effect was thought to involve surface reactions, rather than the disruption of metabolically dependent transport processes. The relatively nonspecific effect of Sc^{3+}, Fe^{3+}, and Al^{3+} in controlling calcium uptake seemingly depends primarily on their polyvalency rather than on their chemical nature (Clarkson and Sanderson, 1971).

Concentrations of 10–100 μM Sc^{3+} added to the nutrient solution significantly reduced the root and shoot elongation, caused leaf and stem reductions, induced chlorosis, and decrease chlorophyll concentration in leaves of loblolly pine (*Pinus taeda*) and honey locust (*Gleditsia triacanthos*) (Fig. 8-19). In the honey locust, the root, leaf, and stem dry weight reduction was measurable

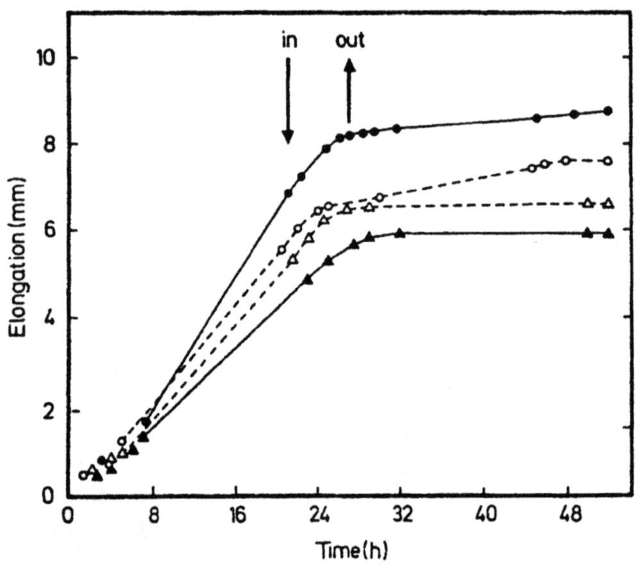

Figure 8-17. Effect of scandium (——) and aluminum (– – –) at two concentrations on root elongation in onion (*Allium cepa*). ●, 200 μM Sc^{3+}; ○, 200 μM Al^{3+}; ▲, 100 μM Sc^{3+}; △, 100 μM Al^{3+}. Duration of treatment, 6 h (shown by arrows). Reprinted from Clarkson and Sanderson (1969), with permission of Springer-Verlag.

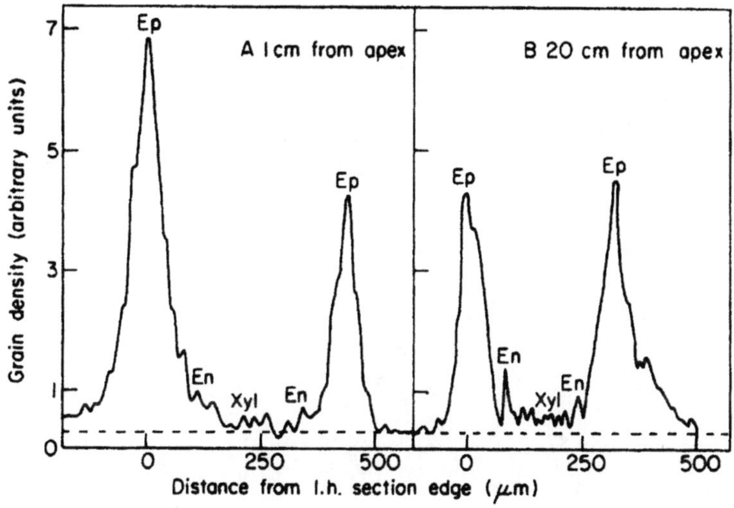

Figure 8-18. Distribution of ^{46}Sc in transverse sections from two zones of barley (*Hordeum vulgare*) seminal axes. Grain-density measurements in microautoradiographs were made along a transect, 10 μm wide, across the diameter of sections. Ep, Epidermis; En, endodermis; Xyl, positions of central xylem vessel. Reprinted from Clarkson and Sanderson (1971), with permission of Oxford University Press.

at 10 μM Sc^{3+}. With loblolly pine, this scandium concentration induced a small, insignificant increase of the elongation of the primary root. Further dry weight increase in old, mid, and new leaves was significantly reduced. Addition of 10 μM Sc^{3+} to the nutrient solution increased the concentrations of calcium and magnesium in the stem and leaves (Table 8-2). At 100 μM Sc^{3+}, loblolly pine developed short roots, and the potassium concentration in the tissues was significantly reduced; neither calcium nor magnesium concentrations in the tissues were affected. Interestingly, in honey locust grown in nutrient solutions containing 10, 25, or 50 μM Sc^{3+}, the roots, stem, and new leaves contained up to 400-, 100-, and 3.6-fold more scandium, respectively, than in control plants. Scandium was bound to the root apoplastic sites. Thus, scandium presumably interfered with potassium uptake and translocation. A direct effect of scandium on the ATPase function, combined with a potassium leakage from exposed tissues, was assumed. The scandium toxicity seemed to be similar to copper toxicity, in that the symptoms of copper toxicity include both chlorosis and change in tissue potassium content. However, large variations between individual plants subjected to the same treatment occurred, such that the differences between the various treatments were not statistically significant (Yang *et al.*, 1989). No explanation was given as to whether this was the result of individual variability of the plant response to the mineral nutrition or the experimental design.

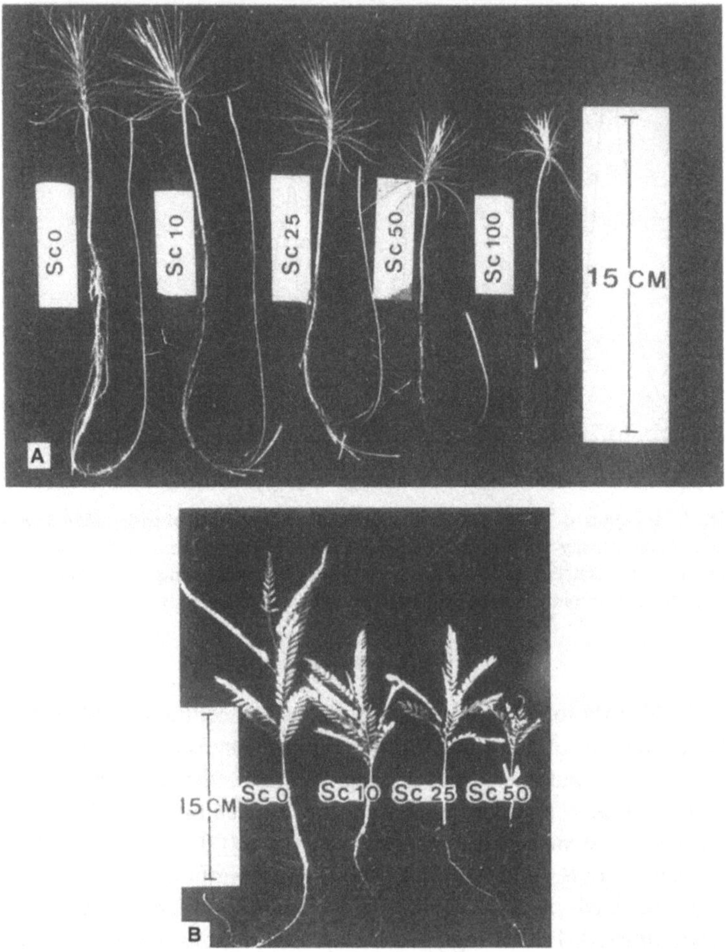

Figure 8-19. Elongation of the primary root of loblolly pine (*Pinus taeda*) (A) and honey locust (*Gleditsia triacanthos*) (B) seedlings, as a function of time and scandium concentration in the nutrient solution. Reprinted from Yang *et al.* (1989), with permission of Elsevier Science.

Accumulation of fission products in plants may be influenced by the concentration of the carrier, plant species, morphological structure of plant organs, and other factors. Although plants absorb more radioactive strontium than yttrium, more yttrium than strontium is transported within plants (Klechkovsky and Tselishcheva, 1957). Barley and pea plants were shown to take up ^{90}Y and other radioisotopes present in trace amounts on the surface of clay or soil particles. The greatest fixating of ^{90}Y was at the root surface (Jacobson and

Table 8-2. Mineral Composition of Roots, Stem, and Leaves of Honey Locust Grown in Solution Culture Containing 10 μM Sc^{3+} [a]

Element	Organs	Control	Scandium, 10 μM	Percent difference from control
K, mg/g	Roots	20.7	18.5	89.3[b]
	Stem	8.3	7.3	87.9
	New leaves	21.0	15.6	74.3
P, mg/g	Roots	2.93	4.0	137.2[b]
	Stem	0.92	0.9	94.6
	New leaves	2.67	2.5	93.6
Mg, mg/g	Roots	0.83	0.7	84.3
	Stem	0.70	0.9	130.0
	New leaves	1.57	2.3	149.1[b]
Ca, mg/g	Roots	1.09	2.4	216.5[b]
	Stem	0.66	0.85	128.8
	New leaves	0.47	0.6	123.4
Sc, μg/g	Roots	8.27	824.7	9972[b]
	Stem	0.52	10.7	2067
	New leaves	0.40	2.1	520

[a] Data recalculated from Yang *et al.* (1989), with permission of Elsevier Science.
[b] Values significantly different from control at $P < 0.01$.

Overstreet, 1948). Essington *et al.* (1963) provided data concerning the uptake of ^{91}Y and other radioisotopes by beans cultivated in pots containing soil. The plant/soil concentration ratio of the metals was highest in roots and leaves and also increased in roots and leaves when chelating agents were added to the soil. The uptake of yttrium and other TE by plants grown on calcareous soil was enhanced when chelating agents were added to the soil. Radiochemical analysis of lettuce (*Lactuca sativa*) seedlings grown in contact with neutron-activated Apollo 11 lunar material showed an uptake of ^{46}Sc, ^{54}Mn, and ^{60}Co (Baur *et al.*, 1974). Although the lunar material has a low solubility, some of the TE released into solution were taken up by the plants and incorporated into the cytoplasm of the stem and leaf. Concentrations of 0.40, 0.27, and 0.05 mg/kg were found for ^{46}Sc, ^{54}Mn, and ^{60}Co, respectively. A cellular incorporation was assumed, on the basis of tracing microscopical autoradiography (Baur *et al.*, 1974). An investigation of the availability of radioisotopes from reactor-activated soils to barley (*Hordeum* sp.) seedlings indicated that the dominant γ-emitting radioisotope in soil was ^{46}Sc; only a small amount of this radioisotope was taken up by plants. Large differences in plant absorption capacity from six types of soil were found in barley, bush bean (*Phaseolus vulgaris*), and alfalfa (*Medicago sativa*) grown on ejecta from a thermonuclear detonation (Nishita *et al.*, 1966). The uptake of ^{46}Sc, ^{124}Sb, and ^{185}W was higher in leaves than in stem, whereas ^{182}Ta

was concentrated more in stems. The plant uptake of ^{46}Sc was influenced by the type of soil. The uptake was less from acidic soil than from neutral or alkaline soils (Romney and Rhoads, 1966). Experiments testing the effectiveness of desorption agents demonstrated the relative inability of Sc^{3+}, and of Ca^{2+}, to desorb the readily exchangeable aluminum from the apoplasm of excised roots of wheat (*Triticum aestivum*) (Zhang and Taylor, 1989).

The calculation of the biological uptake coefficient factor, K_b, for chemical elements in land plants and marine phytoplankton was performed by comparing the Clarke index in soil with that in river water. Values of K_b of 0.50 and 0.44 were calculated for scandium and yttrium in land plants; K_b was in the range of 12–50 for silver, boron, bromine, copper, iodine, and zinc, but values as low as 0.15–0.20 were calculated for titanium and uranium (Zenina, 1986). A soil-stressor-sensitive cultivar of sorghum (*Sorghum bicolor*) was compared to two acid-soil-tolerant cultivars, concerning their capacity to extract scandium from soil solution at various pHs (Wilkinson and Duncan, 1992; Fig. 8-20). A 1000-fold difference was found between the quantities of scandium washed from the roots with water and the nonextractable scandium retained by roots. The roots of the three sorghum cultivars varied in their capacity to retain nonextractable scandium, and this capacity was strongly influenced by pH. The genotypic differences of the cultivars were explained as being due to different isozymes (Wilkinson and Duncan, 1992). No significant changes in the concentration of scandium, aluminum, and other elements were found in rice plants from soils irrigated with water of pH varying between 2.5 and 6.5 (Maeno and Tanizaki, 1996).

The effect of scandium and other TE on growth of aluminum-tolerant and aluminum-sensitive lines of wheat (*Triticum aestivum*) was measured in plants grown in solution culture (Wheeler *et al.*, 1993). None of the tested metal ions could be used as an analog for aluminum. A solution containing 1.1 μM Sc^{3+} reduced by 50% the yield of the roots from aluminum-tolerant plants. The following order of increasing tolerance to other TE was found for aluminum-tolerant lines: Cu < Sc < La < Ga < Al < Zn < Fe < B < Mn. Varying the scandium or gallium concentration in a low-ionic-strength solution produced significant changes in the concentrations of phosphorus, sulfur, calcium, and especially magnesium in wheat plants (Wheeler and Power, 1994). No significant differences in the uptake of various elements between aluminum-resistant and aluminum-tolerant plants were found (Table 8-3). The higher concentration of the metals in the root than in aerial organs was thought to be caused by the precipitation in the root, rather than by plant physiological differences.

Scandium did not affect the net Ca^{2+} uptake at various pHs by *Amaranthus tricolor* protoplast, whereas aluminum, cerium, gadolinium, and lanthanum inhibited the net Ca^{2+} uptake to various degrees (Rengel, 1994; Fig. 8-21). No clear relationship was found between the effects of various metal ions on net $^{45}Ca^{2+}$ uptake and rhizotoxicity of these metals. The absence of inhibitory effects

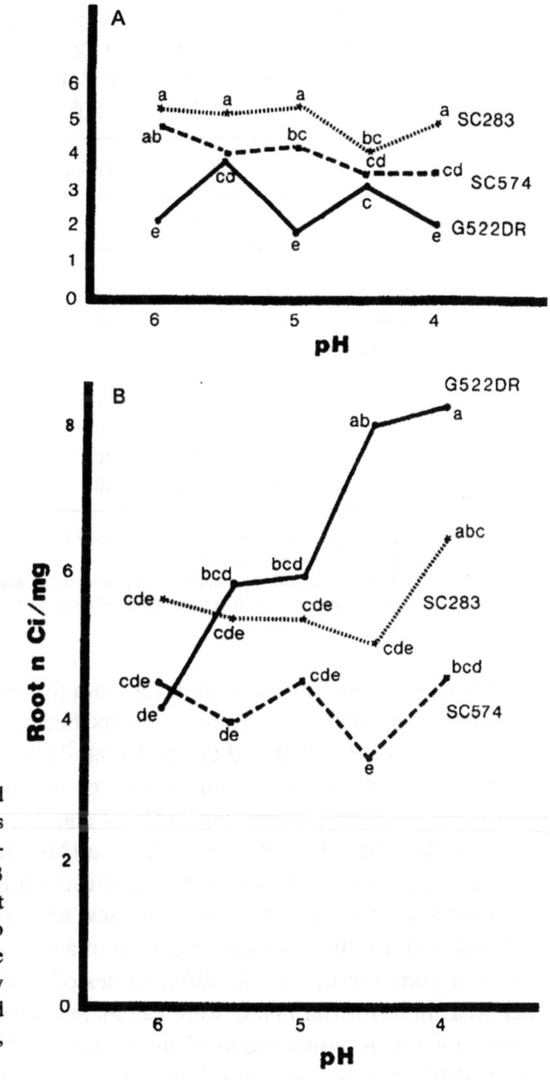

Figure 8-20. Extractable (A) and nonextractable Sc^{3+} (B) in root tips of sorghum (*Sorghum bicolor*) cultivars grown at various pHs. SC283 and SC574 are aluminum-resistant cultivars; GS22DR is nonresistant to aluminum. Points followed by the same letter are not significantly different at the 5% level. Reprinted from Wilkinson and Duncan (1992), by courtesy of Marcel Dekker, Inc.

of Sc^{3+} and Cr^{3+} on net Ca^{2+} uptake was thought to be result of relatively low ionic activity and a different ionic speciation of the metal ions (Rengel, 1994). Al^{3+}, Y^{3+}, Yb^{3+}, and Gd^{3+} also induced extracellular polysaccharide (EPS) formation in mesophile protoplasts of oat (*Avena sativa*), whereas Sc^{3+}, Fe^{3+}, and Cr^{3+} did not have such an effect (Schaeffer and Walton, 1990; Table 8-4). No obvious correlation was found between the physical and chemical characteristics of these metal ions and their ability to induce EPS formation.

Table 8-3. Percent Change Induced by Scandium and Aluminum in the Chemical Composition of Wheat Plants, Grown in a Nutrient Solution That Reduced Yield by 50%[a]

	Percent change in:					
Metal	N	P	S	Mg	Ca	K
			Tops			
Sc	−3	37	13	115	19	3
Al(R)[b]	−6	−13	−13	−39	−18	−18
Al(S)[b]	−6	−21	−12	−26	−18	−17
			Roots			
Sc	−8	23	−27	−51	−8	−13
Al(T)	−5	53	−36	−58	5	−20
Al(S)	−16	−35	−30	−54	−4	−24

[a] Data reprinted from Wheeler and Power (1994), with permission from Kluwer Academic Publishers.
[b] T, Aluminum-tolerant plant near isogeneic line; S, aluminum-sensitive plant near isogeneic line; R, aluminum- resistant plant near isogeneic line.

Numerous elements are complexed with proteins or humic complexes in the humic horizons of the soil. Humic layers become biogenic when enriched in scandium and other TE. Based on the Irving–Williams rules (which state that the stabilities of metal–organic complexes are independent of the nature of the ligand), the following decreasing order of stability for trivalent ions was found: Fe > Ga > Al > Sc > In > Y > Pr > Ce > La (Moeller *et al.*, 1965). The cumulative effect of the complexation is the significant enrichment of certain elements in the topmost layer of the soil. The biological absorption coefficient (BAC) is the plant/soil ratio for the plant concentration of a given element plotted as a function of its concentration in the soil. BAC values of 0.003 for scandium, yttrium, and other REE in terrestrial plants were calculated (Welch, 1984). In a review of the extensive Russian biogeochemical investigations, Dobrovolsky (1994) indicated higher BAC values for several TE, i.e., 0.42, 0.33, and 0.15 for yttrium, lanthanum, and uranium, respectively. It was speculated that this index of dispersed elements is independent of their crustal abundance.

8.5 Biochemistry in Lower Animals (Mollusca, Arthropoda)

Martoja and Martoja (1984) reviewed the accumulation of 34 TE in various groups of invertebrates. These authors emphasized the influence of the chemical

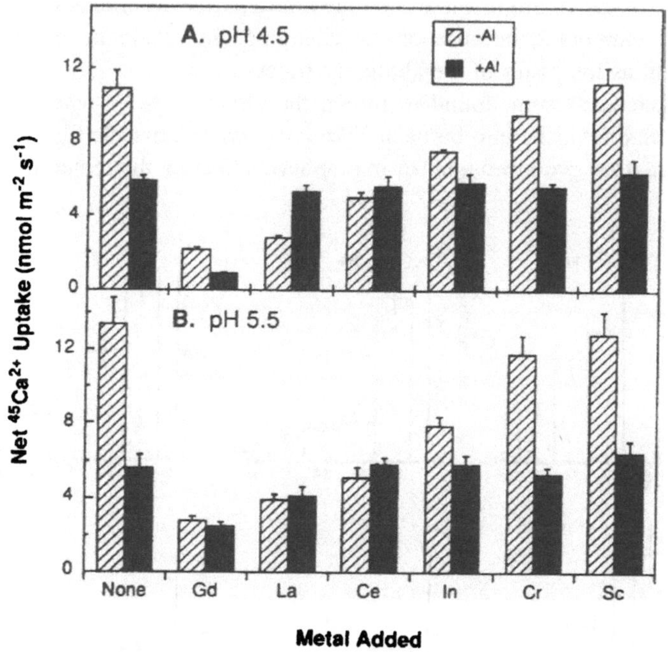

Figure 8-21. Effects of various trivalent cations, added alone or in combination with aluminum, on net ^{45}Ca uptake by *Amaranthus* protoplasts incubated in uptake medium of pH 4.5 (A) or 5.5 (B). Reprinted from Rengel (1994), with permission of the author and Gustav Fischer Verlag.

Table 8-4. Ability of Various Metal Ions to Produce Extracellular Polysaccharide in Oat Protoplasts[a]

	$[^{14}C]$Glucose incorporation mCi	Fold stimulation over control
Control	42.1 ± 3.0	—
$AlCl_3$	156.0 ± 3.1	3.9
YCl_3	111.2 ± 8.3	2.6
$ScCl_3$	59.0 ± 7.0	1.4

[a] Protoplasts were incubated with 200 μM of each metal ion for 24 h, and the radioactivity incorporated into ethanol-insoluble material was measured. Data reprinted from Schaeffer and Walton (1990), with permission of the authors and the American Society of Plant Physiologists.

composition of the environment and pointed out that the accumulation of several TE may be temporary, because of the changing environmental conditions or accumulation as the result of metabolic dysfunction.

Sc^{3+} and Ga^{3+} were found to mimic the effect of Al^{3+}; when they were injected as chloride salts into barnacle (*Balanus nubilus*) muscle fibers poisoned with ouabain, they produced a bi- or monophasic effect on the remaining sodium

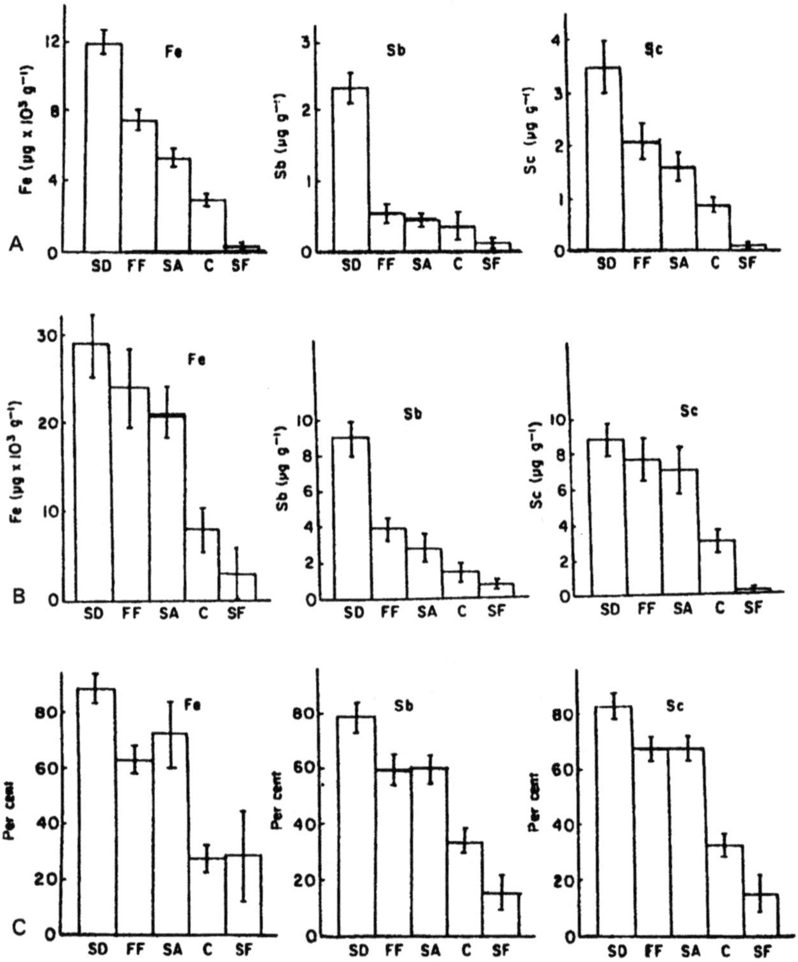

Figure 8-22. Mean concentrations (A) and mean gut content concentrations (B) of Sc, Sb, and Fe in insects, grouped according to the five-feeding-category classification of insects, and percentages of the whole-body concentration of each metal associated with the gut content material, separated according to the feeding category (C). Feeding categories: SD, sediment-dependent; FF, filter feeder; SA, sediment-associated; C, carnivore; SF, surface feeder. Reprinted from Smock (1983), with permission of the author and IUPAC.

efflux (Bittar and Huang, 1990). The stimulated efflux was followed by inhibition or only a stimulation effect alone because the response was dose-dependent. Injection of 0.5 M Sc^{3+} into nonpoisoned and ouabain-poisoned fibers caused a stimulation on the order of 21% and 60%, respectively. Fibers injected with Sc^{3+} were significantly less sensitive to 10^{-4} M ouabain than companion controls. The stimulatory response elicited by Al^{3+}, Ga^{3+}, and Sc^{3+} was caused by a fall in myoplasmic pCa, resulting from the activation of voltage-dependent Ca^{2+} channels; it involved the operation of the Na^+–Ca^{2+} exchanger in the reverse mode (Bittar and Huang, 1990).

Smock (1983) determined whole-body TE concentrations in 40 taxa of aquatic insects. The species were grouped into five feeding categories, which reflected both the type of material ingested and their contact with sediments

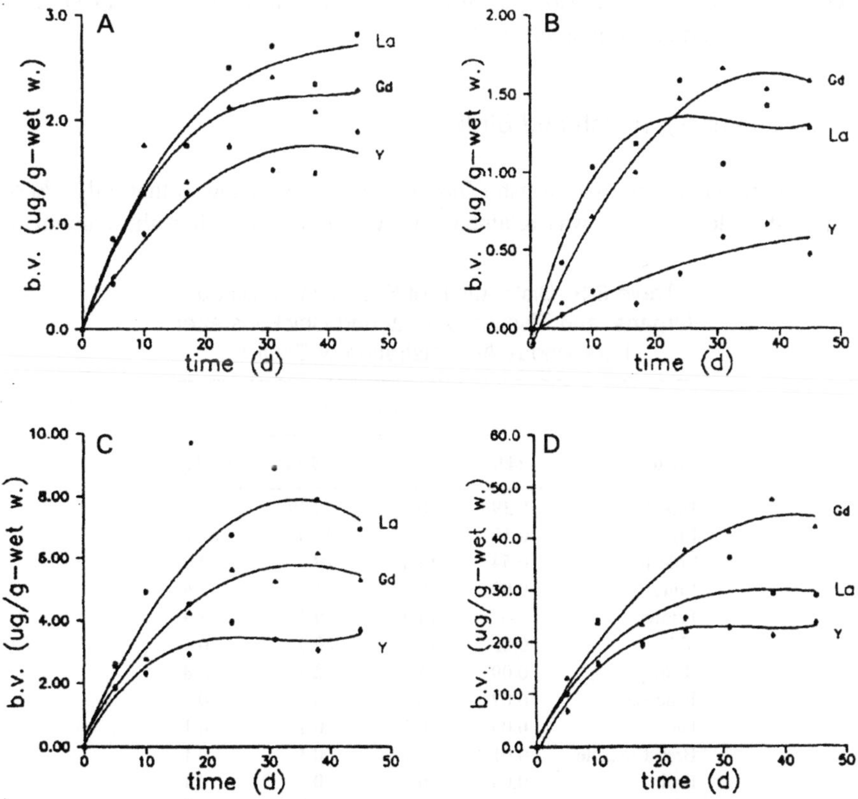

Figure 8-23. Variation of bioaccumulation value (b.v.) of yttrium, lanthanum, and gallium during 45-day exposure of carp (*Cyprinus carpio*) to solutions containing 0.50 mg/l of each element at pH 6.0. (A) Skeleton; (B) muscle; (C) gills; (D) internal organs. Reprinted from Tu *et al.* (1994), with permission of Elsevier Science.

during feeding. Species such as *Ephemeridae* and *Chironomidae* showed the highest concentrations of most metals. Analysis of both gut content and bodies of individuals indicated an exponential decrease in concentration of scandium, as well as of cobalt, chromium, iron, and antimony, with increasing organism size. Both the concentration of metals in the gut material and the proportion of the whole-body burden of the metal that was associated with gut contents varied with the feeding habit of the species (Fig. 8-22).

A radioecological study of marine and freshwater organisms indicated that the CFs of radioisotopes in plants are much higher than those in invertebrates and fish (Polikarpov, 1966). The mean CF value of ^{91}Y in freshwater plants was 14,400, with maximum values as high as 120,000. A CF of 10,000 in egg membranes of marine fish was recorded. In mollusks a CF of 2020 was found, whereas in insects lower figures (120 for ^{91}Y) were estimated. An relatively high accumulation of both radioisotopes of the $^{90}Sr/^{90}Y$ pair in brine shrimp *Artemia salina* was found (Boroughs *et al.*, 1958).

8.6 Biochemistry in Fish and Birds

Experiments with marine fish (*Tilapia mosambica*) showed that only 2% of an ingested dose of ^{91}Y was retained; this was much less than the amount of

Table 8-5. Distribution of Scandium in Blood, Organs, and Tissues of Fattened Chickens after Intravenous Administration of $ScCl_3$ [a]

	mg/kg d.w.			
Tissue	24 h	3 days	7 days	15 days
Blood	0.39	0.01	0.19	0.3
Liver	0.30	49.7	49.3	24.7
Spleen	0.54	33.4	29.3	27.6
Lung	0.09	58.3	21.2	33.6
Femur	—	15.6	9.3	4.3
Yolk	0.05	11.5	20.9	0.4
Kidney	0.09	4.5	2.5	2.8
Pancreas	0.02	0.3	0.2	0.3
Fat	0.03	0.1	0.1	0.1
Breast muscle	0.01	0.1	0.1	0.1
Brain	0.03	0.1	0.1	0.1
White of egg	0.0	0.01	0.01	0.0

[a] Administered dose was 1 mg $ScCl_3$/kg body weight. Data reprinted from Teherani and Badawi (1988), with permission of the authors and Akademiai Kiado.

strontium retained (Tu et al, 1994). After 14 days, about 40% of the radioisotope was found in the visceral organs; the muscles, skeleton, and the gills retained 30, 20, and 5%, respectively. The retention of ^{91}Y and other short-lived fission products in marine organisms is known to be harmful to the biota. Carp (*Cyprinus carpio*) exposed continuously to solutions containing 0.50 mg/l of each of yttrium, gadolinium, and lanthanum for 45 days at pH 6.0 exhibited a relatively low ability to accumulate these REE (Tu *et al.*, 1994). The order of maximum bioaccumulation in organs was internal organs > gills > skeleton > muscle. This order was opposite to that found by Boroughs *et al.* (1956). As seen from Fig. 8-23, the accumulation of yttrium was the smallest. Gadolinium and lanthanum accumulation showed slight differences with the mixed REE solutions, neither synergistic nor antagonistic effects on carp were observed (Tu *et al.*, 1994).

A review of the effects of TE on birds reported that teratogenic, delayed toxic effects on the central nervous system and decreased hatchability were caused by arsenic, nickel, selenium, and a few other TE that may have high environmental burdens (Williams, 1985). At present, no such data concerning the REE are available.

Table 8-6. Distribution of ^{91}Y in Tissues of Hens 96 h after Intramuscular and Oral Dosing[a]

	10^{-3} Percent dose/g	
Tissue	Intramuscular administration	Oral administration
Sternal crest	1700 ± 1300	—
Yolk	190 ± 20	0.09
Liver	110 ± 20	0.03 ± 0.01
Ovary	60 ± 18	—
Spleen	37 ± 13	—
Tibia shaft	27 ± 9	0.06 ± 0.07
Kidney	22 ± 6	—
Gallbladder	18 ± 8	—
Lung	16 ± 3	—
Whole blood	14 ± 4	0.02 ± 0.01
Muscle, pectoral	10 ± 13	0.05 ± 0.05
Heart	7.3 ± 1.5	—
Uterus and vagina	4.3 ± 1.2	—
Brain	1.3 ± 0.5	—
Egg shell	0.93 ± 0.88	0.06
Albumen	0.02 ± 0.03	0.0

[a] Data reprinted from Mraz *et al.* (1964), with permission of Williams and Wilkins.

Higher concentrations of scandium and other TE were found in egg white than in egg yolk of farm and domestic chickens (Waheed *et al.*, 1985; see Table 8-20). Intravenous injection of 1 mg/kg $ScCl_3$ into fattened chickens over a period of 15 days caused high scandium accumulations in liver, spleen, and lungs, and also in the egg yolk (Table 8-5). Maximal accumulation occurred on the 3rd day; the concentration subsequently decreased steadily in all organs. In egg yolk, scandium accumulation reached a maximum on the 7th day, but even at 36 days, 0.5 mg/kg scandium was found in yolk (Teherani and Badawi, 1988). Richards and Steele (1987) proposed a model which describes the interrelated processes governing the uptake and metabolism of TE in the newly formed egg, by tissues of the developing embryo and its associated membrane. The distribution of ^{91}Y and other fission products administered to hens and their transfer to eggs was investigated by Mraz *et al.* (1964). Highest concentrations of ^{91}Y were found in several organs of hens following intramuscular (i.m.) injection (Table 8-6). Trace amounts of yttrium were detected in the tissues and eggs from hens that received these fission products orally.

8.7 Biochemistry in Mammals

8.7.1 Metabolism of Extravascular Entrance

Small laboratory mammals (mice, rats, and guinea pigs) and a few other mammals are the preferred subjects for biochemical and toxicological investigations. When extrapolated to medical problems in humans, the findings from animal models should be interpreted with care. The metabolism of colloidal $^{90}YF_3$, ^{90}Y lactate, and $H^{32}PO_4$ administered by i.m., i.p., and intratumor injection to various mammal species has been studied. Intraperitoneal injection of colloidal $^{90}YF_3$ resulted in the widespread accumulation of radioactivity throughout the body, with a high concentration in spleen and liver (Table 8-7). Apparent gross differences in this response among species merely reflected differences in weight. After i.p. injection of $^{90}YF_3$ in mice with carcinomatous ascites, 95% of the radioisotope was retained within the peritoneal cavity and deposited on serous surfaces (Mayer and Morton, 1956). Injection of $^{91}YCl_3$ into mice resulted in different organ distributions of yttrium, depending on the mode of administration (Table 8-8). After subcutaneous injection, activity was found in liver and kidney and, after a few days, was high in bone, but most remained at the site of injection. Following i.p. injection, high concentrations appeared in bone. The very high retention by pancreas was ascribed to be the result of adsorption on the surface of this organ (Spode and Gensicke, 1957). After i.p. injection of $^{90}YCl_3$ with a specific activity of 1 mCi/mg in dogs (*Canis* sp.), one-third of the radioactivity appeared in the prostate, with only small amounts in distant organs

Table 8-7. Percent Distribution of Injected Dose Recovered per Gram of Tissue after Intraperitoneal Injection of $^{90}YF_3$ in Various Animal Species[a]

	Percent dose/g			
Tissue	Rat	Mouse	Rabbit	Ascites mouse
Spleen	45	470	1.4	1.3
Liver	2.5	26	0.075	0.3
Lung	0.12	0.2	0.005	0.2
Bone	0.046	0.8	0.0025	0.3
Heart	0.004	0.02	0.0002	0.4
Blood	0.002	0.09	0.0005	0.9
Muscle	0.001	0.02	0.0008	0.1

[a] Data reprinted from Mayer and Morton (1956), with permission of the U.S. Atomic Energy Commission, Oak Ridge.

(Cooper *et al.*, 1956). Radiation effects were characterized by hemorrhaging and necrosis followed by healing with fibrous replacement of glandular tissue.

The oral administration of 10 μCi ^{91}Y and ^{46}Sc of high specific activity in mice resulted in a very low uptake in tissues outside the gastrointestinal (g.i.) tract. There was no detectable absorption of ^{46}Sc from the intestine, because none

Table 8-8. Specific Activities Blood and Tissues of Mice as a Result of Subcutaneous or Intraperitoneal Injection[a]

	Specific activity (10^{-3} mCi/100 mg tissue)					
Tissue	30 min	1 h	6 h	24 h	7 days	30 days
			Subcutaneous injection			
Blood	0.1	0.2	0.2	0.1	0.1	≈ 0
Liver	≈ 0	0.2	0.3	0.3	1.2	1.2
Kidney	≈ 0	0.2	1.2	1.1	1.9	0.9
Spleen	≈ 0	0.1	0.1	0.1	0.2	0.4
Pancreas	0.1	0.1	0.3	0.1	0.1	0.2
Femur	0.1	0.6	0.9	2.6	11.4	12.2
			Intraperitoneal injection			
Blood	0.5	0.5	0.1	0.1	≈ 0	≈ 0
Liver	1.6	1.0	1.9	1.5	2.2	1.6
Kidney	1.3	1.3	1.5	2.8	2.8	0.6
Spleen	3.9	6.4	6.1	1.8	1.5	0.5
Pancreas	17.5	6.1	34.7	16.9	21.2	0.7
Femur	0.5	0.7	2.0	3.0	12.7	17.1

[a] Data reprinted from Spode and Gensicke (1957), with permission of Springer-Verlag.

of the organs examined showed any radioactivity, except for the different segments of the g.i. tract (Menczel *et al.*, 1982). Similar findings were obtained with ^{91}Y (Table 8-9). Oral administration of REE chlorides to rats caused a 7–17 times higher accumulation of yttrium than of lanthanum in femur; lanthanum had a stronger effect than yttrium on the accumulation of essential elements (Suzuki *et al.*, 1991). Providing mice a mixture of ytterbium and yttrium in their drinking water for 73 days resulted in the deposition of these elements in stomach, small intestine, gallbladder, and teeth; small amounts were also detected in bone, in spleen, heart, lung, and other organs, in embryos, and in fur (Zhang *et al.*, 1988).

8.7.2 Metabolism of Soluble Intravascular Administration

Forty-eight hours after an i.p. injection of ^{91}Y into the desert pocket mouse (*Perognathus penicillatus*), large quantities of ^{91}Y were found in the kidney, and lesser amounts in teeth and bones; very little radioisotope was found in either blood or muscle (Slater *et al.*, 1961). Considerable variation between individual male and female animals was found in the vascular pathways. Thomassen and Leicester (1964) found that yttrium injected intraperitoneally into rats was initially deposited in liver and spleen and apparently transferred to bone at a later time; teeth also retained yttrium, with the molars having an almost constant amount up to 150 days (Table 8-10).

Because orally ingested Ln are poorly absorbed and transported by mammals, most metabolic studies have used parenteral administration. Scandium-46 has

Table 8-9. Organ and Tissue Distribution of ^{91}Y and ^{46}Sc in Mice after Oral Administration[a]

	Percent dose/g of tissue			
	^{91}Y			^{46}Sc
Tissue	4 h	24 h	48 h	24 h
Muscle	0.007	—[b]	—[b]	—[b]
Liver	0.01	0.01	0.006	—[b]
Kidney	0.02	0.006	0.004	—[b]
Bone	0.009	0.005	0.01	—[b]
Stomach mucosa	2.8	1.2	0.1	0.3
Stomach	1.4	0.5	0.04	0.05
Small intestine	1.0	0.08	0.02	0.05
Large intestine	1.9	0.3	0.08	0.8

[a] Data reprinted from Menczel *et al.* (1982), with permission of the authors and Williams and Wilkins.
[b] $< 2\times$ background.

Table 8-10. Yttrium Retention in Rat Tissues after Intraperitoneal Injection[a]

	Percent dose/g of tissue				
Tissue	3 days	30 days	90 days	120 days	150 days
Spleen	9.11	1.04	1.17	2.98	0.63
Liver	8.89	0.47	0.45	0.96	0.22
Long bone	1.40	0.58	0.41	0.50	0.42
Molars	1.24	0.90	0.77	0.76	0.79
Incisors	0.75	0.38	0.13	0.14	0.10
Small intestine	0.57	0.19	0.15	0.12	0.04
Kidney	0.55	0.18	0.05	0.07	0.05
Lung	0.22	0.04	0.34	0.34	0.75
Heart	0.07	0.02	0.008	0.01	0.007
Striated muscle	0.03	0.02	0.001	0.003	—
Brain	0.008	0.005	0.003	0.003	0.002
Blood	0.01	0.001	—[b]		

[a] Data reprinted from Thomassen and Leicester (1964), with permission of the American Society for Dental Research.
[b] Not significantly above background.

been i.p. injected (Balasiewicz and Kierylowicz, 1961) or i.v. injected (Achor *et al.*, 1962) as a tracer to study calcium metabolism in many experiments with mammals. In rats, stable scandium caused increased accumulation of i.v. administered ^{46}Sc in liver, marrow, intestine, and brain and increased long-term whole-body retention of ^{46}Sc (Byrd *et al.*, 1975; Fig. 8-24). Three compartments of "carrier-free" ^{46}Sc were identified: (a) 21% was in a compartment with a biological half-time (BHT) of 1.0 day; (b) 22% was in a compartment with a BHT of 7.1 days; and (c) 56% was in a compartment with a BHT of 485 days. When ^{46}Sc was administered with stable scandium, three compartments also were identified: (a) one with a BHT of 1.3 days and containing 23% of scandium; (b) another with a BHT of 7.3 days and containing 16% of scandium; and (c) a third with a BHT of 1200 days and containing 62% of scandium. Thus, stable scandium had a pronounced effect on the long-term whole-body retention of ^{46}Sc. The effective half-times for the retention of ^{47}Sc in liver and spleen were 2.6 and 3.4 days, respectively (Byrd *et al.*, 1975).

Freed (1977) investigated the *in vivo* distribution of ^{47}Sc generated by decay of i.p. injected ^{47}Ca in several soft tissues and bones and found that, following an initial decrease, the relative concentration of ^{47}Sc in liver, spleen, kidneys, and skin stabilized at about 9 days and remained practically constant for the next 30 days. In tibia and femur, the ^{47}Sc did not change significantly between the 1st and the 9th day after ^{47}Ca injection (Fig. 8-25). The radiation dose of internally

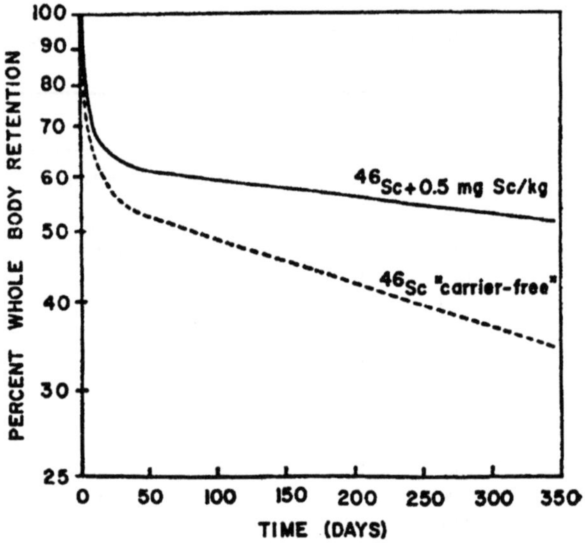

Figure 8-24. Effect of stable scandium on the long-term whole-body retention of ^{46}Sc in rats. Reprinted from Byrd *et al.* (1975), with permission of the Health Physics Society.

generated ^{47}Sc was delivered almost exclusively to bone (Freed, 1977). When a $^{47}Ca/^{47}Sc$ mixture was administered orally, more than 98% of the ^{47}Sc remained in the g.i. tract as unabsorbed scandium; i.e., the deposition of ^{47}Sc in liver and spleen was very small (Pearson, 1966).

Comparison of the metabolic behavior of calcium and scandium administered intravenously into rats as a $^{47}Ca/^{47}Sc$ mixture indicated that 40% of ^{47}Sc was rapidly taken up into the liver and spleen 6 h after i.v. injection (Taylor, 1966). In the subsequent 18 days, the ^{47}Sc activity in both organs decreased exponentially. More than 90% of ^{47}Sc arising from the decay of ^{47}Ca was retained in the skeleton (Fig. 8-26). The i.v. administration of scandium before or with ^{67}Ga in rats and mice increased ^{67}Ga excretion and bone deposition; the uptake of ^{67}Ga in soft tissues markedly decreased (Hayes *et al.*, 1980). Apparently, scandium interfered with the binding of ^{67}Ga to plasma proteins, forcing ^{67}Ga into an unbound or loosely bound state. It was speculated that ^{67}Ga may be distributed between two plasma compartments (PC). PC 1 (free ^{67}Ga) routes the metal mainly to bone, excreta, and nonosseous tumors, whereas PC 2 (protein-bound ^{67}Ga) routes it to normal soft tissues (Hayes *et al.*, 1982).

Intravenous administration of ^{91}Y to cattle provided evidence that bone is the most important tissue accumulator of yttrium; liver, kidney, spleen, lymph nodes, lung, pituitary, adrenals, thyroid, heart, rumen, and intestine also retained significant quantities (Hood and Comar, 1956). Bones, the sternal marrow, and

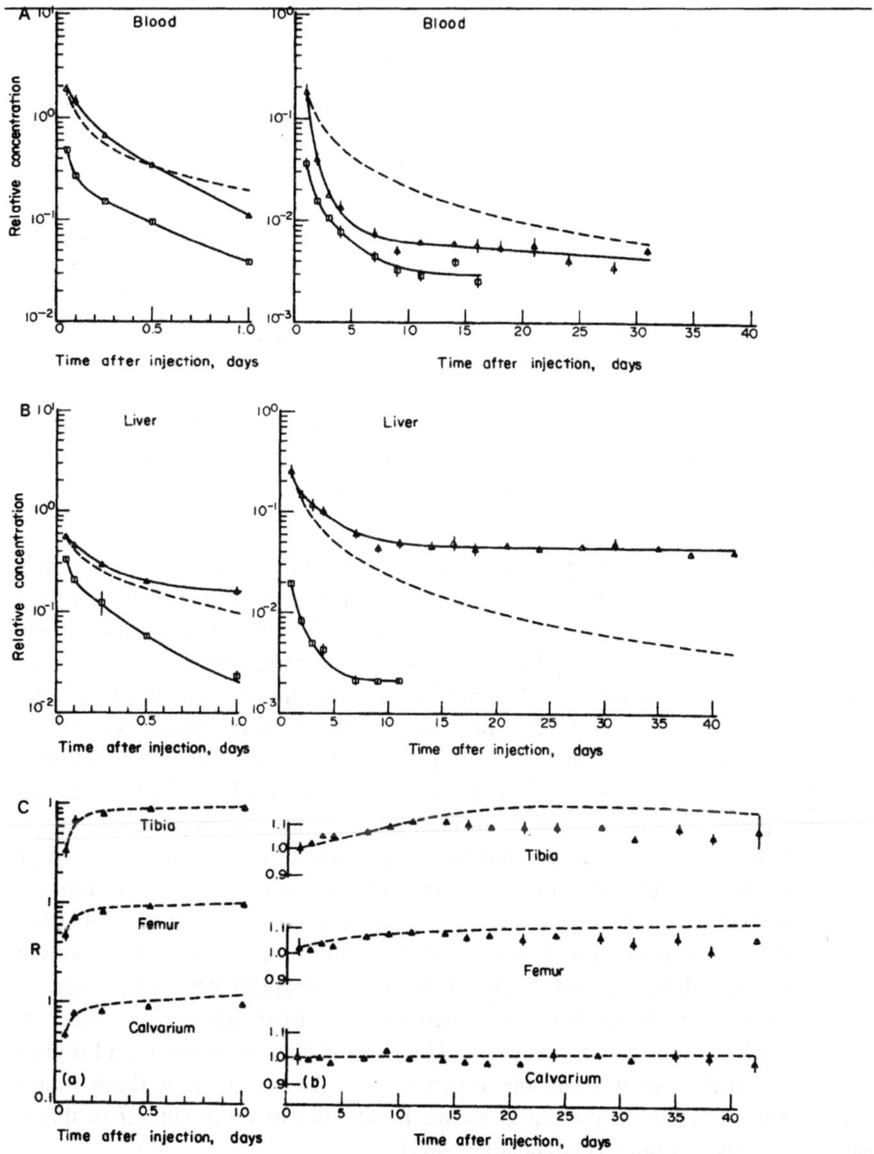

Figure 8-25. Distribution of ^{47}Sc generated by decay of i.p. injected ^{47}Ca *in vivo* in blood (A), liver (B), and bone (C) of mice, during a period of 42 days. By 1 day after injection, the skeleton contained most of the whole-body content of both parent and daughter radionuclides. The value of *R*, the logarithm of the $^{47}Sc/^{47}Ca$ activity ratio, for bone was 37 times that for the liver at 1 day after injection and 188 times that for the liver at 42 days (B. R. Freed, personal communication, 1998, with permission). Reprinted from Freed (1977), with permission of the author and the Health Physics Society.

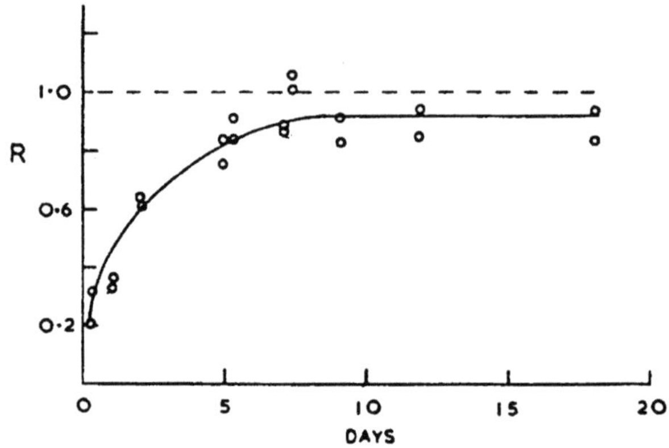

Figure 8-26. Bone retention of ^{47}Sc produced by the decay of ^{47}Ca after i.v. injection of a $^{47}Ca/^{47}Sc$ mixture into rats. Reprinted from Taylor (1966), with permission of the British Institute of Radiology. R = The $^{47}Sc/^{47}Ca$ ratio in the femur and in the injection solution.

the adrenals lost yttrium very slowly, whereas other soft tissues lost 97% of their yttrium content in 5 months. During the first 3 h after i.v. administration of several radioisotopes to rats, the blood concentration of ^{91}Y decreased most rapidly, whereas that of ^{166}Ho and ^{169}Yb decreased the slowest (Magnusson, 1963). In the kidneys, the concentrations of ^{91}Y and ^{160}Tb were the highest, and those of ^{166}Ho and ^{169}Yb were the lowest. The following order of liver uptake was found: $Ce \approx Pm \geq Ho > Tb > Yb > Y$. During the first 3 h, the urinary excretion of ^{91}Y was greatest, and that of ^{166}Ho and ^{169}Yb was the least. Fecal excretion occurred partly via bile and partly through the g.i. wall. A comparative study of the distribution of carrier-free yttrium administered intravenously to various mammal species found that in rhesus monkey (*Macaca mulatta*), yttrium was deposited mainly in bone and tissues with phagocytic activity (Daigneault, 1963; Table 8-11). Treatment with zymosan, for the purpose of stimulating the activity of the reticuloendothelial system (RES), produced a hypertrophy of the liver and spleen but did not stimulate the RES activity. This treatment did not affect the excretion of yttrium, as was expected. The lack of activity in brain and thymus indicated that the ions were unable to penetrate the blood barrier in these organs. The yttrium distribution pattern in rhesus monkeys was more similar to that in rats than to that in mice (Daigneault, 1963).

The percent of administered dose of yttrium and Ln increased several times in the organs of rats which were i.v. injected with 10–50-mg/kg doses. A significant correlation coefficient between the amounts of yttrium and calcium in liver, spleen, and lung was calculated. The i.v. administration of yttrium induced an increased accumulation of calcium in liver and spleen (Nakamura *et al.*, 1993). Following i.v. injection of 10 mg/kg REE chlorides (Y, Dy, Eu, and Yb) to rats,

Table 8-11. Comparison of the Distribution and Excretion of Intravenously Administered Yttrium Chloride in Mammalian Species, without and with Zymosan Treatment[a]

	Without Zymosan-treatment			Zymosan-treated[b]		
Tissue	Rhesus monkey 4 h	Rat 24 h	Mouse 24 h	Rhesus monkey 4 h	Rat 4 h	Mouse 24 h
Liver	8.9	4.1	8.0	2.4	9.1	12.3
Femur	5.1	7.2	3.3	3.4	8.2	4.0
Kidney	2.8	1.8	1.1	3.8	7.5	0.9
Spleen	1.0	0.7	4.3	0.9	1.4	0.02
Adrenal	0.9	1.6	—	0.8	0.5	—
Lung	0.6	0.5	2.3	0.6	1.1	0.6
Blood	0.5	0.1	0.1	0.3	0.1	0.1
Marrow, femur	0.4	0.01	0.1	0.5	1.0	0.01
Muscle	0.3	0.2	0.2	0.1	0.1	0.2
Brain	0.01	0	0.03	0.02	0.02	0
Urine, %[c]	4.7	17.2	9.6	6.6	12.9	5.3
Feces, %[c]	0	3	0.3	0	0.5	1.2

[a] Values were calculated as total counts administered/total weight (in grams). Data reprinted from Daigneault (1963), with permission of Academic Press.
[b] Zymosan (25 mg/kg) was administered intravenously 3 days after injection of yttrium chloride.
[c] Excretion was expressed as the percentage of the administered dose.

these elements were still detected in the blood 4 h later (Nakamura *et al.*, 1991b). The REE accumulated mainly in the liver, spleen, and bone, where they were retained for a long time (Fig. 8-27). High amounts of yttrium did not significantly affect the concentrations of magnesium and iron. The calcium concentration in the liver and spleen increased markedly (Table 8-12). Administration of yttrium and other REE at a dose of 25 mg element/kg to rats and mice increased the calcium concentration in spleen, lungs, and liver (Shinohara and Chiba, 1990). The exposure of pregnant mice to yttrium and ytterbium led to a rapid placental transfer; 14.1% of the total amount administered was detected in newborn mice (Zhang *et al.*, 1988). Young and developing organs were especially susceptible to the accumulating metals. Ando *et al.* (1989a) investigated the uptake rates after i.v. injection to rats of 54 elements and 65 radioactive compounds. Hard acids of multivalent metals, accumulated into soft tissue organs but were eliminated more slowly from these organs than the bivalent and monovalent metals. Administration to rats of 48Sc labeled scandium citrate caused a higher urinary excretion rate, and a lower accumulation of the metal ions in liver and spleen, as compared to the administration of labeled scandium and yttrium chlorides (Table 8-13).

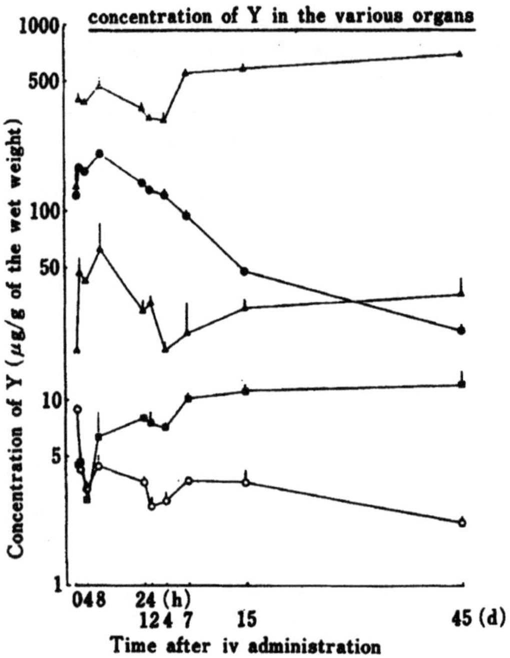

Figure 8-27. Changes in the accumulation of yttrium in various organs of rats during a period of 45 days after i.v. injection of 10 mg YCl_3/kg. △, spleen; ●, liver; ▲, lung; ■, femur; ○, kidney. Reprinted from Nakamura *et al.* (1991b), with permission of the authors and the Pharmaceutical Society of Japan.

Table 8-12. Concentration of Calcium, Magnesium, and Iron in Various Organs of Rats after 8 Days of Intravenous Administration of Yttrium[a]

Tissue	Element	Concentration of element (mg/g)		
		Control	Y, 10 mg/kg	Y, 50 mg/kg
Liver	Ca	36	207	108
	Mg	238	237	240
	Fe	101	79	67
Kidney	Ca	86	80	114[b]
	Mg	243	215	215
	Fe	50	67	66
Spleen	Ca	46	593	7806[b]
	Mg	261	226	358[b]
	Fe	102	126	83
Femur	Ca	82	76	98
	Mg	1.8	1.8	2.3
	Fe	53	37	66
Lung	Ca	91	107	—
	Mg	171	175	—
	Fe	60	63	—

[a] Solutions of $YCl_3 \cdot 6H_2O$ in 5% glucose were administered to male rats at doses of 10 or 50 mg/kg. Data adapted from Nakamura *et al.* (1993), with permission of the authors and the Pharmaceutical Society of Japan.
[b] Statistically different from control at $P < 0.05$.

Table 8-13. Blood Retention Values, Urinary Excretion Rates, and Uptake of ^{46}Sc and ^{90}Y Compounds in Various Organs of Rats[a]

	$^{46}ScCl_3$		^{46}Sc citrate		$^{90}YCl_3$	
	3 h	48 h	3 h	48 h	3 h	48 h
Blood retention value	2.3	0.1	2.8	0.1	5.1	0.02
Urinary excretion rate	0.7	—	10.2	—	0.03	—
Uptake in organs						
Liver	10	10	2	5.4	9.9	11.4
Spleen	2	2.6	1.1	2.3	2.9	3.3
Kidney	1.6	1.3	1.2	1.5	1.0	1.5
Cardiac muscle	—	—	0.8	0.3	0.7	0.2
Skeletal muscle	0.1	0.1	0.4	0.2	0.1	0.03
Pancreas	0.4	0.3	0.7	0.5	0.4	0.1
Bone	—	—	0.7	0.7	—	—
Lung	1.2	0.5	1.6	0.5	3.1	0.9
Brain	—	—	0.1	0.1	—	—
Adrenals	2.6	2.2	1	1.1	4.5	4.5
Thymus	0.2	0.3	0.4	0.7	0.3	0.2

[a] The blood retention value at 3 and 48 h, the urinary excretion rate at 3 h, and the uptake rates in various organs at 3 and 48 h after i.v. injection of radioactive compounds are expressed in percent dose/g. Data reprinted from Ando *et al.* (1987a), with permission of the authors and Elsevier Science.

8.7.3 Metabolism of Insoluble Forms

The consequences of inhalation of radioactive particles, as a result of atmospheric nuclear explosions or by accident, are of major concern, because following their inhalation, they remain in the lungs for long periods of time. These particles may be rapidly transformed into nontransportable forms by precipitation or by binding to macromolecules and adsorption on cellular surfaces. The radioisotopes of concern in this regard include ^{90}Y, ^{90}Sr, and ^{144}Ce.

One of the early attempted medical uses of ^{90}Y and other REE radioisotopes involved preparing them in colloidal form, which would be biochemically inert. Colloidal radioactive yttrium hydroxide and yttrium phosphate were administered intravenously in carrier-free forms to rats (Speer *et al.*, 1961). The composition of the inorganic solvent, containing citrate or celloidin, influenced the distribution of the radioisotope in urine and plasma. An immediate excretion of the injected inorganic yttrium was followed by excretion of an yttrium–amino acid complex formed in the plasma and tissues (Kawin, 1953). Studies of the metabolism of a mixture of ^{89}Sr, ^{90}Sr, and ^{90}Y in rabbits fed diets containing different amounts of calcium showed that the amount of ^{90}Y present in bone 9 days after injection was roughly equal to the amount of ^{90}Sr. The hazard of any intake of ^{90}Y is enhanced by the retention of high β-emitting doses. Analysis of radioactivity loss through excretion suggested that the excretion of ^{90}Y is greater in old than in young animals. Dietary calcium did not affect the excretion of yttrium by young animals (Kawin *et al.*, 1950).

Superselective embolization was used to administer ^{90}Y to pigs as resin particles (Schubiger *et al.*, 1991). Less than 1% of the injected activity accumulated in bone, bone marrow, liver, and lung, as compared to $>99\%$ retention by the kidneys. Complications due to acute necrosis or inflammatory reactions and alterations in the liver and lung were not seen. However, severe arterial wall changes and fibrotic necrosis, due to radiation, were observed in the ^{90}Y-treated kidneys.

Precipitation *in vitro* of 0.01 mM ^{90}Y with several solvents at pH 7.3–7.4 in dog serum was due to the *in vivo* formation of complexes with phosphate, especially when the carrier concentration and pH were low (Kanapilly, 1980). In the presence of chelating agents, such as citrate, or higher carrier concentrations, carbonate was also capable of precipitating the Ln^{3+}.The radioisotope in the colloidal form was incorporated into rat sperm. Whereas guinea pig and rat sperm incorporated similar amounts of ^{90}Y, sperm from bulls and rabbits incorporated notably less. With carrier-free $^{89}YCl_3$, the maximum yttrium uptake was 3×10^7 yttrium atoms per rat sperm. Sperm from all investigated species incorporated relatively greater amounts of ^{90}Y than did erythrocytes. These findings reflect an apparent dissimilarity in cellular surface characteristics of various tissues and of tissues from diverse species (Coomes, 1968). Further toxicological aspects of insoluble forms of scandium and yttrium that may enter into organisms are presented in Chapter 9.

8.7.4 Metabolism of Chelated Forms

The use of natural and synthetic chelators of scandium and yttrium permits their study in soluble form, because it prevents the formation of colloids and precipitates in a biological milieu. Moreover, chelation offers many possibilities for medical applications, as well as for solving environmental problems.

A forced excretion of yttrium was achieved when ^{90}Y chelated by EDTA was i.v. injected at 48 h, 19 days, and 49 days. Calcium-EDTA was more efficient in promoting yttrium excretion than the excretion of cerium and promethium (Foreman and Finnegan, 1956); it caused a threefold increase in urinary excretion and a twofold reduction in yttrium accumulation in long bones and carcass, following i.v. administration to rats (Table 8-14). Studies of the tissue distribution and excretion of scandium and yttrium administered to mice in ionic and chelated forms showed that the strong chelates (Sc-EDTA and Y-EDTA) were well excreted; very little was retained in the tissues (Rosoff *et al.*, 1963). The weak chelates were largely deposited in liver and spleen, whereas chelates with intermediate stability (Y-NTA and Sc-NTA; NTA, nitrilotriacetic acid) appeared in relatively high concentrations in bone. A quantitative 100% urinary excretion of scandium and yttrium chelates with EDTA was noticed, in contrast to the low urinary excretion of the scandium ionic and citrate forms (Fig. 8-28). Y-EDTA has been demonstrated to leave the body quite rapidly when injected intravenously (Hart, 1956). Organ distribution was slightly changed as a result of administration of 0.01 mg yttrium citrate per kilogram body weight (Kutzner *et al.*, 1986). Up to 1 mg of inactive yttrium per kilogram body weight caused a considerable reduction in bone uptake and an elevated accumulation of fluorine in liver and spleen by transforming this element to colloidal form. The transformation into colloids varied with changes in the citrate concentration. A small amount of yttrium carrier caused a considerable reduction of ^{90}Y in bone, when the radioisotope was i.v. administered as the citrate complex (Fig. 8-29). This treatment caused an increased accumulation of ^{90}Y in liver and spleen, where it was attached in colloidal form; its uptake in lung occurred in macroparticle form (Kutzner *et al.*, 1986).

The rate of fluorine clearance was enhanced by scandium, and the extent of accumulation in bone was markedly reduced (Anbar and Ernst, 1962). Apparently, scandium fluoride has a smaller affinity for osteoid tissue than the Sc^{3+} ion. Increasing doses of yttrium had a high affinity for bone, decreasing the relative concentration of fluorine in bone. The decrease in the rate of fluorine clearance from blood in the presence of high yttrium concentrations indicates a

Table 8-14. Effect of Calcium-EDTA on the Distribution and Excretion of Yttrium Administered Intravenously to Rats[a]

	Percent of injected dose	
Tissue	Ca-EDTA treated	Control
Urine	51.3 ± 3.2	18.1 ± 1.0
Feces	3.9 ± 1.1	2.8 ± 0.04
Liver	1.4 ± 0.8	2.8 ± 0.5
Spleen	0.6 ± 0.1	0.15 ± 0.03
Long bone	5.4 ± 0.4	11.8 ± 0.4
Carcass[b]	31.4 ± 0.5	60.7 ± 1.3

[a] Data reprinted from Foreman and Finnegan (1956), with permission of the U.S. Atomic Energy Commission, Oak Ridge.
[b] Carcass includes the remainder of the viscera, the skeleton, and skeletal muscle.

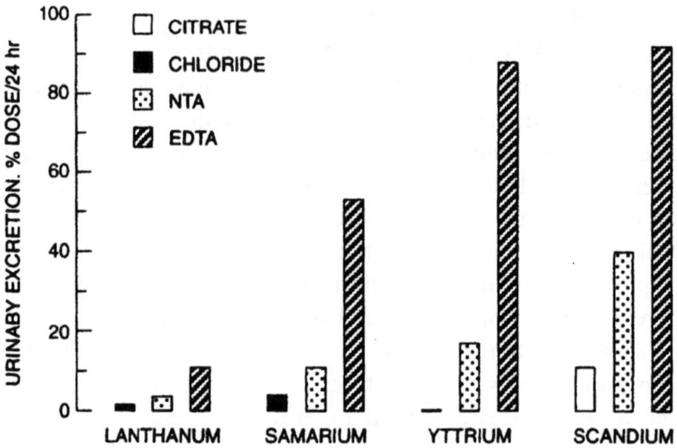

Figure 8-28. Urinary excretion of ionic and chelated lanthanum, samarium, yttrium, and scandium by mice during a period of 24 h following i.v. injection. From Rosoff *et al.* (1963), with permission of the authors and Elsevier Science.

parallelism between the distributions of fluorine and yttrium (Anbar and Ernst, 1962). Moskalev (1963) demonstrated the efficiency of hexametaphosphate in increasing the excretion of ^{91}Y and other radioisotopes administered intravenously to various mammalian species.Other aspects of the biochemical effects, as well as medical and environmental uses, of chelators are presented in Section 6.8 of Part 1, Sections 7.4 and 8.8, and Chapter 10.

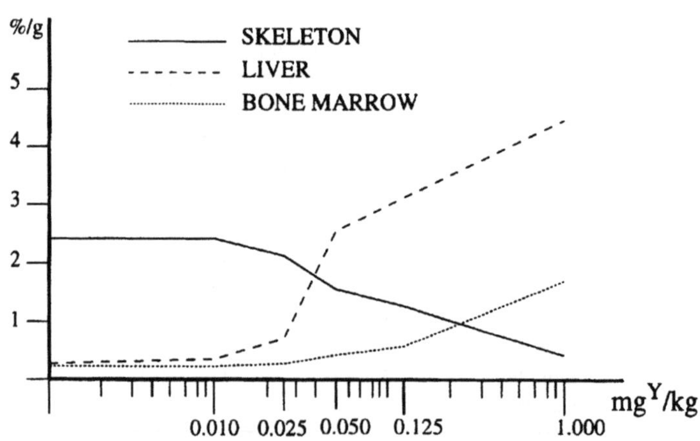

Figure 8-29. Influence of the amount of an yttrium carrier on the specific activity (%/g) in various organs of rats. The yttrium carrier was administered as the citrate complex, and the organ activity was measured 24 h after i.v. injection of ^{88}Y. Modified from Kutzner *et al.* (1986), with permission of the authors and F. K. Schattauer Verlagsgesellschaft mbH.

8.8 Biochemistry in Humans

Many polyvalent cations do not diffuse readily across the gut wall and therefore are mostly excreted in the feces. In a review of the metabolism of the elements in the human body (Spector, 1974), scandium and yttrium were classified as elements that are very poorly (less than 5%) absorbed across the gut wall. Scandium as well as larger Ln are accumulated in liver, as well as in bone, whereas yttrium and the smaller Ln are taken up mainly by bone (Spector, 1974).

Less than 0.4% of an i.v. dose of 85 mg $^{90}YCl_3$ in humans was excreted via urine over a 3-day period, whereas at a low dose (4.2 μg), over 12% was recovered in the urine during the same period. Administration of ^{90}Y with a carrier resulted in greater retention in liver; in the absence of a carrier, more of the radioisotope was excreted and was partly localized in nonreticuloendothelial organs (Andrews, 1956). After i.p. injection of colloidal $^{90}YF_3$ to a patient with extensive carcinomatous ascites, less than 1% of the radioactivity was found in blood at 10 min as well as up to 48 h after injection (Mayer and Morton, 1956). High amounts of several REEs were found in fecal specimens of Apollo astronauts, in contrast to low amounts of scandium (Brodzinski, 1972).

Intrapleural administration of $^{90}YCl_3$ into cancer patients with bilateral effusion and ascites resulted in a low radioactivity in the plasma, bone, and liver (Laszlo, 1956). Trace amounts of yttrium complexes with cyclohexane-*trans*-1,2-diaminetetraacetic acid (CDTA) and DTPA were excreted within 8 h in amounts approaching 100% of the administered dose (Table 8-15). A positive correlation between urinary excretion of chelated yttrium and the stability constant of the metal chelate was noted (Kroll *et al.*, 1957).

Detailed studies on the metabolism of REE in humans were made in the 1950s and the early 1960s by Spencer and her many colleagues at the U.S. Veterans Administration Hospital in Hines, Illinois, and at the Montefiore Hospital in New York (Spencer, 1960; Spencer and Rosoff, 1963). Patients diagnosed with cancer or with other diseases were treated with radioactive chelated scandium, yttrium, and other metals and compounds. In addition to the expected therapeutic results, these studies provided data about the metabolism of scandium and yttrium and their distribution, retention, and long-term turnover in humans (Stannard, 1988).

The excretion of ^{46}Sc in stool was considerably higher than in urine. The fecal ^{46}Sc excretion was thought to reflect the transfer of scandium from the liver through the bile into the intestine. In a study performed on a patient with a biliary fistula, the excretion of ^{46}Sc in the bile approximated fecal excretion (Spencer and Rosoff, 1963). Following administration of ^{46}Sc strongly chelated by EDTA or DTPA in various organs of 12 patients, plasma ^{46}Sc decreased rapidly; up to 82% of the dose was excreted in urine within 24 h (Fig. 8-30). The concentration of ^{46}Sc in tissues, determined 5, 6, and 7 months after i.v. administration of ^{46}Sc-

Table 8-15. Plasma Levels and Cumulative Urinary Excretion of Yttrium Chelates in One Patient of Groups of 6–13 Patients[a]

	8 h		24 h		
Yttrium chelate	Plasma[b]	Urine[c]	Plasma[b]	Urine[c]	Percent urinary excretion[d]
Y-EDTA	5.3	36.5	0.5	46.7	55.9
Y-CDTA	5.7	68.4	0	97.6	87.9
Y-DTPA	5.0	75.0	0	101.3	96.7

[a] Data reprinted from Kroll *et al.* (1957), with permission of *Nature* (London).
[b] Percentage of the administered dose in total plasma volume.
[c] Percentage of the administered dose.
[d] Average cumulative urinary excretion in groups of 6–13 patients, as percentage of the administered dose.

NTA, showed that the highest uptake occured in spleen, averaging five times as high as in the uptake liver. Vertebrae and adrenals also accumulated substantial amounts of scandium (Rosoff *et al.*, 1961). After several months, radioactivity could be detected in vertebrae, while the concentration in kidney was very low, but still detectable (Fig. 8-31). The whole-body count studies indicated that a very small fraction of an i.v. administered dose continue to be excreted for a long period. Long biological half-lives of 1300 and 1557 days for ^{46}Sc-NTA in two patients were calculated (Rosoff *et al.*, 1965). The removal of ^{46}Sc administered intravenously as a single tracer dose in the form of its weak NTA chelate was very slow from the vascular space; it was excreted principally via the intestine. Infusions of either EDTA or DTPA, used as enhancing agents, at 24 h on the 6th day after ^{46}Sc tracer injection induced an increased excretion of ^{46}Sc in urine. Moreover, the urinary ^{46}Sc excretion remained elevated after the discontinuation of the infusions of DTPA (Spencer and Rosoff, 1965).

Scandium citrate was fed to eight volunteer patients. Some patients displayed no adverse effects from the scandium; however, in one patient, a hemolytic effect was observed. *In vitro* autohemolysis testing indicated that this effect of scandium was unique to human red blood cells (Hayes and Edwards, 1973).

Complexation with DTPA was the most efficient method of quantitative excretion through urine in 48 h. Following oral administration of yttrium, scandium, and lanthanum in ionic form, practically the entire dose could be recovered in stool; apparently, these elements pass unabsorbed through the intestine (Fig. 8-32). A small-scale study of the biokinetics of yttrium in human volunteers, to whom ^{88}Y citrate was injected intravenously, indicated that about 22% of the ^{88}Y was excreted with a half-time of 16 h, and 78% was retained with a half-time of several years (Etherington *et al.*, 1989). The distribution and retention in an animal model were similar to those obtained in this experiment. It was calculated that of the element absorbed, 25% goes directly to excreta, 50% is

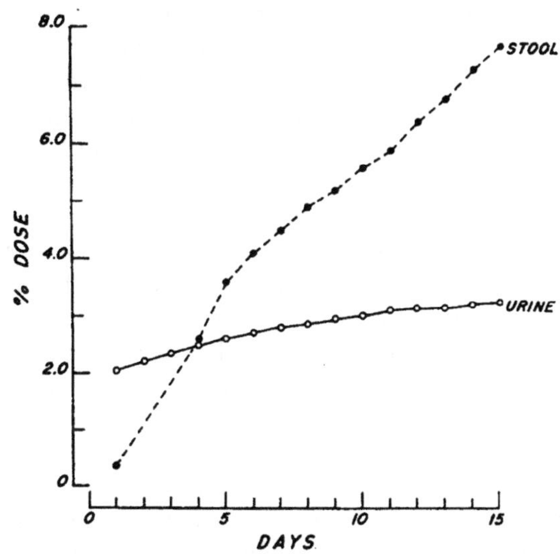

Figure 8-30. Excretion of ^{46}Sc-NTA in humans over a period of 15 days. A single tracer dose of ^{46}Sc-NTA was injected intravenously, and the amounts of ^{46}Sc excreted in urine and stool were determined. Reprinted from Rosoff *et al.* (1965), with permission of the authors and Elsevier Science.

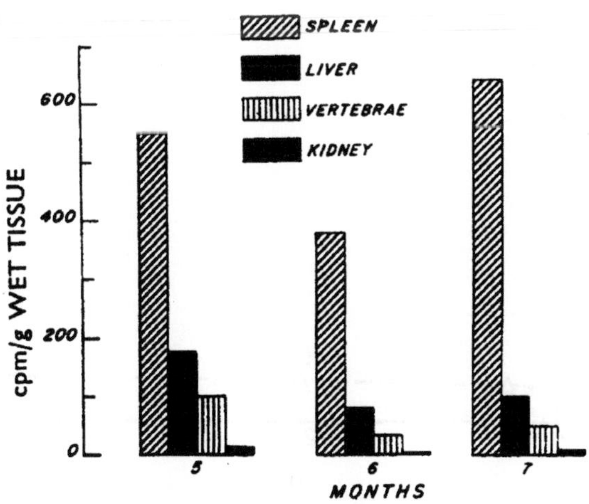

Figure 8-31. Long-term distribution of ^{46}Sc-NTA in several organs of humans. Tissue samples were analyzed at autopsy of three patients, 5, 6, and 7 months after i.v. injection of a single dose of ^{46}Sc-NTA. Reprinted from Rosoff *et al.* (1965), with permission of the authors and Elsevier Science.

transported to the skeleton, 15% is transported to liver, and 10% is distributed uniformly through all other organs and tissues. Subsequent to i.v. injection to a patient, only 29% of ^{90}Y-EDTA was found in the blood, whereas 95% of $^{90}YCl_3$ was retained in the body (Hart *et al.*, 1955). Parenteral application of amino acids to humans caused a parallel displacement of scandium and silver between intra- and extracellular space in blood; an antiparallel displacement of iron and antimony, and selenium and cobalt was reported (Bernhard *et al.*, 1977). These changes of the TE concentrations in blood serum (Fig. 8-33) were assumed to occur as a consequence of stimulation of metabolic processes by the amino acids. The concentrations of scandium and other TE increased in serum of 21 patients during an oral glucose test (abnormal glucose tolerance). A significant correlation between some TE, glucose, insulin, and growth hormones was found (Kvicala *et al.*, 1991).

Intra-articular ^{90}Y treatment of persistent synovitis of the knee induced a marked deposition of ^{90}Y in Baker's cysts, which may have diminished the therapeutic effect (Smith *et al.*, 1988). Deposition ranged from little or no measurable uptake of ^{90}Y to a considerable proportion of the administered activity. The poor relation between the treatment and clinical response may be indicative of biochemical (pathologic) and physiological individuality. *In vitro* study of Nd : YAG laser irradiation of articular cartilage indicated stimulation of

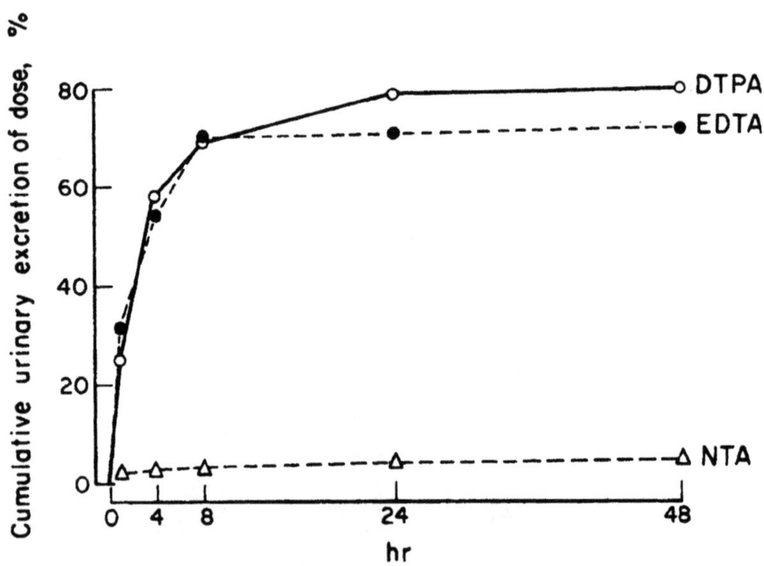

Figure 8-32. Urinary excretion of scandium administered in chelated form with EDTA, NTA, or DTPA in humans. Reprinted from Rosoff et al. (1965), with permission of the authors and Elsevier Science.

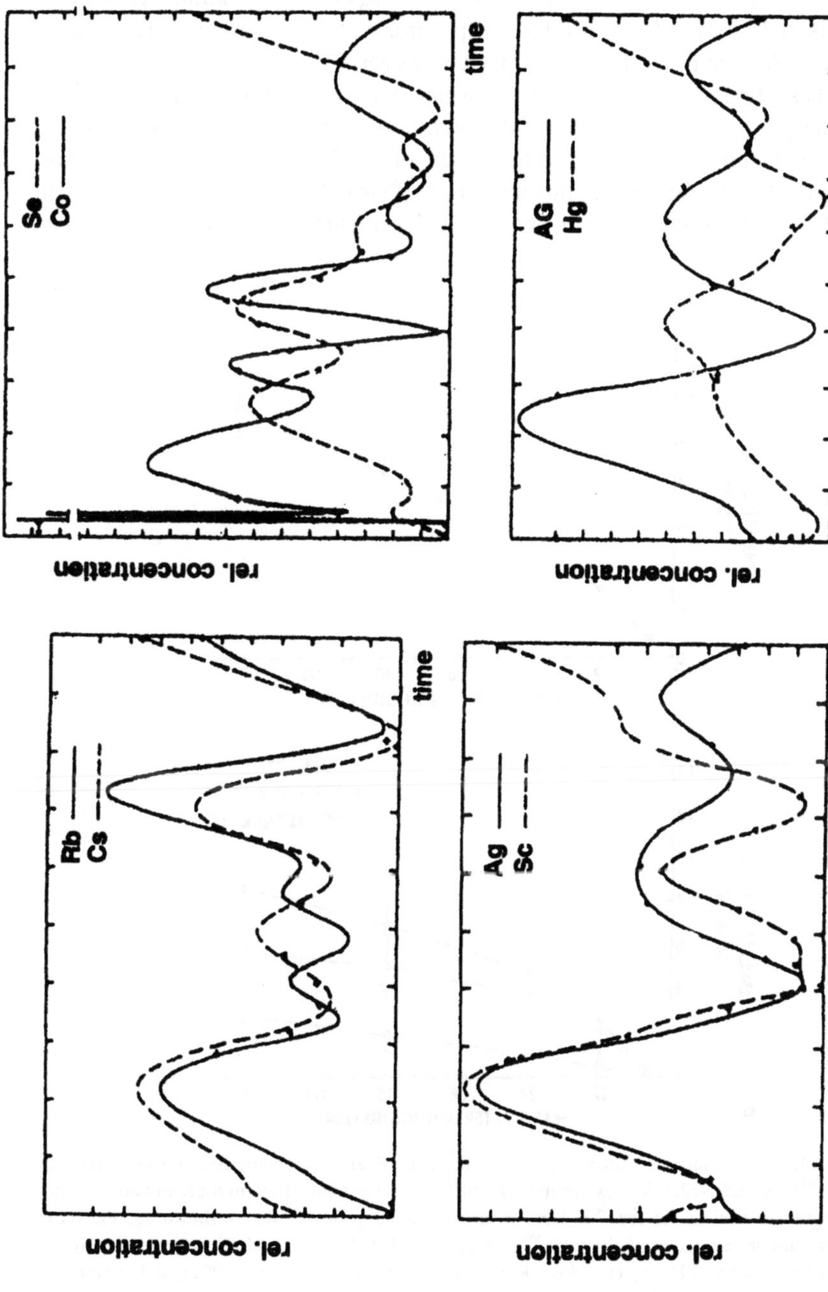

Figure 8-33. Comparison of scandium concentrations with concentrations of other trace elements in human serum before and three hours after parenteral infusion of amino acids. From Bernhard *et al.* (1977), with permission.

both matrix and DNA synthesis (Spivak *et al.*, 1992). Irradiation of both canine and bovine cartilage explants maintained in organ culture stimulated matrix synthesis at 6–7 days following laser exposure, with a concomitant decrease in baseline DNA synthesis. By 12–14 days, however, these effects were no longer seen. These findings indicate that exposure to low-level noncontact Nd:YAG laser energy promotes a significant stimulation of cartilage matrix synthesis.

Several studies have analyzed the uptake of administered radiolabeled MAbs by tumors. Following i.p. administration of ^{90}Y-labeled high molecular-weight MAb B72.3 to one of three investigated patients with epithelial ovarian

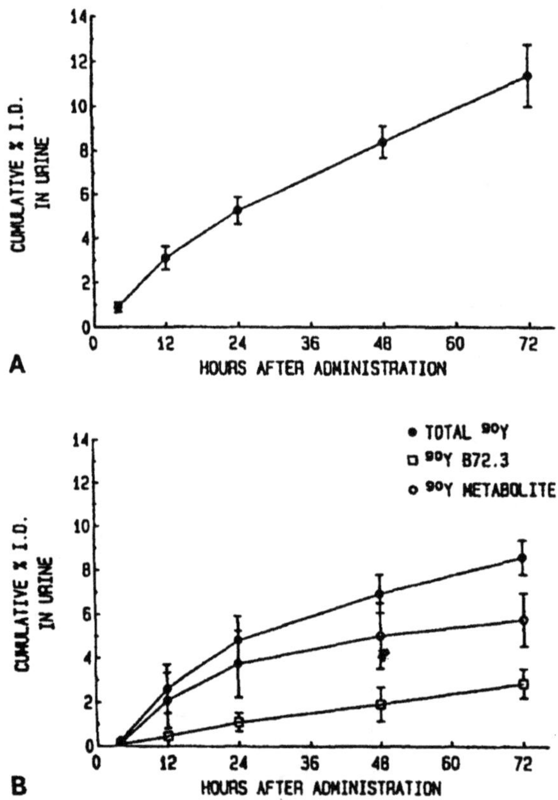

Figure 8-34. (A) Cumulative urinary excretion of ^{90}Y label after i.p. administration to patients of 1 mCi of ^{90}Y-labeled B72.3. Values are means ± SEM for all patients. (B) High-performance liquid-chromatographic analysis of the ^{90}Y label in urine from 3 patients. Values shown are the mean cumulative urinary excretion of the total ^{90}Y label, ^{90}Y-labeled B72.3, and a low-molecular-weight ^{90}Y-labeled metabolite. I.D., Injected dose. Reprinted from Rosenblum *et al.* (1991), with permission of the authors and Oxford University Press.

carcinoma, a ^{90}Y metabolite with a molecular weight lower than 1000 was found in plasma, peritoneal fluid, and urine (Rosenblum *et al.*, 1991). This metabolite was considered to be either a ^{90}Y-DTPA complex or free ^{90}Y (Fig. 8-34). Both the ^{90}Y-labeled metabolite and the ^{90}Y-labeled B72.3 were found in plasma, with biological half-lives of 16 ± 2.2 h and 25 ± 5 h, respectively. About 70% of the label was present as the ^{90}Y-labeled metabolite; 30% of the ^{90}Y label in urine was identical to the authentic ^{90}Y-labeled B72.3 standard. The tissue distribution showed that normal tumor tissue and omentum contained the highest concentration of ^{90}Y (Fig. 8-35). Large heterogeneity existed in the uptake of ^{90}Y by various sites in the same patient, as well as among different patients (Rosenblum *et al.*, 1991). Hird *et al.* (1990) investigated a prospective treatment of ovarian cancer with i.p. administration of ^{90}Y-labeled MAb. The ^{90}Y accumulation in bone, liver, or spleen following the treatment was considered to be due to the release of ^{90}Y from the complex as a result of liver metabolism. The synergistic action between Taxol (paclitaxel) and the ^{90}Y-labeled chimeric L6 antibody was found to be relevant to therapy for breast cancer. This effect may relate to the p53 mutant status and BCL2 expression in HBT 3477 cells (DeNardo *et al.*, 1997).

The role of chemical elements in aging processes has been intensively studied in recent years. Calculation of the correlation coefficients (CC) between age and amounts of TE in human brain found significant positive CC for cadmium, molybdenum, lead, and tin and negative CC in medulla, pons, midbrain, and thalamus for potassium, magnesium, scandium, and zinc (Saito *et*

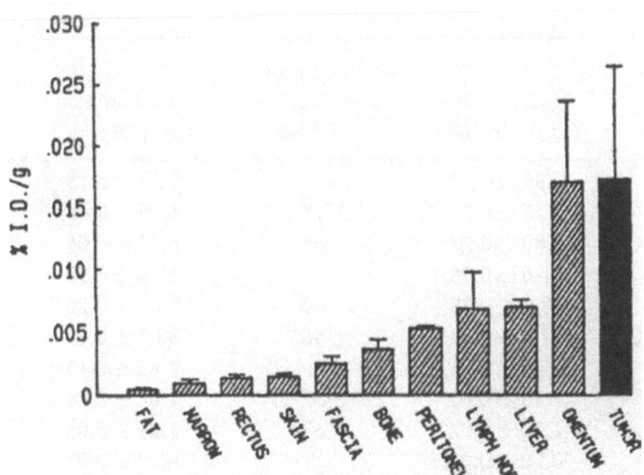

Figure 8-35. Tissue distribution of ^{90}Y-labeled B72.3 in patients. Values shown are mean ± SEM for samples obtained from laparotomy of 9 patients who had received 1 mCi of ^{90}Y. I.D., Injected doses. Reprinted from Rosenblum *et al.* (1991), with permission of the authors and Oxford University Press.

al., 1994). Further investigation of the dynamics of TE in regions of the rat brain revealed that, except for the steady increase of copper during 52 weeks of ontogeny, most of the TE investigated showed a marked decrease with age. A pronounced decrease was found for strontium, and also for manganese, rubidium, scandium, and zinc (Table 8-16). These changes differed markedly among the eight brain regions that were analyzed (Saito *et al.*, 1994). Investigation of the dynamics of TE in regions of the human brain over the life span found that scandium, silver, copper, and iron concentrations increased up to the age of 59 years and then declined; in infant brain, the concentrations of scandium and other TE were lower than in adult brain (see Part 1, Fig. 5-10). Markesbery *et al.* (1984) suggested that essential elements remain within narrow concentration limits throughout adult life. It was speculated that the increased concentrations of nonessential elements may reflect accumulation from the environment and impaired removal or altered balance with other elements.

8.9 Nutrition and Diet

The relative abundances and correlations between several chemical elements in biological, geochemical, and hydrochemical systems were assessed in Chapters

Table 8-16. Changes with Age in the Concentrations of Scandium, Zinc, Manganese, and Strontium in Rat Brain Regions[a]

		Medulla and pons		Midbrain and thalamus	
Metal	Age	Concentration (μg/g dry wt.)	Percent of concentration at 3 wk	Concentration (μg/g dry wt.)	Percent of concentration at 3 wk
Scandium	3	9.85 ± 0.34		7.51 ± 0.15	
	26	7.58 ± 0.21	77	6.75 ± 0.06	90
	52	6.80 ± 0.22	68	6.43 ± 0.05	86
Zinc	3	31.6 ± 0.78		48.4 ± 0.90	
	26	18.9 ± 0.38	60	34.7 ± 0.20	72
	52	17.7 ± 0.34	56	33.2 ± 0.19	69
Manganese	3	2.40 ± 0.06		1.85 ± 0.13	
	26	1.26 ± 0.02	52	1.04 ± 0.06	56
	52	1.27 ± 0.02	53	1.07 ± 0.05	58
Strontium	3	0.20 ± 0.01		0.13 ± 0.01	
	26	0.07 ± 0.01	35	0.038 ± 0.01	31
	52	0.08 ± 0.01	40	0.036 ± 0.01	28

[a] Data recalculated from Saito *et al.* (1994), with permission of the authors.

4 and 5 of Part 1. Hamilton (1987) published a survey of the correlation coefficients between the abundances of elements in biotic and abiotic systems, such as diet and the Earth's crust or seawater, obtained by least-squares linear regression analysis. High correlations existed between diet versus human blood ($r = 0.95$); poorer correlations were found between diet g.i. transfer versus human blood. Limited data were available for dietary intake of scandium and yttrium.

A literature survey of the g.i. absorption of selected TE by human subjects indicated a value of 5 for total absorption of scandium from the ingested dose by adult subjects. Similar or even lower values were assigned for aluminum, cadmium, manganese, niobium, nickel, tin, and zirconium; values higher than 20% were assigned for copper, selenium, zinc, and other TE (ICRP, 1993). Total parenteral nutrition administered to four patients, which included a fat emulsion, as well as calcium, copper, fluoride iodine, manganese, and zinc, caused a substantial retention of iron, scandium, and several other elements (Jacobson and Wester, 1977).

Duffield (1991) compiled a database on the total ligand and metal concentrations in saliva, stomach juice, intestinal fluid, milk, and blood plasma and the physical and chemical constants for all feasible low-molecular-mass complexes. Computer models were used to calculate the distribution of chemical species at equilibrium. Several aluminum complexes may exist as net-neutral charged species, which are potentially bioavailable, whereas others are charged and highly solvated and are removed by the renal system. It may be speculated that similar chemical species of REE might occur in biofluids because several forms of chemical species of scandium and other REE occur in environmental systems (see Part 1, Fig. 4-19).

A study which investigated the pathways of 15 TE from the environment to humans found that the average daily intake of scandium in Italy through air, drinking water, and diet was 0.04, < 0.02 and 0.2 μg, respectively (Clemente, 1976). Calculation of the metabolic balance of scandium gave values of <0.01 μg for daily urinary excretion and 0.2 μg for total daily fecal excretion. An alternative approach employed by Cohen (1985) to analyze the transport of elements from soil through plants and animals to the human diet took into account the average daily human ingestion of each element and its abundance in soil. The relative dietary intakes of 35 chemical elements in various types of food (plant-derived, milk, and meat) were compared. The transport of elements from soil to the average American human diet was estimated by assuming that 0.53 kg of each kilogram of food consumed is plant-derived. For yttrium, the value of the ratio of quantity actually ingested to what would be ingested if diet consisted entirely of 1 kg/day of plant-derived food was equal to 0.5. This value was close to those of several TE, whereas values of 1.4–2.0 were found for sodium, phosphorus, and zinc (Cohen, 1985).

Experiments performed in the 1960s with several animal species indicated that ^{46}Sc is a suitable marker in nutrient studies; depending on the mode of administration, it is a very poorly absorbed reference material. An amount of 0.12 mg Sc^{3+}/kg administered together with REE to mice was considered safe for use as a nutritive marker, even when used in multiple forms for accurate determination of nutrition intake and utilization by animals (Hutcheson *et al.*, 1975). Tracer amounts of ^{46}Sc given to calves as a single i.v. injection resulted in a high concentration of ^{46}Sc in liver; urinary excretion of ^{46}Sc was negligible, and it cleared from blood at a much slower rate than ^{144}Ce (Miller and Byrne, 1970). Analysis of tissue and feces indicated that orally administered ^{46}Sc is essentially not absorbed (Table 8-17). In a study reported by Hutcheson *et al.* (1983), scandium and other metal markers were pulse-fed to pigs with different parts of a meal. The passage of scandium through the transverse and descending colon was maximal at 20–24 h, whereas the passage through the stomach was maximal at 4 h following the pulse dose. The rate of passage of scandium through the entire g.i. tract was 3.8% per hour. Table 8-18 presents the ^{46}Sc distribution in tissues of pigs as percent of the dose, which was administered intravenously or orally. Excretion and tissue distribution of scandium, expressed as percentage of concentration found at 5 min after i.v. administration, decreased, respectively, from 88% to 33% at 10 h. Urinary scandium averaged 0.1% of the i.v. dose; scandium was not detected in blood and urine after oral dosage. Within 8 days, 98% of the orally administered dose was found in feces. Scandium had mostly disappeared by 39 h; the transverse and descending colon and the sigmoid rectum retained 7.4 and 4.9%, respectively, of the administered dose.

Studies on the metabolism of ^{90}Sr and ^{90}Y in rabbits fed diets containing different amounts of calcium found that ^{90}Y excretion was greater in old than in young animals (Kidman *et al.*, 1950). Diet had no effect on excretion in young animals; in 6-month-old rabbits fed on a low-calcium diet, the percentage of yttrium excreted was low, but in rabbits fed a normal calcium diet, the yttrium excretion was higher (Fig. 8-36). When mice were provided with scandium or yttrium in their diet, the retention of several essential elements was enhanced or changed (Table 8-19). Such interactions may alleviate the toxic effects of other TE (Schroeder and Nason, 1976). A balance study of intake and excretion of several TE, including scandium, did not find any regular influence of an increased intake of calcium on the investigated TE (Wester, 1974).

Multiple Ln markers were used by Hutcheson *et al.* (1979) to determine the uptake and apparent utilization of nutrients in humans. The scandium marker was given as a capsule at each of three meals per day, as an internal marker at the rate of 8.5 μg Sc^{3+}/g of fat. The scandium concentration in feces, obtained by averaging the daily concentration for the entire 5-day period, was 38–48 mg Sc^{3+}/kg per day. When direct and marker methods of determining nutrient intake and apparent utilization were compared, no significant differences were found for

Table 8-17. Percent of ^{46}Sc Recovery in Various Tissues of Calves and Percent Gastrointestinal Distribution within 72 h after Intravenous Administration or 8 Days after Oral Administration[a]

Tissue	Single i.v. administration	Single oral administration	Daily oral administration
	Percent of Sc recovery		
Liver	26.9	0.003	0.02
Lung	7.6	0.005	0.01
Spleen	2.8	< 0.001	0.001
Heart	0.9	< 0.001	0.004
Kidney	0.8	< 0.001	0.003
Femur	0.8	< 0.0005	0.002
Adrenal	0.1	< 0.001	0.002
Rib	0.1	—	—
	Percent of the dose		
Total digestive tract	2.1	0.3	75.3
Fecal excretion	1.3	98	101.3
Urinary excretion	0.1	< 0.01	—
	Percent of total digestive tract content		
Small intestine 1	19.3	2.2	0.4
Small intestine 2	35.6	1	2.6
Small intestine 3	18.7	4	8.4
Large intestine	13.9	15	31
Rumen	4.3	47	44
Cecum	4.2	3.3	4.5
Abomasum	3.2	4	2.7
Omasum	0.8	24.3	6.3

[a] Data reprinted from Miller and Byrne (1970), with permission of the American Society for Nutrition Sciences.

gross energy, fat, protein, and calcium (Hutcheson *et al.*, 1979). Luckey *et al.* (1979) fed nonradioactive scandium and Ln oxides with a standard diet to human volunteers to ascertain rates of flow and passage of TE from three meals. The findings indicated that the human alimentary tract has three physiological compartments. The food remains in the digestion–absorption compartment for 10–15 h. Presumably, this is the stomach and small intestine. An intermediate compartment which provides time for considerable fermentation was observed in some subjects. In the first 24 h, some individuals excreted only 10% of the markers, while others excreted as much as 60%.

Adequately mixed diets may include practically all the inorganic mineral elements existing in nature. Problems of mineral deficiency or excess may arise in

Table 8-18. Passage of Scandium through the Gastrointestinal Tract of the Pig, Expressed as Percent of the Marker Given[a]

Segment tissue	Sacrifice time (h) following pulse dose				
	4	12	20	24	39
Stomach	20.9	5.0	9.9	0.2	0.2
Jejunum	14.3	1.7	0.8	0.1	0.1
Ileum	20.1	0.6	3.1	0.2	—
Cecum	6.8	0.9	1.3	0.4	0.2
Ascending colon	30.0	8.0	1.8	14.6	1.8
Transverse and descending colon	7.6	11.3	16.7	17.7	7.4
Sigmoid rectum	0.1	2.1	4.9	7.8	4.9

[a] Data reprinted from Hutcheson *et al.* (1983), with permission of Elsevier Science.

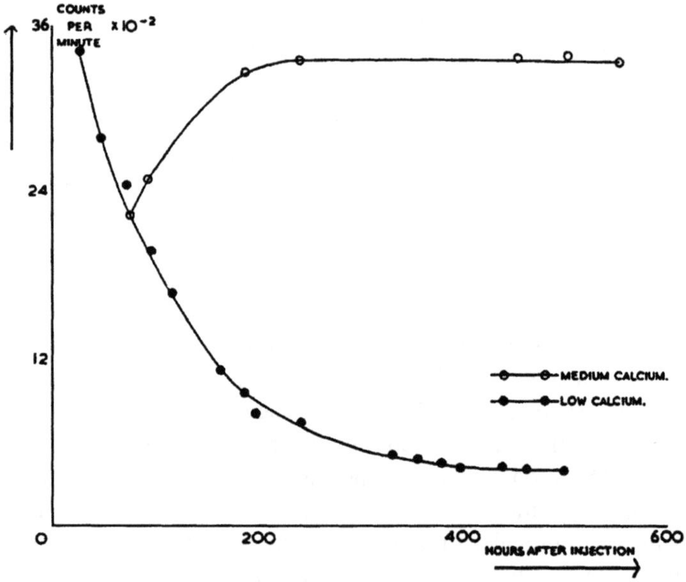

Figure 8-36. Curves showing the restoration of equilibrium concentration of ^{90}Y in urine excreted by rabbits aged 6 months fed on whole medium and on a low-calcium diet. Reprinted from Kidman *et al.* (1950). We made efforts without results to trace the copyright holders.

Table 8-19. Concentrations of Chromium, Copper, Manganese, and Zinc in Tissues of Mice Fed with Scandium or Yttrium[a]

	Concentration (mg/kg)				
Metal fed	Liver	Lung	Heart	Kidney	Spleen
			Chromium		
Control	0.50 ± 0.19	1.84 ± 1.43	3.82 ± 1.41	1.79 ± 0.56	4.03 ± 1.48
Sc	0.93 ± 0.31	1.70 ± 0.56	8.43 ± 4.01	5.91 ± 2.27	7.63 ± 3.63
Y	0.46 ± 0.19	1.84 ± 0.93	1.77 ± 0.51	1.84 ± 0.54	2.68 ± 0.73
			Copper		
Control	19.8 ± 0.90	19.4 ± 1.68	32.5 ± 3.38	18.6 ± 1.67	23.8 ± 3.69
Sc	25.1 ± 4.28	17.5 ± 3.23	42.7 ± 8.27	30.7 ± 6.37	26.2 ± 9.44
Y	21.6 ± 2.16	25.7 ± 1.82	33.2 ± 2.88	20.9 ± 2.73	26.3 ± 3.37
			Manganese		
Control	4.34 ± 0.31	2.71 ± 0.26	3.95 ± 0.69	3.68 ± 0.27	4.08 ± 0.42
Sc	4.73 ± 0.53	3.95 ± 0.91	5.97 ± 0.76	6.09 ± 0.83	5.46 ± 1.63
Y	4.29 ± 0.38	3.52 ± 0.31	3.60 ± 0.39	4.84 ± 0.47	4.41 ± 0.68
			Zinc		
Control	146 ± 8.5	100 ± 5.0	104 ± 5.8	100 ± 7.5	156 ± 14.5
Sc	198 ± 31.2	146 ± 31.0	161 ± 33.0	157 ± 38.8	158 ± 47.8
Y	163 ± 14.7	136 ± 12.3	122 ± 10.0	173 ± 26.9	173 ± 10.5

[a] Scandium or yttrium (10 mg/kg) was added to the drinking water of rats throughout their lifetime; the diet was relatively low in TE. Data reprinted from Schroeder and Nason (1976), with permission of the American Society for Nutritional Science.

regions where there is an inadequate or excessive content of TE in soil and water (Bender, 1993). The 10th edition of the Recommended Dietary Allowances (RDA, 1989) for humans examined the dietary sources, bioavailability, usual intakes, signs of deficiency or excessive intake, and toxicity of minerals (calcium, phosphorus, and magnesium) and electrolytes (potassium and sodium); only nine TE (chromium, copper, fluorine, iron, iodine, manganese, molybdenum, selenium, and zinc) were considered. Limited information is presented in the RDA about the other 80 chemical elements, despite their presence in diets and interactions with elements of major concern, as well as with compounds of biological interest. The concentrations of 24 elements in milk quality control materials have been published by the World Health Organization (WHO, 1989). Scandium concentration in breast milk ranging between 7 and 38 ng/l, and a concentration of yttrium less than 4 ng/l were found by Durrant and Ward (1989). The concentrations of scandium in food products and in reference dietary products of various countries are given in Table 8-20. Only isolated data are

Table 8-20. Concentrations of Scandium in Various Food Products

Product	Country	Sc conc (μg/kg or μg/l)	Reference
Milk powder	Austria	0.18	Suschny *et al.* (1982)
Dairy products	USA	0–1.2	Cunningham and Stroube (1987)
Meat products	Japan	0.2–15	Inarida *et al.* (1987)
Egg yolk	Pakistan	0.2–1.4	Waheed *et al.* (1985)
Soups, mushrooms	Greece	0.9–5.5	Kanias (1991)
Rice	Brazil	1.1–2.1	Maihara and Vasconcellos (1989)
Bread	Brazil	1.6–2.6	Maihara and Vasconcellos (1989)
Wheat	USA	< 1	Iskander and Morad (1986)
Oils and fats	USA	< 0.4	Cunningham and Stroube (1987)
Vegetables	USA	0.23–5.7	Cunningham and Stroube (1987)
Fruits	USA	0.73–1.1	Cunningham and Stroube (1987)
Mustard seeds	Canada	5	Ila and Jagam (1980)
Onions	Pakistan	22	Zaidi *et al.* (1992)
Tea leaves	Pakistan	49	Ahmad *et al.* (1983)

available for yttrium. Improvements in analytical chemistry have allowed a preliminary estimation of the dietary intake of scandium (Table 8-21). The daily intake of scandium through wheat, pulses, and rice in Pakistan was reported to be is 2.71, 0.69, and 0.073 μg/day, respectively (Qureshi *et al.*, 1993).

Analysis of Japanese commercial laboratory animal diets found that the scandium concentration in seven brands varied between 6 and 90 μg/kg (Sato, 1986). Tea and coffee are a source of many TE. A concentration of 49 μg scandium/kg was found in tea leaves; 28% was transferred into the drinkable

Table 8-21. Dietary Daily Intake of Selenium, Chromium, and Selenium in Various Countries, as Determined by Analytical Techniques

			Dietary daily intake (μg/day)		
Country	Year	Author(s)	Sc	Cr	Se
Germany	1977	Schelenz	1.0	62	59
Italy	1978	Clemente *et al.*	0.17	40	14
Germany	1983	Schelenz	0.38	62	150
Norway	1984	Bibov *et al.*	0.2	62	31
Canada	1984	Gibson and Scythes	0.38	47	130
Switzerland	1987	Wyttenbach *et al.*	0.30	50	70
Turkey	1988	Mumcu *et al.*	2.2	49	52
USA	1989	Iyengar	< 0.5	37	130
Worldwide	1996	WHO	0.38	50	60

portion, a percentage that is lower than those for other TE (Ahmad *et al.*, 1983). Interindividual variability is manifested in the uptake and utilization of TE in the daily diet. A study of the food and beverage intake by a group of 26 women, 25–45 years old, revealed a relatively large variation in daily TE intake (Schelenz, 1983; Fig. 8-37). Analysis of 87 foodstuffs (dairy, fish, vegetable, grain, and fruit) found that all of the samples contained amounts of scandium, lanthanum, and several other TE in the range of 9 ng–0.1 μg/g ash (Braunstein *et al.*, 1982). Modern-day processing of foods results in changed intakes of minerals, especially in industrial societies. The TE concentration in many foods is reduced during refining processes (Hamilton, 1987). Refining of sugar and flour and polishing of rice deprive humans of TE, which remain in the residue; this may create TE deficiency problems. An analysis of brown sugar found values of 2 and 0.07 μg/kg for scandium and yttrium, respectively. In refined sugar, the values for these elements were 0.001 and < 0.005 μg/kg, respectively. Up to 85% of bulk elements and TE, as well as vitamins, are lost in the refining of whole wheat to produce flour for white bread (Jones, 1994).

8.10 Involvement in Enzymes and in Main and Secondary Metabolism

Central metabolism is crucial to all activities within a cell. Identical or closely similar metabolic pathways occur in archaea bacteria, eubacteria, and eukaryotes, although the precise details may vary between various organisms (Danson, 1993). On the other hand, it is well documented that metal ions play a critical role at every level in biological systems. These systems contain a great number of ions in solution, which may be attached to macromolecules and electroneutral polyelectrolytes (Bergethon and Simons, 1990). The concentrations of various ions and other species in cells are regulated by homeostatic processes. The need for regulation arises from the imbalance between supply and demand and reflects changes in intake, utilization, and excretion from cells (Vaidhyanathan, 1993). A general chart of intermediate metabolism in animals shows the interactions of about 700 to 1,000 small organic molecules (Kauffman, 1995). When the involvement of inorganic molecules in the cellular traffic is included, the picture becomes substantially more complex.

Transition metals, which possess unpaired electrons, can form complexes with oxygen; this results in toxic reduced oxygen species. The free radicals (FR) are important in health and disease. Investigation of the influence of 14 metal ions on FR formation in subcellular fractions of the lung mitochondria and brain synaptosomes of rats found that iron and manganese were associated with a highly significant increase in FR formation in lung. The response to the other metal ions investigated showed a slight, insignificant decrease in FR. Sc^{3+} slightly enhanced FR formation at the low concentration of 0.01 μM but

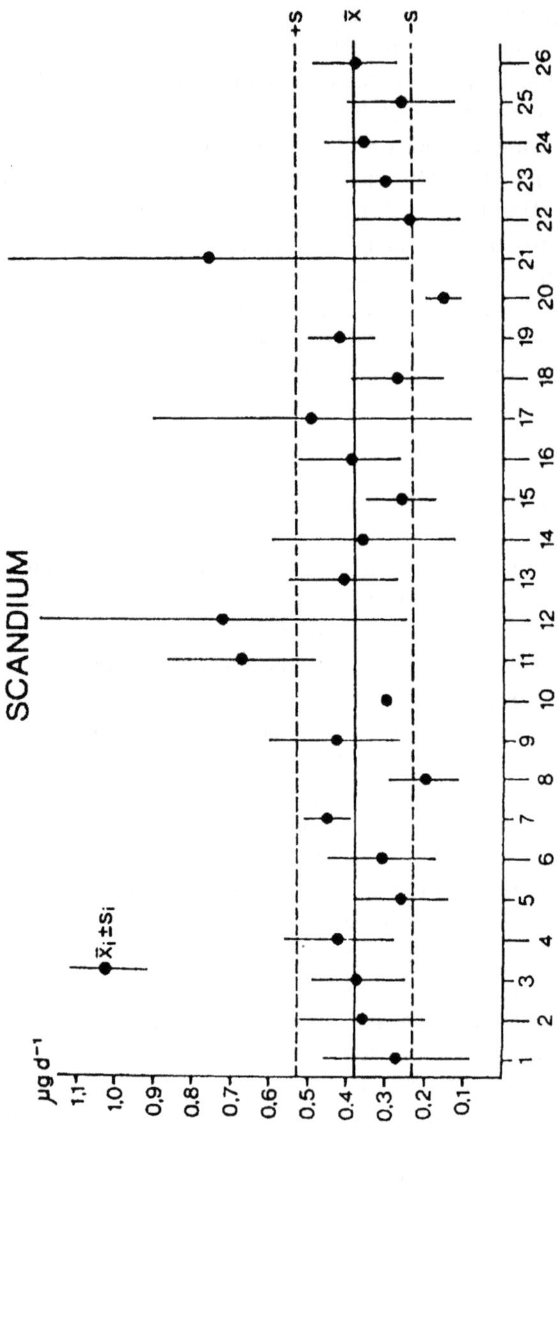

Figure 8-37. Individual variability of daily dietary intake of scandium among 26 German women. Reprinted from Schelenz (1983), with permission of the author and Bundesforschungsanstalt für Ernährung.

decreased FR formation in lung at higher concentrations. In brain cerebellum, only iron caused a significant increase in FR; manganese, as well as aluminum, lanthanum, and scandium, caused only minor changes (Fig. 8-38). The discrepancy between the responses of pulmonary and cerebellar tissues was suggested to be the result of the different biochemical functions and chemical composition of these organs (Horovitz and Bondy, 1994). The reason for the enhancement of FR formation in lung tissue by low concentrations of scandium was apparently related to the greater electron density, increased net charge, and higher coordination number of Sc^{3+}, compared to divalent metal ions (Scheer, 1989). When suspensions of the pigment melanin, which is regarded as a large FR, were incubated with various metals, mainly Sc^{3+} and La^{3+} induced the highest amount of FR formation, which was presumed to be due to the formation of radical–metal-ion complexes (Felix *et al.*, 1978). Metal-induced FR may act in various ways in tissues having distinctive functions. For example, aluminum hydroxide administered to rats increased brain lipid peroxidation, while such an effect did not occur in lung, kidney, liver, or spleen (Ohtawa *et al.*, 1983).

The presence of scandium and other TE in the sarcoplasmic reticulum vesicle isolated from skeletal muscle of rabbit apparently induced the modification of SH groups and increased lipid peroxidation (Papp *et al.*, 1985). The following decreasing order of changes in membrane physical properties, i.e., release of 5(6)-carboxyfluorescein from liposomes, was found: $Al \approx In > Ga \approx Sc > La \approx Y$. At concentrations between 10 and 200 μM, beryllium, gallium, indium, and scandium caused liposome aggregation and fusion; yttrium and lanthanum were maximally effective at 20 μM. Metal-induced fusion of liposomes, was dependent on the negative charge density of the liposomes (Fig. 8-39). These TE stimulated Fe^{2+}-initiated lipid peroxidation in a time- and dose-dependent manner. At 50 μM concentration, the effect of metals on liposome fusion was positively correlated with their capacity to promote Fe^{2+}-supported lipid peroxidation. The changes in the organization of the bilayer that facilitate the initiation or propagation of the lipid peroxidation chain reaction may be a common mechanism whereby Al^{3+}, Ga^{3+}, In^{3+}, Sc^{3+}, and Y^{3+} promote lipid peroxidation in the presence of an external source of FR without redox capacity (Verstraeten and Oteiza, 1995).

Phospatidylinositol (PI) and its phosphate are cell-membrane components responsible for the transduction of extracellular signals into cells. PI is hydrolyzed by chlorides of yttrium and some other REE. This activity, which mimics phospholipase activity, was optimal at pH 7.5–8.5 and 30°C (see Part 1, Fig. 6-15). In contrast Fe^{2+}, Zn^{2+}, and Cu^{2+} chlorides were totally inactive. The catalytic activity of Y^{3+} was the highest, with a conversion of 32 mol %. However, Y^{3+} showed only a marginal hydrolytic effect on ATP. PI can be hydrolyzed nonenzymatically with the use of REE as catalysts (Matsumura and Komiyama, 1994).

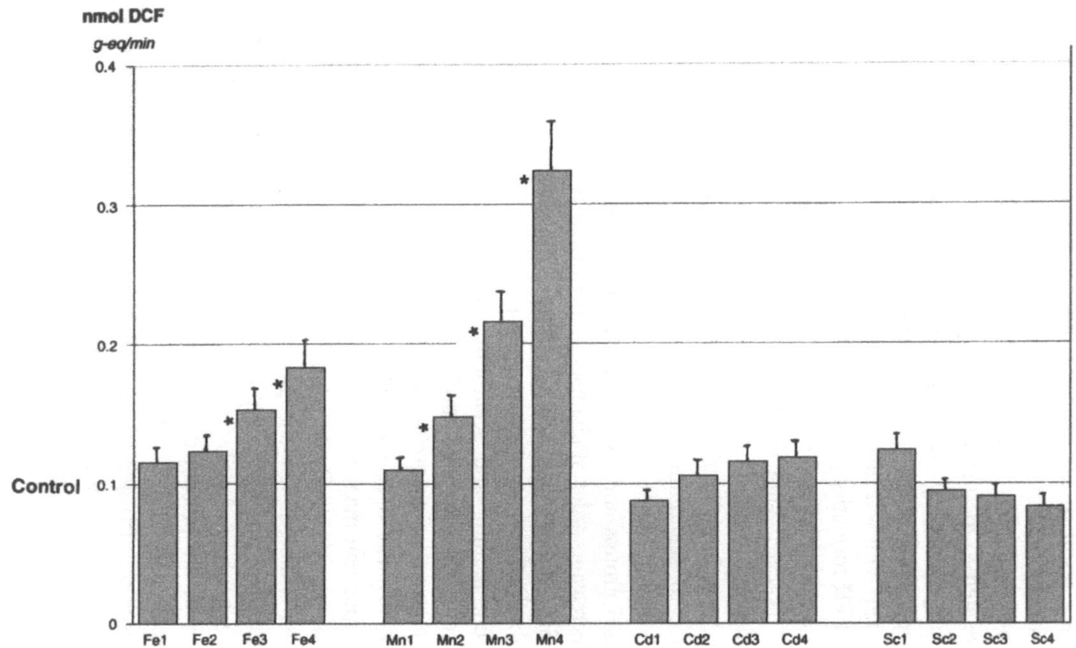

Figure 8-38. Influence of increasing concentrations of Fe^{3+}, Mn^{2+}, Cd^{2+}, and Sc^{3+} on free radical formation in a rat lung. Metal concentrations: 1, 0.01; 2, 0.1; 3, 1; 4, 10 μM. *Value differs significantly from that of the controls ($P < 0.05$). Reprinted from Horovitz and Bondy (1994), with permission of the authors and Gustav Fischer Verlag.

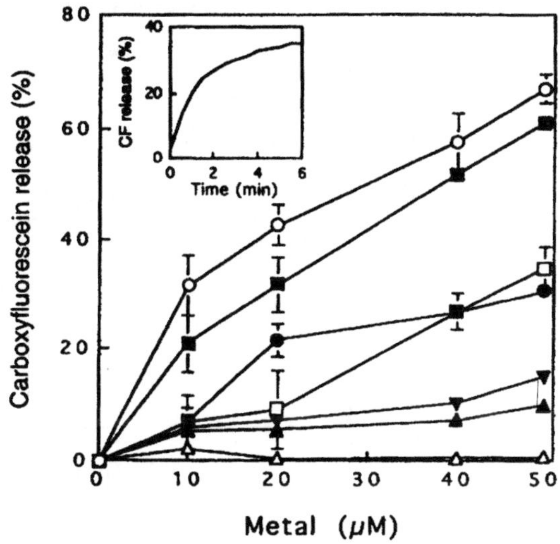

Figure 8-39. Effect of metal ions on 5(6)-caboxyfluorescein release from liposomes. Liposomes of phosphatidylcholine : phosphatidylserine (60 : 40) containing 0.1 M 5(6)-carboxyfluorescein were incubated in the presence of 10–50 μM Al^{3+} (○), Sc^{3+} (●), Ga^{3+} (□), In^{3+} (■), Be^{2+} (△), Y^{3+} (▲), or La^{3+} (▼). Reprinted from Verstraeten and Oteiza (1995), with permission of the authors and Academic Press.

Calmodulin (CaM) is an acid metal-binding protein, with a multitude of regulatory functions. CaM-stimulated cGMP-dependent cyclic nucleotide phosphodiesterase was stimulated by the addition of one Y^{3+} ion per CaM molecule (Suhayda and Huag, 1987). The initially observed conformational changes in CaM upon Y^{3+} addition resembled those induced by Ca^2, but the stimulatory effect was 20% greater with Y^{3+}-CaM than with Ca^{2+}-CaM (Fig. 8-40). An obvious correlation was found between the high charge-to-radius ratio of the metal and its ability to induce conformational changes in CaM.

Enzymes are biocatalysts of myriad reactions and chemical pathways involved in the metabolism of proteins, fats, and carbohydrates and the intricate energetics of the cell. Metals have three main types of chemical effects on enzymes: they may be required for activity, stimulate activity, or inhibit activity. Another type of catalysis of enzymatic reactions which is neither chemical nor physical, has been proposed. This type of reaction is characterized by an adsorbate that (a) changes in frequencies which may range from small to very large, and (b) bonds weakly to the adsorbent. A computer program was written to obtain simultaneous solutions for various reactants. An optimum fraction of 0.0889 for the reaction of a hydrogen atom across the bond of CO was found, which corresponds to a tentative value of a 0.0888 unstable fraction of Sc^{2+} in the reaction of oxidation or hydrogenation of CO. The hydroxyl group, rather than water, acts as a hydrating reactant for carbonic anhydrase (CE 4.2.1.1) (Gardner and Willard, 1986).

Electrophilic catalysis is a major mechanism by which group IIIA metal ions (Sc^{3+}, Y^{3+}, La^{3+}, and Ln^{3+}) act in enzyme-catalyzed reactions. The optimal size

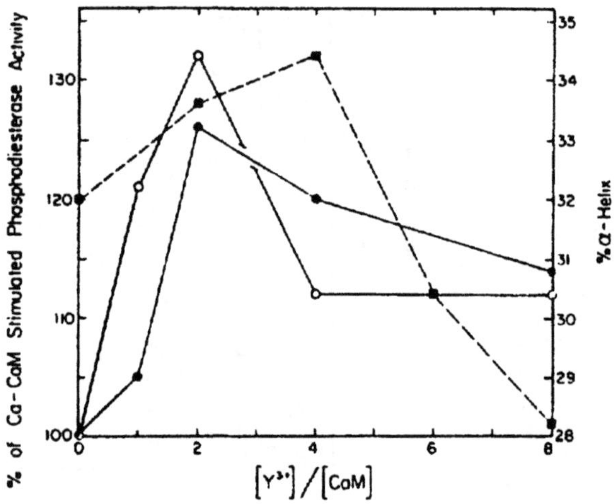

Figure 8-40. Stimulation of CGMP-dependent cyclic nucleotide phosphodiesterase activity by the yttrium–calmodulin (CaM) complex and its relationship to the yttrium-induced change in alpha-helix content of (CaM). The yttrium-CaM stimulation of the enzyme was measured in the absence (○) and the presence (●) of 50 μM Ca^{2+}. The concentration of the bovine brain CaM was 0.2 μM, and Y^{3+} were added at the respective molar ratios shown on the abscissa. The yttrium-induced change in the helix content (■) of 10 μM CaM was measured in Tris buffer. Reprinted from Suhayda and Haug (1987), with permission of Springer Verlag.

for optimal binding affinity of metal ions in this group lies between 0.81 A (Sc^{3+}) and 0.925 A (Yb^{3+}) (Villafranca and Nowak, 1992). The Ln series elements are site-specific metal substituents in some calcium-containing enzymes, such as phospholipase A2. Such replacement of Ca^{2+} results in loss of enzymatic activity by some, but not all, of these enzymes; some do not become inactive because their coordination ligands may contain 7–9 oxygen donors (Williams, 1991). Enzymes, such as enolase possess, in addition to the "catalytic" sites for REE ions, an "inhibitory" site with a high affinity for transition metals. Examples of enzymes inhibited by Ln^{3+} ions as well as of enzymes whose activities are not affected by these ions were outlined by Evans (1990). A summary of the effects of scandium and yttrium salts on oxidoreductases, transferases, hydrolases, and lyases is given in Table 8-22. The enzymes are classified according to the enzyme nomenclature of the International Union of Biochemistry and Molecular Biology (Webb, 1992). The minimum effective concentration of the metal ion may vary from 1 μM up to 3 mM. No similar data for isomerases and ligases are available. An exhaustive compilation of inorganic and organic enzyme inhibitors of all six classes of enzymes describes seven types of inhibition: competitive, noncompetitive, uncompetitive, mixed type, partial competitive, irreversible inhibition, and suicide substrate (Zollner, 1993). The predominant inorganic inhibitors that have

Table 8-22. Effects of Scandium and Yttrium on Enzyme Activities

Enzyme and its source	EC no.	Effective conc	Effect[a] of: Sc	Y	Reference
Alcohol dehydrogenase, yeast	1.1.1.1	3 mM	—	−82	Clayton (1959)
L-Lactate dehydrogenase, rabbit muscle	1.1.1.27	830 μM	—	−100	Holten *et al.* (1966)
		400 μM	—	−30	Clayton (1959)
Malate dehydrogenase, rabbit muscle	1.1.1.37	830 μM	—	−80	Holten *et al.* (1966)
Glucose-6-phosphate dehydrogenase, crystalline	1.1.1.49	830 μM	—	−100	Holten *et al.* (1966)
		400 μM	—	−57	Clayton (1959)
Glutamate dehydrogenase, crystalline	1.4.1.2	830 μM	—	−10F, −91R[b]	Holten *et al.* (1966)
Dihydropteridine reductase, rat blood	1.6.99.7	—[c]	−60	—	Cutler and Blair (1987)
Catechol oxidase, crystalline	1.10.3.1	2 mM	—	−25	Clayton (1959)
Laccase, crystalline	1.10.3.2	7.3 μM	−54	—	Cai *et al.* (1994)
		3.7 μM	—	−40	
Catalase	1.11.1.6	2 mM	—	−58	Clayton (1959)
Ferroxidase	1.16.3.1	100 μM	−90	−55	Huber and Frieden (1970)
Hexokinase, crystalline	2.7.1.1	200 μM	—	0	Clayton (1959)
Choline-phosphate cytidyl transferase, mouse tongue	2.7.7.15	1 mM	+ 5	—	Izumi (1974)
Acetylcholinesterase, bovine	3.1.1.7	100 μM	−30	−34	Marquis and Lerrick (1982)
		10 μM	+ 42	+ 9	Marquis and Black (1985)
Cholinesterase, horse serum	3.1.1.8	1 mM	−80	—	Kamaric (1975)
Deoxyribonuclease I	3.1.21.1	200 μM	—	−50	Clayton (1959)
α-Amylase, *Bacillus subtilis*	3.2.1.1	100 mM	—	+ 80[d]	Smolka *et al.* (1971)
β-Fructofuranosidase	3.2.1.26	2 mM	—	0	Clayton (1959)
Plasmin, human serum	3.4.21.7	37 μM	—	−50	Dano and Reich (1979)
Plasminogen activator, human serum	3.4.21.68/73	31 μM	—	−50	Dano and Reich (1979)
ATP pyrophosphatase, crystalline	3.6.1.8	400 μM	—	−67	Clayton (1959)
Cyanide hydratase, *Bacillus pumilus*	4.2.1.66	200 μM	+ 90	—	Meyers *et al.* (1993)
Adenylate cyclase, murine S49 cells	4.6.1.1	5 μM	−34	—	Maguire (1982)

[a] Negative values represent percent enzyme inhibition, whereas positive values represent percent activation of enzyme activity.
[b] F, Forward direction; R, reverse direction.
[c] Enzyme activity was measured in blood of rats, into which 0.3 mg Sc^{3+}/kg was i.p. injected.
[d] Reactivation of enzyme preparate from which calcium had been removed.

been investigated are Ca^{2+} and the divalent heavy metal ions. Fewer data exist for Al^{3+}, Cr^{3+}, Fe^{3+}, and the Ln^{3+} (Kuby, 1991; Zollner, 1993).

Various degrees of *in vitro* inhibition or activation of selected enzymes by yttrium and REE chlorides have been reported. Even at concentrations of 2 mM, Y^{3+} and other REE did not inhibit invertase (CE 3.2.1.26) and hexokinase (CE 2.7.1.1). The *in vitro* inhibitory effects of yttrium and other REE on dehydrogenases varied, depending upon the enzyme system, and the yeast alcohol dehydrogenase was the least sensitive (Clayton, 1959). Inhibition of enzymes by yttrium and other TE was reversed to different extents by the addition of EDTA. The glutamate dehydrogenase reaction was inhibited by the metal ions only in the reverse direction; the forward direction was less inhibited (Holten *et al.*, 1966). At a concentration of 100 μM, Sc^{3+} and Y^{3+} caused a 90% and 55% inhibition of ferroxidase (CE 1.16.3.1), respectively (Huber and Frieden, 1970). The reversibility of the inhibition by Sc^{3+} was indicated by linear plots of velocity v, against enzyme concentration. The plot of $1/v$ against inhibitor concentration showed some nonlinearity for Sc^{3+} (Fig. 8-41), which may indicate a complicated inhibition scheme similar to that for Al^{3+}. The inhibition by Sc^{3+} and In^{3+} was reversed upon dilution. No effects on the visible or UV spectrum of ferroxidase were observed with Sc^{3+}, In^{3+}, and other metal-ion inhibitors. A

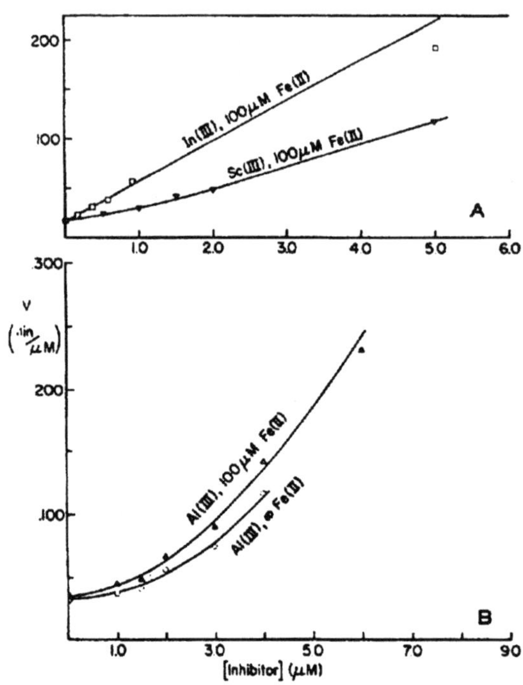

Figure 8-41. (A) Inhibition of the ferroxidase reaction by Sc^{3+}, In^{3+}, and Al^{3+}. (B) The reciprocal of the velocity of the ferroxidase reaction at 30°C is plotted against inhibitor concentration. The final concentration of the reaction mixture was 83 mM Chelex-treated acetate buffer, pH 6.0, 100 M Chelex-treated ascorbate, and 100 μM Fe^{2+}. Reprinted from Huber and Frieden (1970), with permission of the American Society for Biochemistry and Molecular Biology.

concentration of 13 μM Sc^{3+} caused a 50% inhibition. The extents of inhibition by Sc^{3+}, Y^{3+}, and La^{3+} indicated that with a gradual increase in atomic number, the ability to inhibit ferroxidase decreases. The approximate order of decreasing effectiveness in inhibition of 120 nM ferroxidase in the presence of 100 μM Fe^{2+} was $In^{3+} > Zr^{2+} > Al^{3+} > Sc^{3+} > Ga^{3+}$. Tests of reversibility indicated a mixed type of inhibition, with both competitive and noncompetitive effects, for Sc^{3+}, Al^{3+}, and In^{3+}. With Sc^{3+}, the inhibition of ferroxidase increased up to 300 s and than remained unchanged up to 600 s (Huber and Frieden, 1970).

At a concentration of 1 mM, Y^{3+} and several Ln^{3+} caused a reversible competitive inhibitory effect on bovine erythrocyte acetylcholinesterase (AChE) (CE 3.1.1.7). Using the criteria of reappearance of enzyme activity and release of [^{32}P]phosphonate (Howells and Coult, 1971) showed that the REE did not catalyze the dephosphonylation of organophosphorus-inhibited AChE. Because the REE did not catalyze this reaction, they were considered unlikely to be a therapy for poisoning by organophosphates. Purified AChE from the electric organ of *Electrophorus electricus* was noncompetitively inhibited by 0.1 mM Y^{3+} and Sc^{3+} (Marquis and Lerrick, 1982). The inhibitory action of these metal ions was similar to the action of Al^{3+} but contrasted with the mixed competitive–noncompetitive inhibition caused by La^{3+}. The Y^{3+}, Sc^{3+}, or Al^{3+} inhibition was not altered by Ca^{2+} or decamethonium. These metals may alter the ligand binding at the catalytic site by interaction at the ψ-peripheral anionic sites (Marquis and Lerrick, 1982). A mixed type of inhibition of horse serum cholinesterase (ChE) (CE 3.1.1.8) was exhibited by Eu^{3+}, Sc^{3+}, and Yb^{3+} (Kamaric, 1975). As seen from Fig. 8-42, the hydrogen ion interfered with the inhibitory effect of Sc^{3+} through a noncompetitive type of reaction. These reactions reflect the allosteric properties of this enzyme, because a mixture of isoenzymes was thought to be present. Sc^{3+} completely inhibited horse serum cholinesterase, but practically all the enzyme activity was restored by exhaustive dialysis (Kamaric, 1975). Prolonged oral administration of yttrium chloride to rats at concentrations of up to 1 g/kg body weight did not affect the activity of ChE in male animals (Ogawa *et al.*, 1994). In females, the enzyme activity was insignificantly changed at lower concentrations but strongly inhibited at highest concentrations (Table 8-23).

Kinetic analysis of the interaction between chlorides of scandium, yttrium, and other trivalent cations and bovine caudate AChE revealed at least three distinct concentration-dependent effects on enzyme activity (Marquis and Black, 1985). At low ionic strength, high-affinity noncompetitive or allosteric activation occurred at very low concentrations of 1–10 μM (Fig. 8-43). At concentrations of 50–200 μM, these cations were noncompetitive inhibitors. At 20–500 μM, they exerted a mixed competitive–noncompetitive inhibition. The enzyme activation observed at low cation concentrations did not occur in high-ionic-strength buffers; however, enzyme inhibition by all the trivalent metal ions was similar at low and high ionic strength. The inhibition exerted by all cations was fully reversed by a 10-fold excess of EDTA or by a 100-fold dilution of the inhibited enzyme. The affinity of the cations for the active or catalytic site of the enzyme decreased in the order $Y^{3+} > Tb^{3+} > La^{3+} > Sc^{3+}$ (Marquis and Black, 1985).

Y^{3+} reduced the activity of succinate dehydrogenase (CE 1.3.99.1) *in vitro*; La^{3+}, Nd^{3+}, and Sm^{3+} stimulated this activity. Also, both YCl_3 and $LaCl_3$ inhibited the ATPase system *in vitro* (Cochran *et al.*, 1950). Y^{3+} and Sm^{3+} can activate the Ca^{2+}-ATPase of erythrocytic membranes (Wen *et al.*, 1995). Y^{3+} and

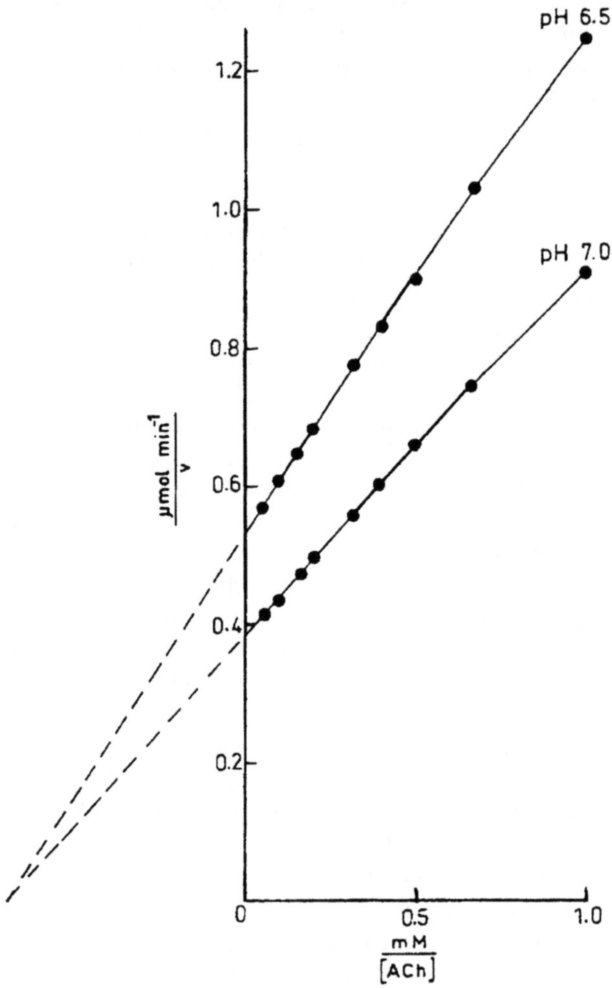

Figure 8-42. Inhibitory effects of 0.5 mM Sc^{3+} on horse serum cholinesterase at two pH values, 25°C. The substrate was acetylcholine. Reprinted from Kamaric (1975), with permission of the Creation Chemical Society.

Table 8-23. Effect of Repeated Oral Administration to Male and Female Rats of Increasing Amounts of Yttrium Chloride for 28 Consecutive Days on Serum Cholinesterase and Transaminase Activities[a]

	Enzyme activity			
	Male		Female	
YCl_3 (mg/kg)	mU/ml	%	mU/ml	%
	Cholinesterase			
0	172 ± 26	100	1022 ± 89	100
40	159 ± 12	92	1257 ± 279	122
200	154 ± 31	89	800 ± 125[b]	78
1000	150 ± 54	87	237 ± 89[c]	23
Recovery 0[a]	147 ± 14	100	1275 ± 149	100
Recovery 1000[a]	169 ± 33	115	901 ± 162	71
	Transaminase			
0	48 ± 2	100	47 ± 10	100
40	57 ± 10	119	49 ± 9	104
200	50 ± 5	104	46 ± 5	98
1000	45 ± 9	94	48 ± 13	102
Recovery 0[a]	52 ± 7	100	43 ± 9	100
Recovery 1000[a]	51 ± 9	98	45 ± 12	105

[a] Data reprinted from Ogawa *et al.* (1994), with permission of the authors and the Pharmaceutical Society of Japan. The results of serum biochemistry tests were, however, considered not to be caused by the direct influence of yttrium, but caused by the decreased consumption of feed (cited from Ogawa, personal communication, 1995, with permission).
[a] Recovery = Animals maintained for a 16 day recovery period after the 28 day administration of 0 on 1000 mg/kg dose of $YCl_3 \cdot 6H_2O$.
[b] Significant at 5% level.
[c] Significant at 1% level.

several Ln^{3+} ions have an ability to reactivate Ca^{2+}-free α-amylase (CE 3.2.1.1) (Darnall and Birnbaum, 1970). The average percent reactivation by the metal ions examined varied between 100% for Ca^{2+} and Lu^{3+} and 10–14% for Mg^{2+} and La^{3+}; Y^{3+} induced 80% reactivation (Fig. 8-44). The effectiveness of various REE in activating apo-amylase decreased in the order $Ca^{2+} \approx Lu^{3+} > Yb^{3+} > Er^{3+} \approx Y^{3+} \approx Dy^{3+} \approx Tb^{3+} > Sm^{3+} > Nd^{3+} > Pr^{3+} > La^{3+}$ (Smolka *et al.*, 1971).

Spraying leaves of grapes (*Vitis* sp.) during the flowering period with scandium sulfate caused an increase of catechol oxidase (CE 1.10.3.1), L-ascorbate oxidase (CE 1.10.3.3), catalase (CE 1.11.1.6) and peroxidase (CE 1.11.1.7) activities in leaves and berries (Dobroliubsky, 1962). Wheat (*Triticum*

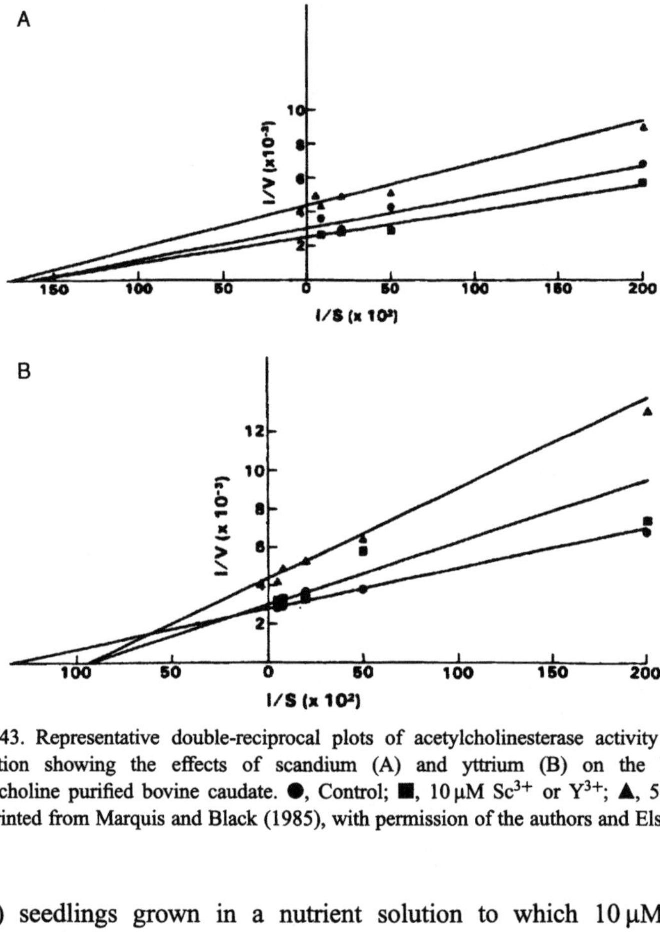

Figure 8-43. Representative double-reciprocal plots of acetylcholinesterase activity vs. substrate concentration showing the effects of scandium (A) and yttrium (B) on the hydrolysis of acetylthiocholine purified bovine caudate. ●, Control; ■, 10 μM Sc^{3+} or Y^{3+}; ▲, 50 μM Sc^{3+} or Y^{3+}. Reprinted from Marquis and Black (1985), with permission of the authors and Elsevier Science.

vulgare) seedlings grown in a nutrient solution to which 10 μM scandium chloride had been added exhibited an increase in acid phosphatase (CE 3.1.3.2) and peroxidase (CE 1.11.1.7) activities in leaves of 20% and 65%, respectively (Horovitz, 1968).

Intraperitoneal administration of 0.3 mg Sc^{3+}/kg to rats resulted in decreased dihydropteridine reductase (CE 1.6.99.7) activity in brain (−20%) and blood (−60%); Al^{3+} had a similar effect (Cutler and Blair, 1987). Sc^{3+} acted as a specific antagonist of Mg^{2+} in the catalytic subunit of the β-adrenergic receptor–adenylate cyclase complex (CE 4.6.1.1) of murine S49 lymphoma cells. Two mechanisms by which Sc^{3+} inhibited the enzyme were proposed: (i) by directly binding at one of the Mg^{2+} ions to coupling proteins or (ii) by chelating ATP. The affinity of ATP for Sc^{3+} at 0.4 μM Sc^{3+} was essentially identical to that for La^{3+} at the same concentration. The apparent complex of Sc^{3+} with F^-, namely ScF_4^-, is considered a potent inhibitor of adenylate cyclase (Maguire,

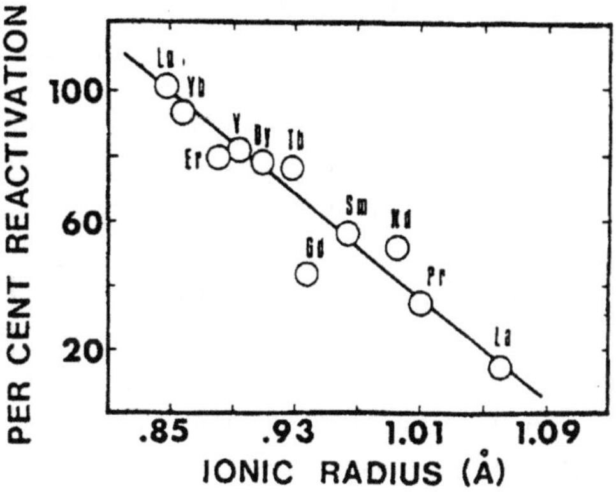

Figure 8-44. Reactivation of apo-α-amylase by Y^{3+} and Ln^{3} ions plotted as a function of the crystal ionic radius of the metal ions. The percent reactivation was determined at a metal–ion concentration of 100 mM in a 20 mM maleate/Tris buffer, pH 6.0. The reactivation by the metal ions was expressed as percent of the reactivation by Ca^{2+}. Reprinted from Smolka *et al.* (1971), with permission of the American Chemical Society.

1982). However, the Sc^{3+}–ATP complex is not regarded as being a competitive substrate inhibitor at Sc^{3+} concentrations of 5–50 μM. These concentrations significantly inhibited adenylate cyclase activity (Fig. 8-45). The dose-dependent decrease in the maximum rate, V_{max}, suggested that free Sc^{3+} itself inhibited noncompetitively and indicated a direct effect of free Sc^{3+}, presumably at the Mg^{2+} site. The activation of the purified guanine nucleotide-binding regulatory

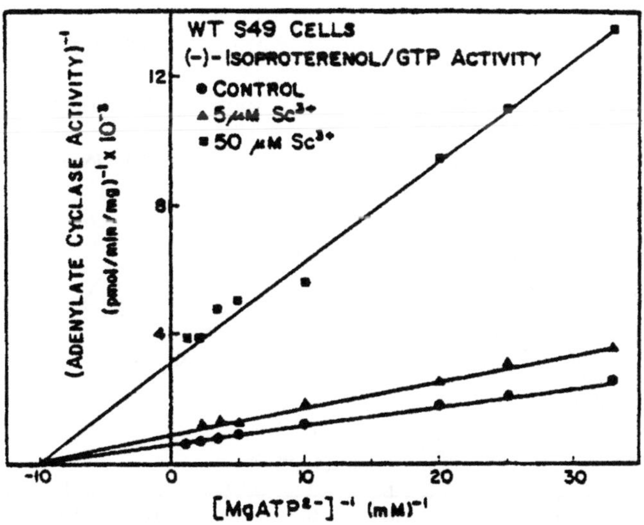

Figure 8-45. Effect of Sc^{3+} on $MgATP^{2-}$ kinetics in membranes from wild-type (WT) S49 lymphoma cells: substrate kinetic parameters for (−)-isoproterenol/GTP-activated adenylate cyclase at various concentrations of $MgATP^{2-}$ in the absence (●) and in the presence of 5 μM (▲) and 50 μM Sc^{3+} (■). Reprinted from Maguire (1982), with permission of the author and Williams and Wilkins.

component of adenylate cyclase by F^- requires the presence of Mg^{2+} and another factor. A highly specific Al^{3+} requirement for the activation of the regulatory component of adenylate cyclase was assumed. Although Be^{2+} partially activated the regulatory component, Sc^{3+} and La^{3+} did not (Sternweis and Gilman, 1982).

Evaluation of the effects of i.v. administered ^{90}Y-labeled MAb B72.3 immunoconjugate GYK-DTPA in beagle dogs revealed an acute toxicity and late side effects in the liver, but not in other organs (Quadri *et al.*, 1993). Higher radiation doses decreased the alkaline phosphatase (CE 3.1.3.1) activity, beginning 30–40 days after injection (Fig. 8-46). At lower doses (1.0 mCi/kg), no change in enzyme activity was observed.

A wide range of biological processes have been associated with cellular secretion of plasminogen activators. Although these activators are serine proteinases of the urokinase type, they induce significant proteolysis only by generating plasmin. The catalytically active product of plasminogen activation, can be nonspecifically inhibited by REE metal ions. Pr^{3+} and Yb^{3+} showed a preference for plasminogen activation, and Eu^{3+} and Cd^{2+} showed a preference for plasmin activation, whereas Y^{3+} displayed a preference for both plasminogen and plasmin activation. At concentrations of 14–37 μM, Y^{3+} equally inhibited (50%) plasmin (CE 3.4.21.7) and human plasminogen, as well as human urinary urokinase (CE 3.4.21.31) (Dano and Reich, 1979). The activity of cyanide

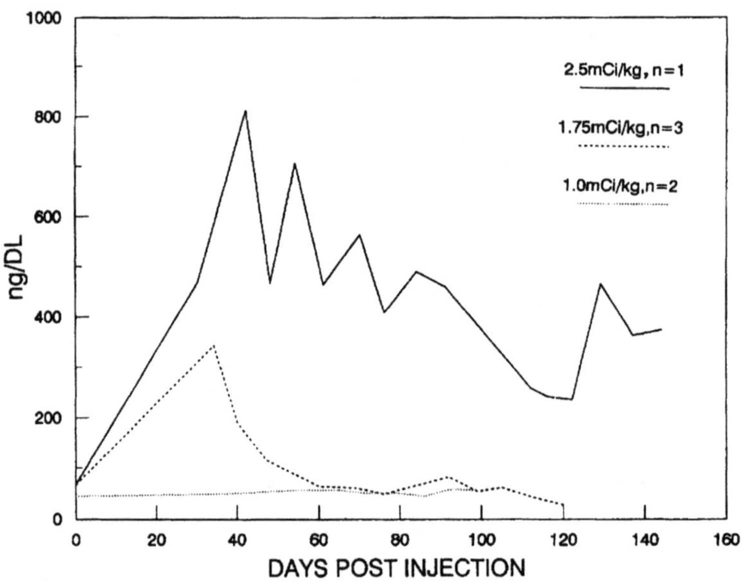

Figure 8-46. Mean alkaline phosphatase activities in dogs after i.v. injection of various doses of ^{90}Y-labeled MAb B72.3 immunoconjugate GYK-DTPA. ———, 2.5 mCi/kg; — — —, 1.75 mCi/kg; ...., 1.0 mCi/kg. Reprinted from Quadri *et al.* (1993), with permission of Elsevier Science.

hydratase (CE 4.2.1.66), which catalyzes the hydration of cyanide to formamide, isolated from *Bacillus pumilus* can be enhanced approximately twofold by Sc^{3+}, Cr^{3+}, Fe^{3+}, and Tb^{3+}, whereas Hg^{2+}, Pb^{2+}, and Ag^{+} are inhibitory (Meyers *et al.*, 1993). The enhancement is independent of metal-ion concentration above 5 μM; it may occur with some metal ions because they act as cyanide carriers. Alternatively, the enhancement of enzyme activity could be the result of changes in the protein structure brought about by the presence of these ions (Meyers *et al.*, 1993). Mn^{2+} and Se^{2+} were the only metal ions that enhanced choline-phosphate cytidylyl transferase (CE 2.7.7.15) activity in the tongue of mice (Izumi, 1974). The enhanced activity of this enzyme was related to changes in the taste threshold for bitter taste. Other metal ions, including Sc^{3+} and Y^{3+}, did not significantly stimulate or inhibit this enzyme (Table 8-24) although they did affect the taste threshold (see Section 8.12).

One of the early biochemical events that occurs in cells exposed to interferon is the synthesis of oligomers of adenylic acid. Sc^{3+}, Y^{3+}, and other trivalent metal ions depolymerize oligoadenylates produced in cells treated with interferon (George *et al.*, 1985). A 94–100% depolarization of 2′,5′-phosphodiester bonds was achieved with Y^{3+}, Sm^{3+}, Nd^{3+}, and Sc^{3+}. Al^{3+}, Fe^{3+}, and other trivalent metal ions caused only a partial depolarization of the oligoadenylates. In these experiments the metals were present at concentrations in the 0.1–1.0 mM range, which exceeds the lethal concentrations of most of these metals. Whether the effects of the metal ions on the interferon response occur *in vivo* or with tissue culture cells has not been established.

Extensive use of irradiation with yttrium and scandium radioisotopes for medical purposes prompted investigations of their effects on metabolic processes.

Table 8-24. Minimum Effective Concentrations (MEC) of Scandium, Yttrium, and Selenium on the Taste Threshold for Bitter Taste and Choline-Phosphate Cytidyltransferase (PCT) Activity in Mice[a]

	Male			Female		
	MEC	Taste threshold[b]	Percent activity of PCT	MEC	Taste threshold[b]	Percent activity of PCT
H_2O	—	No. 13	100	—	No. 13	100
Sc	10 μM	No. > 13	98	1 mM	No. > 13	105
Y	1 μM	No. 12	90	1 μM	No. 12	96
Se	2 mM	No. 11	117	200 μM	No. 12	139

[a] From Izumi (1974), with permission of the author.
[b] The taste threshold was evaluated by applying solutions of 6-propyl-2-thiouracil (PROP) on the tongue epithelium was expressed by the solution number according to an arbitrary scale; a higher solution number represents a higher threshold for PROP; No. > 13 means that the threshold is No. 14 or more.

Irradiation with ^{90}Y implanted for the treatment of acromegaly, a condition of generalized soft tissue hypertrophy, resulted in a substantial reduction in plasma insulin-like growth hormone (MacSweeney *et al.*, 1990).

Biomineralization processes are extremely heterogeneous in lower organisms, plants, and animals. Almost 60 different biogenic minerals that contain several metal ions have been identified in numerous species. Skeletal parts of different phyla contain mineralized macromolecules of carbonate and phosphate, including acidic glycoproteins and/or proteoglycans (Lowenstam and Weiner, 1989). Whereas physiological mineralization involves the formation of carbonated hydroxyapatite crystals in bone and teeth, pathological mineralization involves several other inorganic and organic biominerals, such as calcium phosphate and oxalate, urate, and others (Cheng, 1991). The possible involvement of scandium, and, in particular, yttrium, which is considered a "bone seeker," in biomineralization processes is unknown.

Secondary metabolism deals with such metabolites as siderophores, sex hormones, and germination effectors. The assumed general functions of secondary products are detoxication of substances accumulating as a result of primary metabolism, physiological action, chemical signaling, coordination activity of organisms of the same species, and mediation of ecological relations. In plants, more than 20,000 secondary compounds are known. These compounds are considered vestiges of ancestral (essential?) biosynthetic pathways that, although having no known essential functions, ensure "biological success" (Anderson and Beardall, 1991). Others consider secondary metabolism as being essential in the expression of the diversity of regulatory systems involved in biosynthetic pathways and in many chemical structures (Kleinkauf *et al.*, 1986). The reactivity of secondary metabolites and the differentiation of cells in the presence of TE, such as iron, zinc, and manganese, has been extensively investigated with several species of microorganisms (Weinberg, 1986). An experiment with tobacco (*Nicotiana* sp.) plants grown in nutrient solution culture to which 54 TE were added separately at concentrations of 1 mg/l showed that scandium and several other TE did not affect nicotine yield; other TE caused a definite increase (beryllium, copper, palladium, platinum, and samarium) or decrease (cobalt, nickel, vanadium, and others) (Tso *et al.*, 1973). No explanation was given for these findings.

8.11 Membranes and Transport

An important aspect of the involvement of TE in physiological and biochemical processes is their transport across membranes, a function common to living organisms. Biological fluids consist of suspensions of a discrete collection of small particles immersed in an essentially continuous fluid. It is estimated that there are 250 million inorganic ions per cell, but this number may vary considerably from bacterial cells to higher organisms (Bohinski, 1987). Metals differ from other transported solutes in that they generally do not exist in a "free" form in biological fluids and tissues. Metals are normally complexed with amino acids, peptides, proteins, phospholipids, and other constituents; their transport is regulated by specific ligands. The "hydrated radius" and the number of water molecules of hydration of polyvalent ions are inversely related to the atomic number. These ions are less mobile than monovalent ions of corresponding size. Experimental calculations of the ionic mobilities at infinite dilution in water at 250°C showed high ionic mobilities for the H^+ and OH^- ions, i.e., 36.2×10^{-4} and 20.5×10^{-4} cm s^{-1}/V cm^{-1}, respectively. A value of 6.16×10^{-4} cm s^{-1}/V cm^{-1} was calculated for Ca^{2+}, and lower values would be expected for REE. The movement of ions in membranes is partly dependent on the rate at which water leaves and enters the coordination sphere of the aquo ion. The rate constant for inner-sphere substitution of cations increases 10^{-10} s for H^+ and K^+ to 10^{-8}–10^{-9} s for Sc^{3+}, Y^{3+}, and La^{3+} and to 10^0 s for the smaller, highly charged Al^{3+} ion (Frey and Stuehr, 1974).

Cell membranes are the origin of many vital functions of living cells. Some prokaryotes have, in addition to the plasma membrane, a cell wall and an outer membrane of the cell. Eukaryotic cells are more complex and contain a variety of membrane-bound organelles, such as the nucleus, mitochondria, endoplasmic reticulum, Golgi apparatus, and lysosomes (Yeagle, 1993). A wide variety of methods are used to monitor membrane transport processes. Nuclear-magnetic resonance (NMR) spectroscopy has great potential in the study of fast or slow membrane transport and of intracellular processes (Kuchel *et al.*, 1994).

At a concentration of 0.05–0.07 nmol/mg protein, Ce^{3+}, Ho^{3+}, and Pr^{3+} specifically inhibited the transport of Ca^{2+} in rat liver mitochondria, whereas Sc^{3+} and Al^{3+} did not (Mela, 1969). The inhibition of Ca^{2+} transport by Ln was assessed as noncompetitive. Al^{3+} and Sc^{3+} showed no effect on the transport of Ca^{2+} at concentrations of 0.06–0.1 nmol/mg protein (Mela, 1970). Ogurusu *et al.* (1991) concluded that the Ln^{3+} ions do not have the ability to replace Ca^{2+} at the calcium-specific transport site.

Several modes of regulation of membrane transport are recognized; they have the different chemical regulatory possibilities, as well as oscillatory and even chaotic behavior in biochemical cycles (Tosteson, 1989). Modern techniques can distinguish channel, primary, and secondary transport, as well as ground translocation classes of transport systems (Sahm, 1993). Four general mechanisms of transport of metals have been described: (a) simple diffusion and convection, (b) facilitated diffusion, (c) primary active transport, and (d) membrane turnover. Whereas the transport of monovalent and divalent metal ions has been well studied, primary active transport systems of heavy metals and polyvalent metals have not been identified in organs and organisms (Ballatori, 1991). However,

Silver and Perry (1982) suggested that microbial cells require highly specific membrane transport systems for every inorganic cation and anion; these are needed for nutritional purposes or for ion transport in common growth milieu. Three mechanisms of transport in microorganisms were described: (a) transport via carriers; (b) specific transport of metals complexed with specifically exuded low-molecular-weight ligands; and (c) nonspecific transport of metal ions complexed with substrates that serve as carrier molecules (Collins and Stotzky, 1989).

The mitochondrial transport system is comprised of a spatially organized respiratory chain. The "average" liver mitochondrion contains some 17,000 units of each respiratory chain assembly and several metabolite and metal-ion transport proteins. About 20 distinct transporting systems have been detected in mitochondria isolated from a variety of tissues (Tyler and Sutton, 1984). Under physiological conditions, the movement of several ions between mitochondria and cytosol is catalyzed by specific metabolite carriers. At lower concentrations, the Ln inhibit the calcium carrier, whereas at high concentrations the Ln inhibit the Na–Ca carrier (Tyler, 1992). Ion channels are elementary excitable macromolecular pores in eukaryotic cell membranes of many tissues. The Na, K, Ca channels are cation-permeable channels with little selectivity among cations. Mitochondria from all eukaryotic kingdoms contain channels that allow ions and small molecules to diffuse between the cytoplasm and the mitochondrial species. These channels are called voltage-dependent anion-selective channels (VDAC) (Colombini, 1991).

Aluminum salts, at a concentration of 10 μM, caused an inhibition of the voltage-dependent closure of VDAC of mitochondrial membranes from a wall-less mutant of *Neurospora crassa* (Zhang and Colombini, 1989). Several metallic and organic ions were tested for a possible aluminumlike effect; only the trivalent metals exhibited a similar ability to reduce the channel's voltage dependence. Like Al^{3+}, Sc^{3+}, as well as Ga^{3+}, In^{3+}, Cr^{3+}, and Fe^{3+}, caused the VDAC to open and stay open. However, trivalency was not a sufficient criterion, because La^{3+} was ineffective (Table 8-25). The inhibitory effects of aluminum and other trivalent metals are postulated to be mediated through groups of the protein that enable VDAC to respond to the membrane potential. It was speculated that the trihydroxides of aluminum and other trivalent metals are able to diffuse through cell membranes without transport systems. This does not exclude the possible existence of specific transport systems, similar to those for iron and other well-studied TE (Colombini, 1994). Characteristics such as ionic radius and ionic surface charge density apparently do not play significant roles in the inhibition of the closure of VDAC. The VDAC seem to interact specifically with the neutral species $M(OH)_3$, but not with $M(OH)_4^-$. A mechanism other than simple electrostatic neutralization must be responsible for the inhibition (Zhang and Colombini, 1989). At concentrations in the range of 10–100 μM, the Ln^{3+}

Table 8-25. Effectiveness of Metals in Inhibiting the Voltage-Dependent Closure of the Mitochondrial Outer Membrane Channel from a Wall-less Mutant of *Neurospora crassa*[a]

Metal	Inibition of voltage-dependent closure	Surface charge density (e/Å)	Lowest conc to cause inhibition	Highest conc used
Sc^{3+}	Yes	0.45	10 μM	500 μM
Al^{3+}	Yes	0.92	1 μM	100 μM
Fe^{3+}	Yes	0.58	500 μM	500 μM
Ca^{2+}	No	0.16	—	5 mM
La^{3+}	No	0.23	—	500 μM

[a] Data reprinted from Zhang and Colombini (1989), with permission of the authors and Elsevier Science.

produce comparable potential shifts of the kinetic parameters in the voltage-clamped axons, slowing the gating kinetics, but have no effect on the flow of ions through an open channel (Starzak, 1984).

Metals are taken up into brain across the blood–brain barrier by a variety of mechanisms, which include both carrier-mediated and receptor-mediated transport systems. Metals differ in their primary site of entry, with some being taken up primarily at brain capillaries and others gaining access through the choroid plexus epithelium. The blood–brain barrier limits the exchange of metals between plasma and the central nervous system but facilitates diffusion and active transport systems. The mechanisms through which the blood–brain barrier regulates the uptake and distribution of aluminum in brain (Smith, 1990) conceivably act similarly with respect to polyvalent metal ions with uncertain functions.

The interaction between α-latrotoxin (α-LT), a potent proteinaceous neurotoxin that stimulates the release of a wide variety of neurotransmitters, and trivalent metal ions was investigated in rat striatal synaptosomal preparations (Scheer, 1989). All the metal ions tested inhibited α-LT-induced [^{3}H]dopamine ([^{3}H]DA) release, the order of potency being $Yb^{3+} > Gd^{3+} \approx Y^{3+} > La^{3+} > Al^{3+}$. The inhibitory effects of trivalent metal ions was reversible, provided that the synaptosomes were washed with buffer containing EDTA (Table 8-26). α-LT-induced synaptosomal membrane depolarization was blocked by La^{3+} but was not affected by Gd^{3+}, Y^{3+}, and Yb^{3+}. The α-LT-stimulated uptake of $^{45}Ca^{2+}$ was inhibited by all trivalent metal ions tested. These results demonstrate that at least three means by which trivalent metal ions can inhibit α-LT in rat striatal synaptosomal preparations: (a) inhibition of α-LT binding (Al^{3+}), (b) inhibition of α-LT-induced depolarization (La^{3+}), and (c) inhibition of α-LT-induced $^{45}Ca^{2+}$ uptake (Gd^{3+}, Y^{3+}, Yb^{3+}, and La^{3+}) (Scheer, 1989). The

Table 8-26. Reversibility of the Trivalent Cation-Induced Inhibition of α-Latrotoxin in [^{3}H]Dopamine-Loaded Rat Striatal Synaptosomes of Lactrodectus mactans[a]

	Percent of [^{3}H]DA released	
	Buffer wash	EDTA wash
Control	73.3 ± 2.8	72.9 ± 2.0
Y^{3+}	30.3 ± 16.0	69.0 ± 6.8
Yb^{3+}	18.7 ± 7.9	63.3 ± 1.4

[a] Data reprinted from Scheer (1989), with permission of Lippincott-Raven Publishers.

potassium-stimulated, voltage-dependent influx of ^{45}Ca in synaptosomes from rat brain was blocked by polyvalent cations (Nachsen, 1984); the half-inhibition constants (K_1) are presented in (Table 8-27). Three distinct groups were found: (a) $K_1 > 1$ mM (Mg^{2+}, Sr^{2+}, and Ba^{2+}); (b) $K_1 = 30$–$100\ \mu$M (Mn^{2+}, Co^{2+}, Ni^{2+}, Ni^{2+}, Cu^{2+}, Zn^{2+}, and Hg^{2+}); and (c) $K_1 < 1\ \mu$M (Cd^{2+}, Y^{3+}, La^{3+}, Ln^{3+}, and Pb^{2+}). The metal ions inhibited calcium influx by competing with Ca^{2+} for the channel site that is involved in the transport of divalent cations. A model of the calcium channel binding site was proposed, in which inhibitory polyvalent cation selectivity is determined primarily by interactions between the binding site and the different cations. The Ln and yttrium may exert an extra positive charge, because these elements may concentrate at the mouth of the channel or in the channel; this effect may be caused by negative surface charge and voltage gradients across the membrane (Nachsen, 1984).

The effects of Y^{3+} and other Ln^{3+} on the T-type voltage-gated calcium channel (VGCC) were characterized by the whole-cell patch clamp technique

Table 8-27. Effect of Yttrium, Manganese, and Magnesium on Induced Fluorescece, as Dependent From Potassium-Stimulated, Voltage-Dependent Calcium Influx in Low-Potassium or Potassium-Rich Solutions, Measured in Synaptosomes of Rat Brain[a]

Metal	Conc (mM)	20 mM K	77.5 mM K	I_{50} [b]	K_1 [c] (mM)
Yttrium	0.0007	49 ± 6	55 ± 4	0.001	0.0007
Manganese	0.1	37 ± 5	32 ± 5	0.07	0.005
Magnesium	3.5	23 ± 4	48 ± 4	3.0	5.7

[a] Data reprinted from Nachsen (1984), with permission of the author and Rockefeller University Press.
[b] I_{50}, Potassium-induced fluorescence.
[c] K_1, Half-inhibition constant for the inhibitory ions.

(Mlinar and Enyeart, 1993). All the trivalent metal ions studied blocked the current through T-type VGCC (I_T) in a concentration-dependent manner. Y^{3+} and smaller Ln^{3+} ions were the most potent inorganic antagonists of T-type VGCC in neural crest-derived rat and human thyroid C cell lines. The half-maximal blocking concentrations (IC_{50} of I_T) yielded the following potency sequence (values of IC_{50}, in μM, are given in parentheses): $Ho^{3+}(0.107) \approx Y^{3+}(0.117) \sim Yb^{3+}(0.124) \geq Er^{3+}(0.153) > Gd^{3+}(0.267) > Nd^{3+}(0.429) > Ce^{3+}(0.728) > La^{3+}(1.015) \approx Ni^{2+}(5.65)$. The blocking of I_T by Y^{3+} and Ln^{3+} was voltage-independent in the range of −200 to +80 mV (Mlinar and Enyeart, 1993). Neither depolarization to +80 mV nor repolarization to −200 mV influenced the potency of I_T (Fig. 8-47).

Plant roots are remarkably efficient in absorbing ions from the soil or from solutions. Two routes were described for the passage of solutes across the root: (a) via the apoplast of the cell walls or intercellular spaces outside of the cell membranes, and (b) via the symplast interconnected through segments of

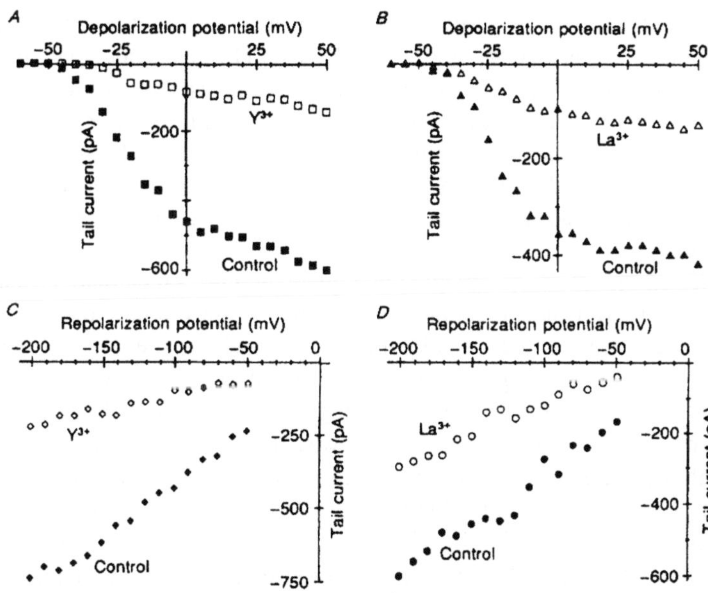

Figure 8-47. Effects of yttrium and lanthanum at extreme voltages on the blocking of the T-type voltage-gated calcium channel (I_T) in rat and human neural cells. (A and B) Tail currents were recorded at a repolarization potential of −80 mV after activation by 0.5-ms steps to various potentials between −60 to +50 mV before and after blocking by 200 nM Y^{3+}. (C and D) Tail currents were recorded at potentials between −50 and −200 mV after activation by 10.5-ms depolarizing steps to 0 mV. Data points represent amplitudes of recorded tail currents before and after blocking by 200 nM Y^{3+} (C) or 1 μM La^{3+} (D). Reprinted from Mlinar and Enyeart (1993), with permission of the authors and the Physiological Society, London.

cytoplasm via the plasmodesmata (Pitman, 1982). The site of metal detoxification or immobilization may either be in the symplasm (internal) or apoplasm (exclusion). An exclusion mechanism is postulated to include (a) immobilization of metals at the cell wall, (b) complexation of metals by chelates excluded from plant roots, and (c) formation of a redox barrier or a pH barrier by the plasma membrane (Taylor, 1987). The damage caused by anthropogenic acid precipitation with a pH 4 or lower to plants and the environment has been extensively investigated in recent years. Smalley *et al.* (1993) examined the effect of cations on the effective permeability of leaf cuticles to sulfuric acid. At concentrations of 1–10 mM, yttrium and gadolinium chlorides decreased the movement of protons through astomatous cuticles isolated from leaves of pear (*Pyrus communis*) and rough lemon (*Citrus lime*) to near zero.

Philips and Rainbow (1989) reviewed the strategies adopted by aquatic organisms and suggested that various species may lie on a continuum between "regulator" organisms, which exhibit low net uptake rates for metals, and "nonregulator" organisms, which may accumulate large amounts of some, or all, elements. Today, no generalization seems possible with respect to accumulation of metals by different species. Marine organisms may employ different strategies for the uptake, sequestration, and excretion of TE. Whereas nonregulatory organisms show an extreme uptake, regulatory organisms show a low uptake of TE. The key determinant of TE concentrations in aquatic invertebrates is their bioavailability from water and from food. Uptake of many TE from seawater conforms to the law of diffusion but may also be influenced by nonspecific binding to ligands in body fluids and cells of aquatic organisms (Depledge and Rainbow, 1990).

Studies on the effects of cationic uncouplers on intact rat cardiac muscle and on Ca^{2+} bound to isolated cardiac plasma membranes found the following selectivity sequence of the sarcolemmal Ca^{2+}-binding site: $Y^{3+} > Nd^{3+} > La^{3+} \approx Cd^{2+} > Co^{2+} \approx Mg^{2+}$ (Fig. 8-48); this order was the same as the effective uncoupling sequence (Bers and Langer, 1979). Further, the sidedness of isolated rabbit cardiac sarcolemmal vesicle was studied by observing the effects of permeability-increasing agents on several biochemical processes. The Na_1^+-dependent Ca^{2+} uptake was inhibited by certain cations in the order $Cd^{2+} > La^{3+} > Y^{3+} > Mn^{2+} > Co^{2+} > Mg^{2+}$. It was noted that the order of effectiveness of La^{3+} and Y^{3+} in the inhibition of Na^+–Ca^{2+} exchange is the opposite of that found for tension depression and Ca^{2+} displacement. The increase in displacing and uncoupling effectiveness of the trivalent ions was correlated with decreasing crystal ionic radius (Bers *et al.*, 1980).

The following order of effectiveness of metal cations in inhibiting the initial rates of Na_1-induced Ca^{2+} uptake in canine cardiac sarcolemmal vesicles was found: $La^{3+} > Nd^{3+} > Tm^{3+} \approx Y^{3+} > Cd^{2+} \approx Sr^{2+} \approx Ba^{2+} \approx Mn^{2+} \approx Mg^{2+}$. Concentrations of 1×10^{-1}–6×10^{-1} μM of trivalent ions caused a slight

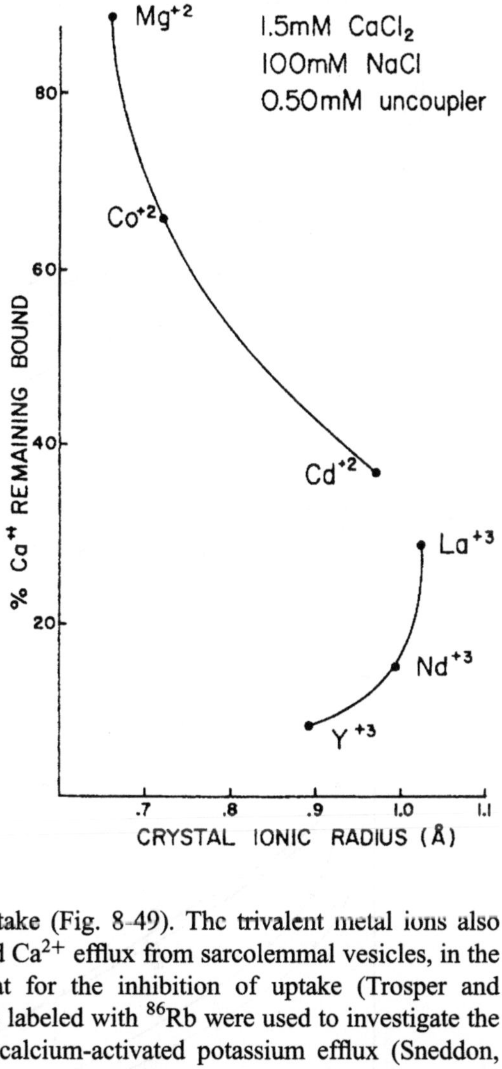

Figure 8-48. Percent of Ca^{2+} remaining bound to sarcolemmal vesicles from neonatal rat hearts in the presence of uncoupling cations (Cd^{2+}, Co^{2+}, Mg^{2+}, La^{3+}, Nd^{3+}, Y^{3+}) as a function of the crystal ionic radius. The incubation medium containing 1.5 mM $CaCl_2$, 100 mM NaCl, 10 mM sucrose, and 500 µM Cd^{2+}, Co^{2+}, Mg^{2+}, La^{3+}, Nd^{3+}, or Y^{3+}, added as chloride salts with 10 µM Tris at pH 7.0. The amount of Ca^{2+} bound without uncoupler present was 84 nmol/mg protein. Reprinted from Bers and Langer (1979), with permission of the American Physiological Society.

stimulation of Ca^{2+}-exchange uptake (Fig. 8-49). The trivalent metal ions also inhibited passive and Na^{+}-induced Ca^{2+} efflux from sarcolemmal vesicles, in the same concentration range as that for the inhibition of uptake (Trosper and Philipson, 1983). Rat erythrocytes labeled with ^{86}Rb were used to investigate the action of various metal ions on calcium-activated potassium efflux (Sneddon, 1987). Yttrium and several divalent and trivalent metal ions, all with an ionic radius of less than 0.090 Å, did not induce efflux of ^{86}Rb and inhibited calcium-activated ^{86}Rb efflux. The inhibitory effect of these metal ions was suggested to be the result of an interaction with Ca^{2+}, because the reduction in ^{86}Rb efflux could be overcome by increasing the extracellular Ca^{2+} concentration (Sneddon, 1987).

Isolated data exist about the possible influence of therapeutic chelating agents or cytotoxic agents (anticancer components, ionophores, tetracycline) on the translocation of scandium and yttrium, as well as of several other TE (May,

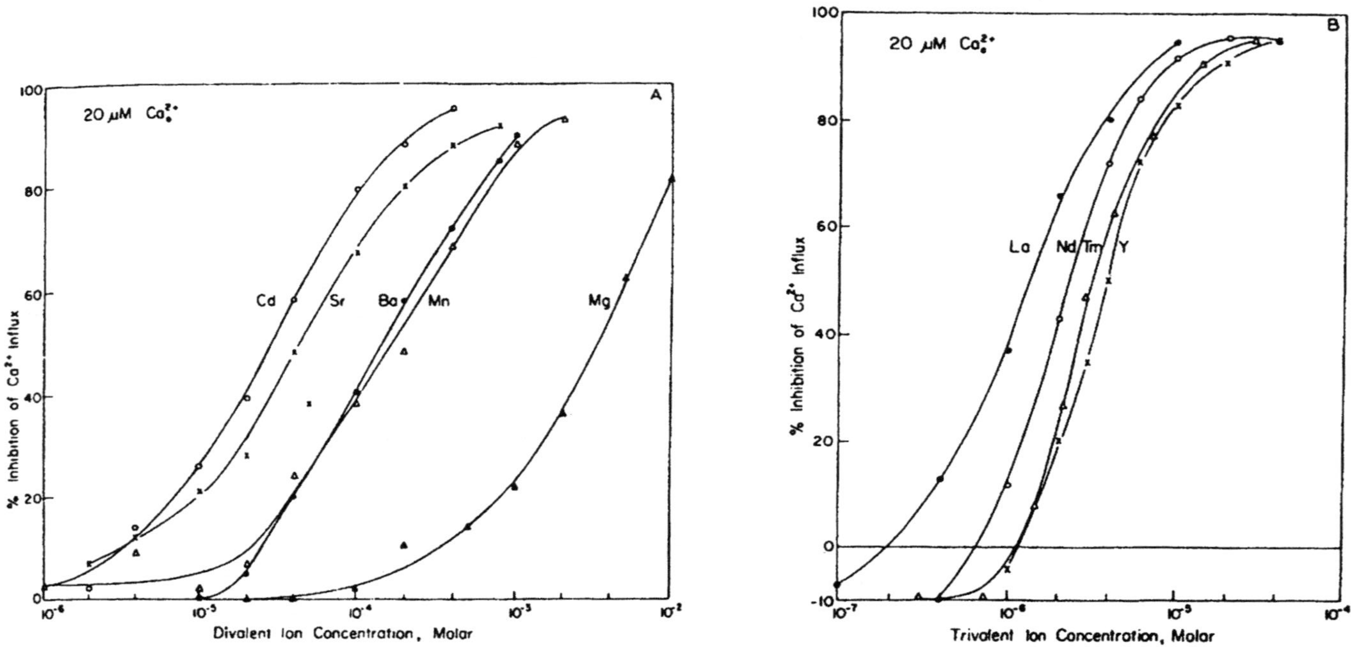

Figure 8-49. Inhibition of Na^+-dependent Ca^{2+} uptake in canine cardiac sarcolemmal vesicles by trivalent metal ions, as a function of the logarithm of cation concentration. In all media 20 μM Ca^{2+} and 0.36 μM valinomycin were present. Inhibition is expressed as percent of uptake in the absence of inhibitors. All data are corrected for nonspecific binding and uptake by mechanisms other than Na^+-induced exchange. Reprinted from Trosper and Philipson (1983), with permission of the authors and Elsevier Science.

1990). The ionophore A23187 induced histamine release from peritoneal rat mast cells in the presence of La^{3+} and Tb^{3+}, as it did in the presence of Ca^{2+} (Amellal and Landry, 1983). The effects of low concentrations of calcium and Ln were additive. These data show that the ionophore A23187 can transport Ln across the plasma membrane of mast cells, allowing them to substitute for calcium in the activation of calmodulin or calmodulin-like proteins (Amellal and Landry, 1983). It is premature to speculate about the ability of ionophores other than A23187 to transport REE in various biological systems.

Sc^{3+} and a few other metal ions (Tl^{+}, Ba^{2+}, Zn^{2+}, Mn^{2+}, and La^{3+}) potently inhibited magnesium influx without causing obvious cell toxicity in murine S49 lymphoma cells (Grubbs *et al.*, 1985; Table 8-28). The trivalent metal ions Y^{3+}, Eu^{3+}, and Tb^{3+} increased magnesium influx by 20–40% at 0.01 mM concentrations; at 1 mM, these metal ions substantially inhibited the influx. A clear dose–response relationship was not apparent. Thus, except for La^{3+} and Sc^{3+}, none of the trivalent metal ions tested inhibited the magnesium influx at concentrations distinctly different from the toxic range. The nature of the inhibition by Sc^{3+}, La^{3+}, and Zn^{2+} was not evaluated. Seventeen other metal ions were ineffective even at maximal nontoxic concentrations (Grubbs *et al.*, 1985). Erdos and Maguire (1983) analyzed the hormone-sensitive transport of magnesium into murine S49 lymphoma cells and its relationship to other metal transport systems. La^{3+} inhibited magnesium transport half-maximally at about 100 μM Sc^{3+}; the divalent metal ions were all less effective than La^{3+} (Fig. 8-50). The inability of Sc^{3+} to inhibit calcium transport, despite its specific and potent inhibition of free Mg^{2+} activation of the catalytic subunit of adenylate cyclase, was not explained. La^{3+} was a more potent inhibitor of magnesium influx than Sc^{3+}, with IC_{50} values of 10–30 and 100–300 μM, respectively (Maguire, 1982).

Table 8-28. Effect of Metals on Mg^{2+} Influx in S49 Lymphoma Cells[a]

Metal	Crystal ionic radius (Å)	Maximal conc tested (mM)	Effect
Al^{3+}	0.51	1.0	NE[b]
Sc^{3+}	0.73	0.1	$IC_{50} = 0.3$ mM[c]
Y^{3+}	0.89	0.1	NE
Ca^{2+}	0.99	30	$K_i = 5$ mM[d]
Mg^{2+}	0.66	3.0	$K_a = 0.33$ mM[e]

[a] Data reprinted from Grubbs *et al.* (1985), with permission of the American Physiological Society.
[b] NE, No effect at highest concentration that was nontoxic to cells.
[c] IC_{50}, Inhibition ion concentration giving 50% inhibition on influx at a stated extracellular concentration of the transported ion.
[d] K_i, Inhibition constant for the inhibitory ions.
[e] K_a, Dissociation constant of an ion.

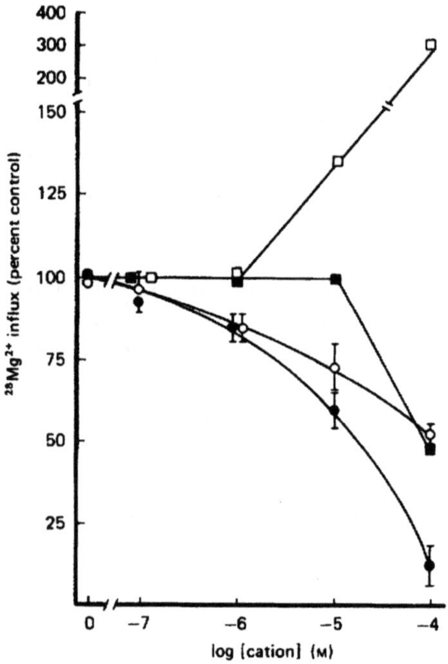

Figure 8-50. Effects of Sc^{3+} (○), La^{3+} (●), Zn^{2+} (■), and Ni^{2+} (□) on $^{28}Mg^{2+}$ influx in murine S49 lymphoma cells, measured after 15 min to reduce the potential toxic effects of the added metal ions. Reprinted from Erdos and Maguire (1983), with permission of the Physiological Society, London.

8.12 Involvement in Physiological Processes

Physiological processes in living organisms represent an extraordinarily complex, self-replicating, self-assembling, dynamic, and open physical and chemical system maintained in or near the steady state by the continual entry and exit of matter and energy (Tosteson, 1989). The presence of 4000–5000 species in a single gram of forest soil (Raven and Wilson, 1992), and a comparably large but different array in a similar volume of a lake, river, or marine sediment, gives an indication of the complexity of life processes.

At the beginning of this century, Bachem (1907), Mines (1910, 1911) and Höber and Spaeth (1914) conducted classical research on the physiological and toxicological effects of REE and other metal ions on animals. Colloidal forms of yttrium and other trivalent metal ions scarcely affected the hearts of frogs and rabbits, even at concentrations higher than those at which the simple metal ions produced diastole cardiac arrest. No toxic effects on heart rate and blood pressure were caused by low concentrations of $Y(NO_3)_3$ (Bachem, 1907). Mines (1910) showed that at concentrations of 1–10 μM, Y^{3+}, as well as Ce^{3+}, Dy^{3+}, La^{3+}, Nd^{3+}, Pr^{3+}, Sm^{3+}, and Tm^{3+}, added to Ringer solution, caused a rapid and reversible diastole arrest of the frog's heart. This effect was reversed by the addition of NaOH to the Ringer solution (Fig. 8-51). Whereas several REE

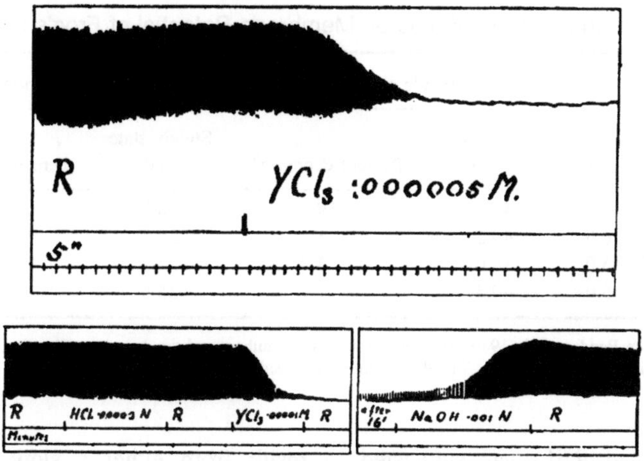

Figure 8-51. Myograms showing the arrest of rabbit's heart by yttrium chloride (*top*) and the arrest of frog's heart by yttrium chloride and recovery after addition of sodium chloride to the Ringer solution (*bottom*). Reprinted from Mines (1910) and (1911), with permission of the Physiological Society, London.

showed similar effects, Sc^{3+} and Al^{3+} produced a smaller inhibitory effect. The addition of Y^{3+}, Ce^{3+}, La^{3+}, and Pr^{3+} at concentrations of 8.7 mM to Ringer solution that was infused into rabbits caused temporary muscle paralysis. This effect was reversed after a few hours. The reversal was much slower at higher concentrations of REE and was concentration-dependent (Höber and Spaeth, 1914).

Results from a study of the effects of polyvalent cations on frog skin membrane potential are given in Table 8-29. Many polyvalent metal ions brought about a surge in the electrical potential (EP) only with respect to the outside solution. A marked decrease in resistivity coupled with an increased voltage indicated an electrokinetic phenomenon. Yttrium and other metal ions produced an EP increase of more than 30%. Scandium apparently fell between the "depressing" and "no-change" categories of metals. An increase of EP was expressed through electrical stimulation and rapid communication originating with the solution and passed from cell to cell (Bender *et al.*, 1980).

The twitch tension of the striated muscle and the neuromuscular junction of the grass frog (*Rana pipiens*) was halved by 0.5–1.1 mM Y^{3+} (Bowen, 1972). At a concentration of 0.0125 mM, Y^{3+} reduced the amplitude of the intracellular recorded miniature end-plate potential (MEPP), which indicates an inhibition of transmitter release. The inhibitory effect was partially to completely reversible during a 30-min recovery period for both indirectly and directly stimulated preparations (Fig. 8-52). The reduction in sensitivity of neuromuscular

Table 8-29. Influence of Metals on Membrane Potential of Frog's Skin[a]

		Outside solution		Inside solution	
Metal salt	pH	Steady state (mV)	Percent decrease[b]	Steady state (mV)	Percent decrease[b]
$ScCl_3·6H_2O$	5.0	45.3	19.8	30.8	3.4
$YCl_3·6H_2O$	6.0	49.2	1.1	30.7	−3.8
$FeSO_4·7H_2O$	5.0	36.6	3.9	34.7	0.8
$Cr(NO_3)_3·9H_2O$	5.0	40.8	−21	25.5	−14.7

[a] Data reprinted from Bender *et al.* (1980), with permission of the authors and the American Chemical Society.
[b] Percent decrease in electrical potential, E (mV), 120 min after steady state.

transmissions to REE was suggested to result from either a slow rate of penetration of yttrium or lanthanum into the muscle or a cumulative depolarization effect of the high MEPP. Concentrations of 0.2 mM Y^{3+} or La^{3+} reduced the amplitude but did not alter the time course of extracellular recorded end-plate potentials (EPP). Analysis of the quantal content of EPP indicated that the principal action of Y^{3+} or La^{3+} was to inhibit transmitter release. No significant differential reaction of the prejunctional membranes against Y^{3+} and La^{3+} was observed.

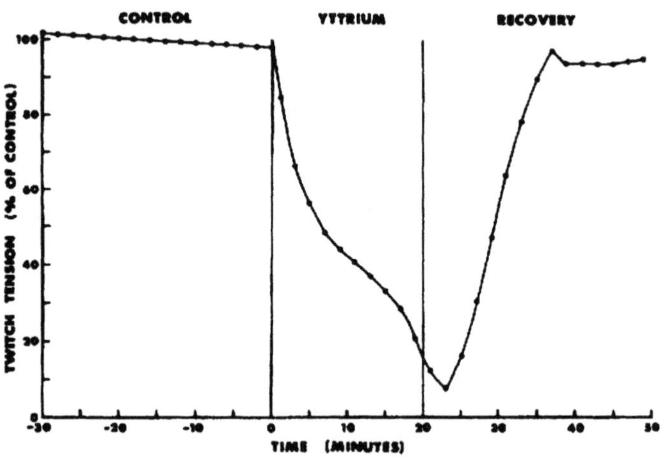

Figure 8-52. Effect of 0.8 mM Y^{3+} on the twitch tension of a directly stimulated sartorius muscle of frog (*Rana pipiens*) in Ringer solution with 20 μg *d*-tubocurarine chloride/ml. The stimulus intensity was increased gradually after addition of yttrium and was supramaximal 15 min after its addition; after that, the stimulus intensity was not changed until 38 min after addition of yttrium, when it was reduced to the level that had been supramaximal during the control phase of the experiment. Reprinted from Bowen (1972), with permission of the National Research Council of Canada, Research Press.

Intravenous injection of REE of low atomic numbers into rats induced fatty infiltration of the liver, whereas yttrium and Ln of higher atomic numbers did not have such an effect (Snyder *et al.*, 1959). The maximal infiltration observed at 48 h was characterized by an increase in neutral fat esters. The differences in response among the seven species that were studied were attributed to biochemical and physiological differences or differences in response time. Fatty liver response *in vivo* was elicited only by Ln with atomic numbers of 57–62, not by Ln with atomic numbers of 63–71. The reason why this effect was unique to the light Ln was not explained (Snyder, 1960).

In a study reported by Schroeder and Darrow (1973), large numbers of rats and mice were fed different elements for 90 days and evaluated for growth, longevity, changes in serum composition, macroscopic pathology, and disease. Data given in Table 8-30 show that feeding yttrium increased life span in both male and female animals. Similar findings were obtained with male and female mice that were fed various concentrations of scandium in drinking water. Haley *et al.* (1962) found that various concentrations of scandium chloride in the diet of rats did not significantly affect growth, except the 1% concentration in males. The lower growth of female animals resulted in minimal inhibitory effects even at high scandium concentrations (Fig. 8-53).

Hoshishima *et al.* (1978) investigated the effects of 24 metals on the sense of taste in mice by applying the bitter-tasting substance 6-propyl-2-thiouracil (PROP) on the tongue epithelium. For $ScCl_3$ and YCl_3, the minimal effective concentrations that altered the taste preference threshold for PROP were 45 and 9 ng, respectively. Differences were found between males and females in the stimulatory or inhibitory actions of the metals on the sense of taste. Scandium and

Table 8-30. Effects of Scandium and Yttrium, as Compared to Palladium and Mercury, Given to Mice in Drinking Water (5 mg/kg) on Median Life Span and Longevity[a]

	Males		Females		
Metal	Median life span (days)	Longevity (days)	Median life span (days)	Longevity (days)	Remarks
Scandium	453	686	484	783	
Yttrium	494	710	574	908[b]	
Palladium	554[b]	815[b]	572[b]	851	Tumors
Mercury	58[c]	—	65[c]	—	Accumulates

[a] Data reprinted from Schroeder and Darrow (1973), with permission of Plenum Press.
[b] Greater than control value, $P < 0.05$–0.005 (control values were not given).
[c] Less than control value, $P < 0.05$–0.005.

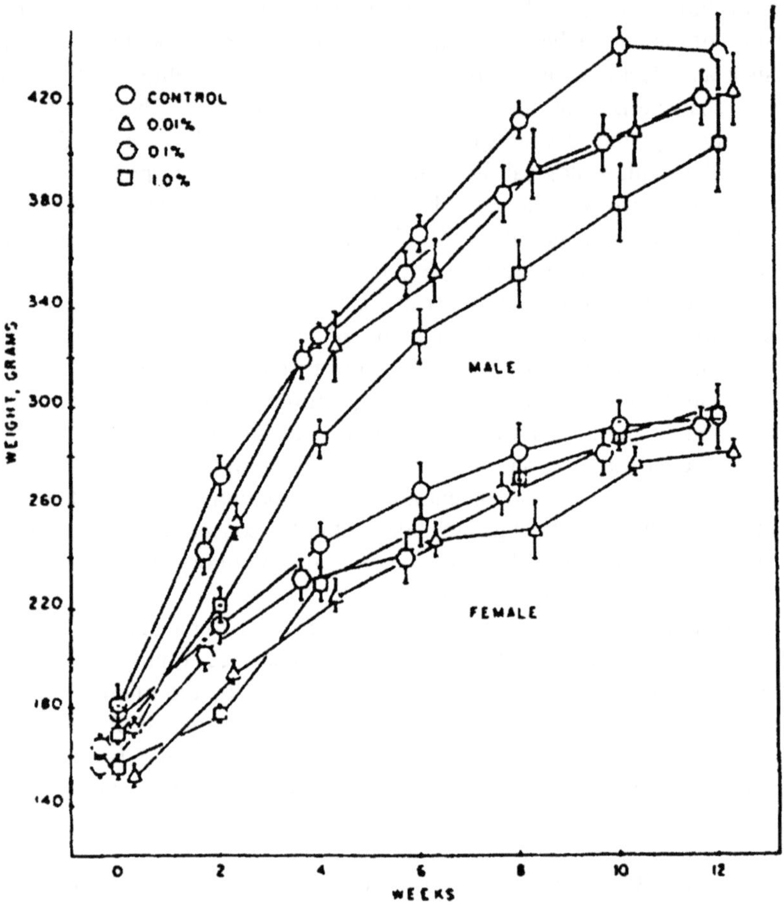

Figure 8-53. Comparison of growth of male and female rats fed diets containing 0.01, 0.1, or 1.0% $ScCl_3$ for 90 days with that of controls. Reprinted from Haley *et al.* (1962), with permission of the American Pharmacological Association.

some other TE increased the threshold for PROP at the minimum effective dose, whereas yttrium decreased the taste threshold in both male and female mice (Table 8-24). With some metals, like selenium and manganese, the change in the taste threshold in mice was mediated through an effect on choline-phosphate cytidyl transferase activity in the surface tissue of the tongue (Izumi, 1974). Scandium, yttrium and other TE injected into mice during pregnancy produced long-lasting behavioral changes, indicating changes in reflex response in suckling mice and in maze learning by the mature pups (Table 8-31). A sex difference was found in the head-turning response in offspring from mothers injected with

Table 8-31. Effect of Repeated Injection of Scandium and Yttrium into Mice on Behavior of Offspring[a]

Behavior test	Scandium	Yttrium
Weaning weight	16–30% increase[b]	16–30% increase[b]
Geotaxis	31–50% delay[b]	1–15% delay
Bar holding, 1 s	Acceleration[b]	16–30%[b]
Upward creeping	Acceleration[b]	None
Straight walking	Acceleration	Acceleration
Play behavior	Prominence	None-maze side

[a] Data reprinted from Tsujii and Hoshishima (1979) and Hoshishima *et al.* (1983), with permission of the authors and Elsevier Science.
[b] Sex difference.

scandium; female pups exhibited an earlier reflex response than male pups, but males showed better maze learning than females. The sex differences were explained as resulting from sexual differentiation of the brain during a critical period between birth and about 10 days of age (Tsujii and Hoshishima, 1979). Observation of the "play" behavior of mice indicated that, whereas yttrium preferentially affected the nonmaze-learning side of the brain during adult life, scandium acted mainly in the early stage of life (Hoshishima, 1983). Combined administration of yttrium and other REE to pregnant mice caused alterations in the behavioral development of the offspring, such as altered swimming ability and water escape behavior (Hoshishima *et al.*, 1983). Also, the prenatal injection into mice of scandium and yttrium caused changes in the circadian drinking rhythm of the offspring (Hoshishima and Shimai, 1985).

Il'enko (1973) reported distinct sex differences in the buildup of radioisotopes in a population of field voles on a territory that had been contaminated by fallout following nuclear tests in Nevada. The concentration of $^{90}Sr/^{90}Y$ detected in females in the reproductive period was higher than that in males, perhaps because of a higher metabolic rate.

Anke (1991) reviewed the role of several TE in animal reproduction, growth, and development; however, limited information exists for REE. No reduction in fertilization capacity or detrimental effects on developing ova were observed as a result of the pretreatment of rabbit sperm with ^{90}Y (Coomes, 1968).

8.13 Patterns in Disease and Pathological Conditions

Several recent books deal with the role of TE in health and disease (Aitio *et al.*, 1990; Tomita, 1990) and their uses in medicine and chelation therapy (Taylor and Williams, 1995). The prevailing viewpoint is that the dynamics of health are ordered and regular and that a variety of pathologies are bifurcations to chaotic states (Degn *et al.*, 1987). However, some pathological conditions manifest regular periodic dynamics, and normal nonpathological conditions may have chaotic dynamics (Goldberger and West, 1987). Biological organisms are considered to be an order/disorder equilibrium, in which order refers to structure and disorder refers to motion (Williams and Frausto da Silva, 1996).

Numerous attempts have been made to find a link between the changes in TE composition of tissues and various diseases (Ward, 1984, 1985, 1986, 1987; Corrigan *et al.*, 1987, 1989, 1990; Huljev *et al.*, 1989). Characteristic changes in blood concentrations of several TE, in particular, iron, copper, and zinc, occur as part of the nonspecific "acute-phase response" to injury and infection. Negative balance in illness is characterized by fever and retention of body water and salt, as well as of nitrogen, potassium, magnesium, phosphorus, zinc, and other TE (Beisel, 1976). Various organs, usually the "target organs", such as the liver, heart, blood, and brain, but also hair, nails, or skin, have been analyzed in disease states. Isolated data exist about changes in yttrium concentration in tissues of animals and humans and in other organisms affected by disease. The changes in the concentration of scandium in various tissues in humans affected by various pathological conditions, ranging from a significant increase or decrease to no change, are shown in Table 8-32. Threefold (Bhandari *et al.*, 1987; Zhuang *et al.*, 1991), 10-fold (Moo and Pillay, 1983), and even up to 20-fold (Kalashnikov *et al.*, 1983) increases in the concentrations of scandium and other TE have been found in pathological tissues, as compared to normal tissues. As compared to healthy persons, 548 patients with various stages and forms of lymphoderma exhibited no changes in the blood concentrations of scandium, iron, and tin concentrations (Bogomolov *et al.*, 1990). No distinct differences were found in urinary concentrations of scandium and several TE in patients before and after an operation for primary hyperparathyroidism (Boström and Wester, 1969). The concentrations of 29 TE in serum from patients with affective disorders at different stages of treatment were compared with those from control groups. Untreated depressive patients exhibited lower scandium concentrations than recovered depressive and recovered manic patients, schizophrenics, and alcoholics (Corrigan *et al.*, 1987).

A generalized metabolism disorder in epileptic patients may affect water transport across erythrocytes and cause TE imbalances in untreated epilepsy. In the serum of epilepsy patients, aluminum, strontium, and zinc were significantly higher, cobalt was significantly lower, and scandium and several other TE were not significantly different than controls (Davidson and Ward, 1988). In schizophrenic patients who had a high cardiac risk, a highly significant correlation between scandium and cesium and between scandium and red blood cell (RBC), phospholipids was found (Corrigan *et al.*, 1990b). This finding and

Table 8-32. Influence of Disease States or Pathological Conditions on Scandium Concentration in Human Tissues and Bodily Fluids

Disease or condition	Tissue or fluid analyzed	Other TE showing similar trend	Reference
		Increase of Sc conc	
Parkinsonism	Cerebrospinal fluid	Ag, Ca, Yb	Shi *et al.* (1991)
Schizophrenia	Brain	Ca, Cs, Mn	Corrigan *et al.* (1990a)
Manic depression	Serum	V	Corrigan *et al.* (1987)
Alzheimer's disease	Hair, nails	As, Br, Zn	Vance *et al.* (1988)
Mental retardation (children)	Hair	Au, Hg	Bhandari *et al.* (1987)
Blackfoot disease	Hair	Br, Fe, Hg	Pan *et al.* (1993)
Leukosis	Liver DNA	Fe	Babenko *et al.* (1980)
Breast cancer	Breast	Fe, Se, Zn	Garg *et al.* (1994)
Bone tumor	Bone	Fe, Se, Zn	Kalashnikov *et al.* (1983)
Malignant tumor	Brain	Cu, Fe, Zn	Zhuang *et al.* (1991)
Uremic failure	Heart	As, Br, Co	Pehrsson and Lins (1983)
Pneumoconiosis	Lung	Th, U	Pietra *et al.* (1985)
Periodontosis	Saliva	Cr, Mn	Zaichick and Bagirov (1993)
		Decrease of Sc conc	
Diabetes	Blood	Mg, Rb, Zn	Ward and Pim (1984)
Amyotrophic lateral sclerosis	Lung	Se	Khare *et al.* (1990)
Colitis	Colon mucosa	Co, Cr, Fe	Draskovic *et al.* (1987)
Leukemia	Blood	Al, Ca, Zn	Andronikashvili and Mosulishvilli (1980)
Undefined cancer	Hair	Al, Ca, Zn	Moo and Pillay (1983)
Gastric cancer	Stomach	Br, Co, Fe	Terai *et al.* (1980)
		No significant change of Sc conc	
Anorexia nervosa	Urine	Cr, Fe, Hg	Ward (1986)
Wilson's disease	Brain	Co, Fe, Zn	Ordögh *et al.* (1985)
Multiple sclerosis	Blood, hair	Mo, Pb, Zn	Ward *et al.* (1985)
Epilepsy	Serum	Ca, Hg, Mg	Davidson and Ward (1988)
Leukemia	Blood	Hg, Zn	Mangal and Sharma (1981)

other interactions between TE and phospholipids may represent different precursors to and states of the RBC membrane. Scandium was detected by thin-layer chromatography in only the 18 : 2ω6 lipid fraction isolated from plasma (Corrigan *et al.*, 1990b). A functional association between calcium, iron, lanthanum, nickel, and scandium and plasma phospholipid fatty acids in alcoholic subjects was found (Corrigan *et al.*, 1989). Findings from a balance study performed on patients with pancreatic insufficiency showed that the metabolism of several TE, including scandium, with known or unknown biological functions, was influenced by pancreatic exocrine function (Wester, 1971). The scandium concentration increased significantly in kidney and spleen and decreased in blood of mice with skin tumors (Gulati and Mangal, 1985). These changes were similar to those found for chromium, iron, selenium, and zinc (Table 8-33). Analysis of the TE in disorders of the breast found that the scandium concentration was higher with fibroadenoma tumor than with fibrocystic disease. The high correlation between cobalt and scandium was presumed to be a characteristic of fibroadenoma tumors (Kanias *et al.*, 1994). The accumulation of REE, including yttrium, in liver and skeletal muscle of lipogenous diabetes model mice was increased by factors of 2–5, as compared to normal mice (Enomoto *et al.*, 1995).

Alzheimer's disease (AD), which causes adult-onset dementia, is the most common degenerative disease of the central nervous system. Several alterations in brain TE concentrations which may be relevant to the etiology and pathogenesis of AD, have been described, such as increased aluminum, silicon, and lead. Analysis of isolated subcellular fractions of the temporal lobe from autopsied brains of AD patients and normal controls revealed that AD was associated with elevated bromine (in whole brain) and mercury (in microsomes) and diminished rubidium (in all fractions), selenium (in microsomes), and zinc (in the cell nucleus). Scandium exhibited only an insignificant decrease in mitochondria (Table 8-34). Whether these changes are a cause or an effect of the degenerative brain disease has not been established (Wenstrup *et al.*, 1990). Scandium and mercury were enriched in amygdala and nucleus basalis of Meynert (tissues which are considered to play important roles in memory function) of AD patients. However, these and other imbalances (cesium, iron, potassium, nitrogen, sodium,

Table 8-33. Scandium Concentrations in Skin Tumor and Tissues of Tumor-Bearing Mice, Compared to Control Animals[a]

Tissue	Sc conc: Control	Sc conc: TBA[b]	Percent difference from control
Liver	0.14 ± 0.001	0.19 ± 0.02	40
Kidney	0.09 ± 0.004	0.20 ± 0.04	122
Spleen	0.14 ± 0.01	0.28 ± 0.10	100
Blood	0.10 ± 0.003	0.07 ± 0.01	−33
Skin	0.16 ± 0.01	0.20 ± 0.02	12
Skin tumor		0.29 ± 0.09	81[c]

[a] Data reprinted from Gulati and Mangal (1985), with permission of the authors and the Executive Editor of the *Indian Journal of Cancer.*
[b] TBA, Tumor-bearing animals.
[c] Percent difference from concentration in skin of control animals.

Table 8-34. Imbalances of Selenium, Mercury, Rubidium, and Scandium in Subcellular Fractions of Autopsied Alzheimer's Disease Brains, Compared to Healthy Ones[a]

	Subcellular fraction			
Metal	Whole tissue	Nuclei	Mitochondria	Microsomes
		Control		
Se, ng/g	901 ± 162	809 ± 212	637 ± 68	513 ± 52
Hg, ng/g	69 ± 9.7	75.3 ± 40	126.7 ± 30	40.0 ± 8.7
Rb, μg/g	10.9 ± 0.8	1.79 ± 0.2	2.55 ± 0.8	2.12 ± 0.3
Sc, ng/g	0.72 ± 0.2	1.8 ± 0.9	7.5 ± 2.3	0.88 ± 1.1
		Alzheimer's disease		
Se, ng/g	865 ± 109	688 ± 165	529 ± 42	379 ± 46[b]
Hg, ng/g	176 ± 129	107 ± 37	174 ± 50	59.8 ± 29[b]
Rb, μg/g	8.30 ± 0.6[c]	1.39 ± 0.2[b]	1.41 ± 0.2	1.64 ± 0.2[b]
Sc, ng/g	0.88 ± 0.6	1.9 ± 0.5	3.9 ± 1.0	1.1 ± 0.4

[a] Data reprinted from Wenstrup *et al.* (1990), with permission of Elsevier Science.
[b] Significantly from control ($P < 0.05$).
[c] Significantly different from control ($P < 0.01$).

phosphorus, rubidium, and selenium) do not directly implicate any TE as causative factors in AD (Thompson *et al.*, (1988). The excretion of some TE (cadmium, molybdenum, and samarium) tended to be decreased, whereas that of others (mercury, lanthanum, and scandium) showed large variations. Based on the action of citrate, gluconate, tartrate, maltol, and acetylacetone complexes of aluminum, Corain *et al.*, (1993) concluded that aluminum speciation may significantly direct the biological effects of this TE. Ward and Mason (1987) stated that the possibility of multielement involvement in AD and other diseases is not clear. The scandium concentration in hair apparently is not affected by AD (Ehmann *et al.*, 1986). The concentrations of calcium, cobalt, manganese, scandium, and zinc in hair of cancer patients were significantly lower, as compared to those in hair of healthy people from the same geographic area (Fig. 8-54). The concentrations of samarium and europium in spleens of alcoholics were higher than in those of healthy subjects, but no significant changes in the concentration of yttrium, lanthanum, and other Ln were found (Erametsa and Sihvonen, 1971). However, because of the large variations in the data obtained from a limited number of subjects, no definite conclusions could be made from this study. A review of other studies revealed meaningful imbalances of Ln in pathologic and disease states (Evans, 1980). Heavy-metal lung disease can be caused by exposure of workers to dust produced during processing of

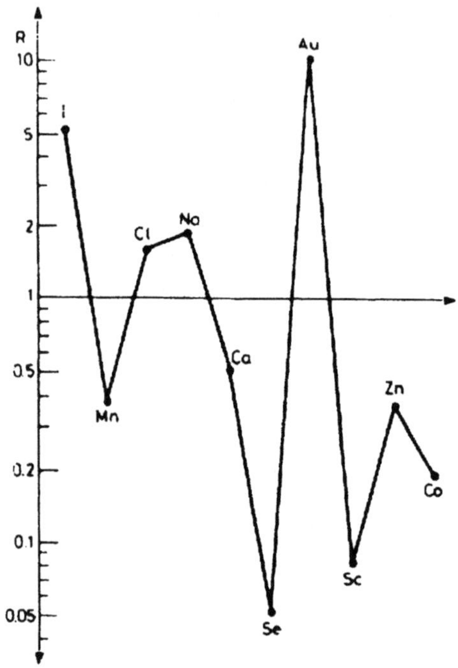

Figure 8-54. Concentration ratios of scandium and other trace elements in hair of cancer patients compared to those of healthy patients. From Moo and Pillay (1983), with permission of the authors and Los Alamos National Laboratory.

Table 8-35. Concentration of Tungsten, Tantalum, Cerium, and Scandium in Lung, Blood, and Urine of a Worker Exposed to Hard Metals, Compared to Healthy Subjects[a]

		Concentration mg/kg or mg/l			
Metal	Subject	Lung	BAL[b]	Blood	Urine
Tungsten	Worker	107,000	60	1.35	12
	Control	1.5	1.5	0.4	0.7
Tantalum	Worker	32,500	1010	1070	3
	Control	10	0.1	0.1	0.02
Cerium	Worker	146	1.9	6.5	45
	Control	70.6	0.5	2.2	0.4
Scandium	Worker	8.5	0.03	0.3	0.02
	Control	4.2	0.004	0.06	0.04

[a] Data reprinted from Rizzato *et al.* (1986), with permission of the authors and adapted from the American College of Chest Physicians.
[b] BAL, Bronchoalveolar lavage fluid.

metals. In a study reported by Rizzato *et al.*, (1986), a male subject exposed for many years to metal dusts from grinding and cutting metals was investigated four years after the exposure ceased. In addition to tremendously high concentrations of tungsten and other heavy metals, an elevated concentration of scandium was found in a lung biopsy specimen of the worker, as compared to control subjects (Table 8-35). The study of these TE in the bronchoalveolar lavage (BAL) fluid, blood, and urine samples of the worker suggested that TE entering the body through inhalation have a poor degree of mobility into organs. The authors of this study pointed out dose-response studies directed toward the understanding of the origin of metal lung disease should consider the multielement aspect of the exposure. Sabbioni *et al.*, (1987) proposed that the examination of BAL fluid of workers suspected to be affected by forms of occupational lung fibrosis be used as an indicator of exposure to TE. Higher concentrations of scandium and some Ln (1.5, 8.2, 757, and 405 μg/g of scandium, thorium, cerium, and neodymium, respectively) were detected in BAL fluid of photoengraver workers with pneumoconiosis disease. When integrated with clinical examinations, BAL may be a useful diagnostic tool (Sabbioni *et al.*, 1987).

In chemotherapy, suitable agents are substances having a strong affinity for the protoplasm of the invading agent and a weak effect on the cells of the host. Gallium-67 has a preferential affinity for soft-tissue tumors, through its interaction with plasma proteins in humans and other mammals (Hayes *et al.*,

Table 8-36. Effect of Scandium on the Four-Hour Tissue Distribution of ^{67}Ga in Male Rats Bearing Hepatomas[a]

	Percent administered $^{67}Ga/g$	
Tissue	Control	0.5 mg Sc^{3+}/kg
Tumor	3.7 ± 0.5	3.8 ± 0.4
Liver	0.7 ± 0.03	0.2 ± 0.01
Spleen	1.3 ± 0.1	0.6 ± 0.07
Lung	0.8 ± 0.06	0.3 ± 0.02
Muscle	0.1 ± 0.02	0.04 ± 0.01
Femur	0.5 ± 0.04	0.8 ± 0.02
Marrow	1.2 ± 0.1	0.2 ± 0.01
Blood	1.4 ± 0.1	0.36 ± 0.02
^{67}Ga retention, %	99 ± 0.9	75 ± 0.7
Tumor weight, g	16 ± 3.3	10 ± 1.2
^{67}Ga in tumor, %	25 ± 3.9	17 ± 1.8

[a] Data reprinted from Hayes *et al.* (1981), with permission of the Society of Nuclear Medicine.

1980). Intravenous administration of scandium prior to or with ^{67}Ga increased ^{67}Ga excretion and bone deposition and markedly decreased and blocked binding of ^{67}Ga by plasma proteins in normal as well as in tumor-bearing rats and mice but had little effect on rodent tumors (Hayes *et al.*, 1981; Table 8-36). The difference between the uptake by normal tissues and tumor uptake was attributed to dissimilar plasma membrane permeability. A different course of accumulation of ^{67}Ga and ^{46}Sc in the lysosomal fraction of Yoshida sarcoma, Ehrlich tumors, and rat liver was observed by Ando *et al.* (1989b). Whereas ^{46}Sc readily accumulated in the lysosomal fraction of tumors, most of the ^{67}Ga accumulated in the cytosol of tumor cells, while only a small amount accumulated in the lysosomes; both ^{67}Ga- and ^{46}Sc-binding substances avidly accumulated in liver lysosomes (Fig. 8-55). Nevertheless, chromatographic separation of ^{67}Ga and ^{46}Sc showed that these two radioisotopes were bound by different substances in the lysosomes of tumor cells (Ando *et al.*, 1989b). In rats, scandium and some Ln accumulated extensively in liver lysosomes, but much less was accumulated in tumor lysosomes; these elements were bound to acid mucopolysacharides with a molecular mass of ca. 10,000 Da (Ando and Ando, 1990; Table 8-37). Tumor uptake of ^{46}Sc and other radioisotopes in a rat hepatoma model was greater in

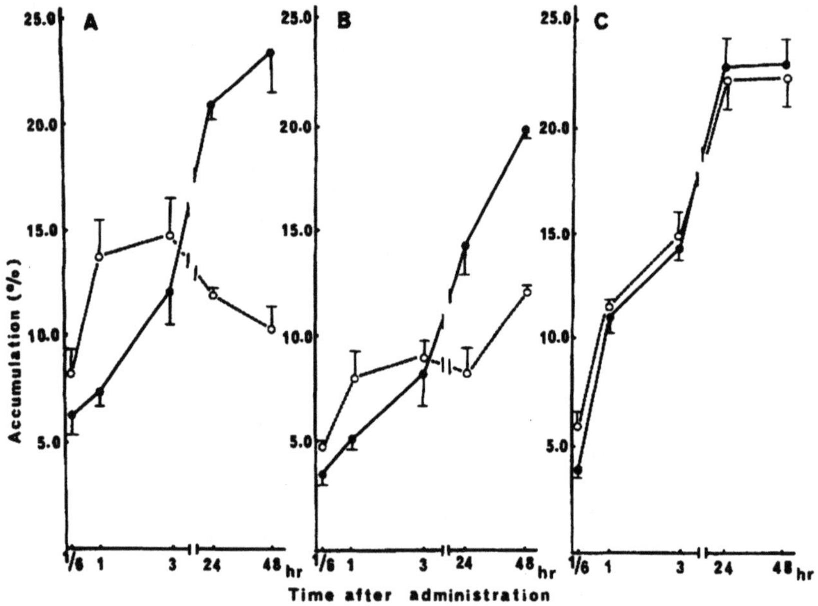

Figure 8-55. Percent accumulation of ^{46}Sc and ^{67}Ga in lysosomal fractions of tumors and livers of rats, after i.v. injection of ^{46}Sc and ^{67}Ga citrate solutions. (A) Yoshida sarcoma; (B) Ehrlich tumors; (C) liver of rats. ●, ^{46}Sc radioactivity; ○, ^{67}Ga radioactivity. Reprinted from Ando *et al.* (1989b), with permission of the authors and Elsevier Science.

Table 8-37. Percent Accumulation of Sc^{3+}, Ru^{3-4+}, and Hg^{2+} in Rat Liver Mitochondrial Fractions Containing Lysosomes and in Lysasomal Fractions of Yoshida Sarcoma[a]

Time after radioisotope administration	Sc^{3+} Liver	Sc^{3+} Yoshida sarcoma	Ru^{3-4+} Liver	Ru^{3-4+} Yoshida sarcoma	Hg^{2+} Liver	Hg^{2+} Yoshida sarcoma
	Percent accumulation					
10 min	9.0	4.0	13.2	5.9	26.4	11.2
60 min	28.0	7.7	22.2	5.9	24.6	10.9
3 h	49.1	15.4	28.9	7.3	20.8	11.5
24 h	53.8	36.2	40.0	15.1	24.2	12.5
48 h	52.5	34.6	43.4	17.4	24.9	12.9
	Percent accumulation of 48 h/percent accumulation at 10 min					
	5.8	8.7	3.3	2.9	0.9	1.2

[a] Data reprinted from Ando and Ando (1990), with permission of the authors and Elsevier Science.

viable than in nonviable tumor tissue at early times (4–24 h) after administration of the radioisotope; after 96 h, the remaining activity in the nonviable tumor tissue exceeded the quantity accumulated in viable tumor tissue (Hagan *et al.*, 1977).

Comparison of the retention of ^{125}I, ^{90}Y, and ^{111}In radioimmunoconjugates by cancer cells found that the antibodies are routed to lysosomes after endocytosis, where they are degraded by proteolysis (Press *et al.*, 1996). Whereas [^{125}I]monoiodotyrosine is rapidly excreted from cells after lysosomal catabolism of MAbs, small-molecular-weight ^{111}In and ^{90}Y catabolites remain trapped in lysosomes. As a consequence, ^{111}In and ^{90}Y conjugates resulted in a superior retention of radioactivity, compared with ^{125}I conjugates, when tumor cells were targeted by using rapidly internalizing antibody–antigen systems (Press *et al.*, 1996).

The use of Ho : YAG laser surgery to repair cartilage damage in horses caused decreased glycosaminoglycan synthesis in treated perilesional cartilage and perilesional zonal cloning of chondrocytes (Pullin *et al.*, 1996). The significance of these biochemical changes which may be influenced also by the chemical composition of the laser, is unknown. In humans, radiosurgery using ^{90}Y injected directly into the cavity of cystic craniopharyngiomas produced a remarkable reduction in tumor size and diminished cyst fluid production (Szeifert *et al.*, 1990; Fig. 8-2). Histological examination before and after ^{90}Y silicate implantation revealed that the lining epithelial cell layer became destroyed and the cyst wall shrunk, following the ^{90}Y irradiation. Large amounts of collagen fibers with focal hyaline degeneration were present, and the proliferation of intimal cells and subendothelial connective tissue narrowing small vessel lumina was observed. The destruction of the squamous epithelium and vascular changes were suggested to be the cause of cyst volume reduction and fluid diminution after ^{90}Y treatment (Szeifert *et al.*, 1990).

Concentrations of cobalt and chromium were elevated by up to two orders of magnitude in several organs of deceased people with CoCr-alloy/polyethylene joint prostheses (Schnabel *et al.*, 1994). A similar trend was found for scandium, even in organs distant from the implants. Because of the toxic, allergenic, and carcinogenic effects of cobalt, chromium, and nickel, and possibly of other TE, corrosion of implanted metallic alloys is considered a severe problem, especially if the frequency and duration of implantation are to be increased (Schnabel *et al.*, 1994).

8.14 Effects on Genetic and Immunologic Processes

Genetic definition of life considers metabolic evolution as some kind of junction between ancient elements (earth, water, air, and the fire of sunlight) and forward extension of metabolic pathways. A variety of colloidal inorganic crystalline minerals, clays, metal oxides, cationic and anionic polymers, and polyphosphates might be most suited to genetic and other functions (Cairns-Smith, 1982). A concept of evolutionary changes in biological functions of mineral elements based on natural selection according to potential for activity has been proposed. After the advent of dioxygen, new roles for molybdenum, selenium, vanadium, and other essential TE arose (Frausto da Silva and Williams, 1991). The concept that the biological functioning of nickel and other "not required" TE virtually disappear from higher organisms may be considered questionable. The importance of gene control via trace metals and the role of metals involved in gene regulation and metal-regulated expression are beginning to be appreciated. Several divalent metal ions and a few toxic metals are known to have roles in positive or negative gene regulation (Cousins, 1994). There are occasional data about the involvement of scandium and yttrium in genetic and immunologic processes.

Levan (1945) reported that treatment of the root meristem of onion (*Allium cepa*) with yttrium nitrate caused colchicine mitosis, an induction of sticky chromosomes, and differential staining of the heterochromatin at metaphase–anaphase. Similar effects were also caused by numerous other metals. The cholchicine-type mitosis was observed at concentrations of 5–50 μM Y^{3+}; these threshold values were considerably lower than those for colchicine.

In mices, survival, growth rate, hematology, morphological development, maturation, reproduction, and lactational performance were not significantly affected when oxides of scandium and other REE were fed with the basal diet for three generations (Hutcheson *et al.*, 1975). No significant differences in the number of mice born were found among the treatment groups; however, different growth rates during differentiation were observed. When a mixture of rare earth nitrates, which included YCl_3 and $ScCl_3$, were administered to rats and mice at concentrations of 1–100 mg/kg, no teratogenic effects were observed (Mu and Shi, 1985). Also, scandium and yttrium did not cause chromosomal aberrations and sister chromatid exchange *in vitro* experiments with human lymphocytes (Table 8-38), but no statistical analysis of these data was given.

Kidman *et al.* (1951) investigated the retention of strontium and yttrium radioisotopes injected into pregnant and lactating rabbits and their offspring. The detection of ^{90}Y and ^{90}Sr in fetuses within 1 h after their injection into pregnant rabbits demonstrated that these radioisotopes crossed the placenta. Lactating rabbits retained slightly less ^{90}Sr and ^{90}Y, produced by the decay of ^{90}Sr, than pregnant or virgin rabbits. The retention of ^{90}Y was roughly proportional to that of ^{90}Sr in pregnant and lactating rabbits and their offspring.

Recent research has found that the immune system is highly susceptible to alterations in nutritional and environmental factors. Deficiencies or excesses of some TE affect cellular and humor immunity, phagocyte function, thymus function, and natural killer cell activity and, as a result, affect infection, tumorigenesis, and autoimmune syndromes. Spallholz and Stewart (1989) presented findings from *in vivo* and *in vitro* experiments that were conducted to ascertain how macroelements (calcium and magnesium) and TE (copper, iron, and selenium) affect various components of the immune system and host resistance to infection. Genetically determined abnormalities of TE metabolism

Table 8-38. Effect of Metals on the Induction of Chromosomal Aberrations (CA) and Sister Chromatid Exchange (SCE) in Human Lymphocytes *In Vitro*[a]

	Concentration (μg/ml)	CA (%)	SCE/cell
Control	—	1.2	5.6
Scandium	5	0.0	—
	20	1.0	—
Yttrium	2.6	—	6.5
	17.0	—	6.9
Nickel	10	5.0	8.5
Mixture of REE	2.5	1.0	5.0
	25	1.5	4.7

[a] Data reprinted from Mu and Shi (1985), with permission of Science Press.

may be primary or secondary consequences of genetic diseases. Zinc has been found to be an essential activator of thymulin, while aluminum and gallium seemed to be as active as zinc (Cousins, 1994). Limited information exists about the effects of several other TE. Various TE acting as stimulating agents or in excess may influence the above-mentioned immune functions and processes; these play important roles in the metabolism of genetic molecules and affect information transfer to the immune system. The list of TE shown to affect the immune system is steadily increasing in length (Ohsawa, 1993). Several studies have been devoted to the effect of Ln on immunocompetent cells *in vitro*; anti-inflammatory effects of several Ln *in vivo* were outlined by Evans (1990).

The effects of yttrium and the Ln salts on antibody formation during primary immunological response has been investigated. In a study reported by Zimakov and Zimakova (1971), CBA mice were injected with ram erythrocytes and, after 24 h were supplied with increasing doses of yttrium or lanthanum propionate in drinking water. Whereas the lanthanum salt was increasingly toxic with increasing dose and was absolutely lethal at 180 mg/kg, the yttrium salt was not toxic. Plaque formation was stimulated by 40 mg of yttrium propionate; the hemolysin and hemagglutinin titers were also increased (Zimakov and Zimakova, 1971). The toxic effects of REE on several functions related to the inflammation caused by bovine alveolar macrophages (BAM) are presented in Section 9.12 (Fig. 9-15). The reasons for these effects are unknown; however, it was postulated that the calcium channel may be involved (Wiener-Schmuck *et al.*, 1990; Polzer *et al.*, 1993).

The relevance of individual variations of biochemical indices is obvious in many biochemical and physiological processes. These processes are sensitive to a wide range of external factors, such as nutrition (Van Toller, 1992). An example of individual variation in daily intake of scandium was presented in Fig. 8-37.

8.15 Comparison to Lanthanides, Aluminum, and Other Trace Elements

The empirical classification of metal ions in biological systems into distinct categories of class A, borderline, and class B elements (Nieboer and Richardson, 1980) was presented in Chapter 2 of Part 1. Some authors have considered this classification to be useful for the clarification of essentiality, toxicity, and therapeutic functions of metal ions (Nieboer and Sanford, 1984). However, no clear definite biochemical functions have emerged for the class A metals, which include Sc^{3+} and Y^{3+}.

The properties of the Ln^{3+} ions are often compared to those of Ca^{2+}, which allows the former to be used to mimic the crucial biochemical functions of Ca^{2+} (Evans, 1990). A comparison of the physical and chemical properties of Sc^{3+} and Y^{3+} (see Part 1, Table 2-4) with those of Ca^{2+} reveals many important characteristics; these may explain the ability of the former to replace Ca^{2+} in a specific, isomorphous manner in many biochemical processes. Sc^{3+} and Y^{3+} show remarkable similarities to the Ln^{3+} in size, bonding, coordination geometry, and donor atom preference. In some conditions, Sc^{3+} can occupy a Mg^{2+} site in a specific fashion. Unlike the Ln^{3+}, Sc^{3+} and Y^{3+} have in common with Ca^{2+} such properties as colorlessness, diamagnetism, and partial lack of luminescence. An difference between Ca^{2+} and Sc^{3+} or Y^{3+} is that the latter form more stable complexes. It was anticipated that Y^{3+} may behave similarly to the Ln^{3+} ions in biological systems (Martin and Richardson, 1979). In the description of the A-group metal ions, a distinct structural separation is made between ions with radius less than 0.75 Å and those with larger radius. This distinction puts a diagonal line in the periodic table, with Li^{+}, Mg^{2+}, Al^{3+}, and Sc^{3+} on one side and Na^{+}, Ca^{2+}, Ln^{3+}, and Y^{3+} on the other. Whereas the complexes formed by the second group of smaller cations have strictly octahedral symmetry, the complexes formed by the second group of larger metal ions have irregular coordination geometries, bond angles, and coordination numbers. These differences may explain the differing biochemical functions of Ca^{2+} and Mg^{2+}, and in particular the exceedingly high stability constants of ligands chelating Ln^{3+}, Y^{3+}, and Sc^{3+}, as compared to the divalent cations (Frausto da Silva and Williams, 1991).

The similarity of the biochemistry of aluminum and other trivalent TE indicates that the isotopes of Sc^{3+}, Ga^{3+}, and Cr^{3+} can be used as substitutes for Al^{3+} in the investigation of its metabolism. A hypothetical model for the metabolism of aluminum proposed by Ganrot (1986) features its main characteristics: low intestinal absorption, rapid urinary excretion, and slow tissue uptake, mostly in the skeleton and reticuloendothelial cells. Because of the lack of a usable isotope of aluminum, a comparative study of the intestinal absorption and tissue distribution of aluminum, gallium, and scandium in rats used the activity of ^{67}Ga and ^{46}Sc as markers (Farrar *et al.*, 1987; Farrar, 1988). The serosal transfer and mucosal uptake of all three elements were linear with time (2–30 min) after an initial rapid binding of the metal ions to the tissue. This was apparently a reflection of the formation of nontransportable metal–ligand complexes, or the food composition may have caused the metals to remain unavailable to the gut surface for transport. The highest concentrations of gallium and scandium were found in femur and spleen. It was speculated that scandium, aluminum, and gallium are transported across the epithelial barrier by a passive energy-independent process; most likely, this process occurs across the tight junctions and down the paracellular pathway between cells. While approximately 1% of aluminum and gallium were taken up by the tissue and transported to the serosal side, only 0.3% of scandium was absorbed in the intestine. These metal ions were taken up by the body more intensively in fasted than in fed animals.

In an extensive study of the biological behavior of high-specific-activity radioisotopes of 70 elements (Durbin, 1960), trace quantities of each were

Table 8-39. Distribution of Radioisotopes of the Elements of Group III in Various Rat Tissues after Intramuscular Injection[a]

Radioisotope	Chemical form	Percentage of absorbed dose			
		Bone	Liver	Muscle	Excreta[b]
^{46}Sc	Sc citrate	15.9	21.2	12.5	31.1 F
^{91}Y	YCl_3	55.6	12.1	2.6	26.3 U
^{140}La	La citrate	18.4	64.2	1.2	11.3 U; F
^{67}Ga	Ga citrate	19.6	7.1	13.2	48.7 F
^{114}In	$InCl_3$	16.8	14.0	14.7	25.8 F

[a] Data reprinted from Durbin (1960), with permission of Williams and Wilkins.
[b] F, Feces; U, urine.

intramuscularly injected into rats. Tissues and excreta were analyzed at autopsy at different intervals after injection. Distribution data were arranged according to the groups of the periodic table of the elements. For the Ln, a continuous transition from mainly hepatic to major skeletal deposition with increasing atomic number was noted. Trivalent metal ions exhibited remarkably similar behavior; liver and bone were the major deposition sites (Table 8-39). Elements which accumulated primarily in bone were excreted through the kidney, whereas elements which were deposited primarily in the liver were eliminated through the g.i. tract. The following properties were described as determinants of the biological behavior: (a) stable oxidation state at body pH, (b) solubility of the stable state, and (c) tendency to be incorporated into organic compounds and to associate with specific proteins (Durbin, 1960). Gensicke *et al.* (1966) compared the metabolism of yttrium radioisotopes with that of Ln radioisotopes. The lung clearance of inhaled ^{91}Y was 19 days, as compared to 21, 23, and 60 days for ^{147}Pm, ^{143}Pr, and ^{144}Ce, respectively. The liver deposition of ^{91}Y was fastest (i.e., 0.14 days, as compared to 0.52–1.09 days for the other three radioisotopes), and the ^{91}Y skeletal deposition was the slowest (2.97 days, as compared to 0.36–0.87 days of ^{147}Pm, ^{143}Pr, and ^{144}Ce).

8.16 Essentiality, Hormesis, and Other Theoretical and Practical Approaches

The free ions of virtually all A-type elements exhibit a variety of biological effects. Because these metal ions exchange relatively readily, their ligands are considered of low biochemical significance, and of lesser importance in terms of risks or benefits, except at relatively high doses (Frausto da Silva and Williams,

1991). On the basis of the similarities in the chemical composition of various groups of organisms, Li (1984) proposed the concept of systematic enrichment of "biophile" elements and depletion of "biophobe" elements. Scandium and yttrium, as well as aluminum, beryllium, chromium, cesium, gallium, silicon, titanium, vanadium, zirconium, and the Ln, were included in the group of "biophobe" elements. The consideration of their physical and chemical properties may not imply essential or a nonessential role for growth (Li, 1984). Several TE, such as barium, antimony, cobalt, chromium, gallium, and yttrium, which are deposited in the skeleton in trace or ultratrace amounts have been characterized as "neutral" cationic elements, because they show relatively low toxicity (Wallach and Chausmer, 1990; Christie and Williams, 1995). Although scarce in comparison with the extensive studies of the well-established essential TE, research on the biochemistry of scandium and yttrium has been marked by tremendous progress in the past three decades. In the previous sections of this chapter, we have seen that various investigations have shown beneficial and stimulatory properties of scandium and yttrium at low and, even very low, concentrations. Even at relatively high concentrations, scandium and yttrium do not have toxic effects in a large number of organisms.

Many experiments on the biological effects of chemical or other agents are concerned primarily with inhibitory or lethal concentrations; investigations dealing with stimulatory (hormetic) concentrations are much less common (Stebbing, 1982). The reason for failing to see hormesis effects, where they apparently exist, seems to be inadequacies in the experimental design. Designs of dose–response experiments to determine no-effect level and threshold (dividing line between no-effect and effect level) (Ottoboni, 1991) or lowest-observed effect level (LOEL) and no-observed-effect level (NOEL) (Hoekstra and Van Ewijk, 1993) have been proposed. The NOEL may be estimated for ultralow concentrations and infinite dilutions, which is the case for many chemical elements in a biological milieu. Induction of beneficial and stimulatory effects by low doses of chemical and physical agents and research on the biological effects of low-level exposure (BELLE) have resulted in the development of a multidisciplinary concept, situated at the crossroads of many sciences. Stimulatory effects of very weak doses of TE and other chemical and physical agents have been observed in animal and human experiments (Doutremepuich, 1991; Ashmarin *et al.*, 1992; Calabrese, 1994).

The concept of the nonspecific beneficial action of low doses of chemical and physical agents, developed by Hippocrates, Hahnemann, Schultz, Cannon, Bernhard, Selye, and others, was further extended to the hormetic action of ionizing radiation (Luckey, 1980). Of the many forms of alternative medicine, homeopathy is the most enigmatic. Research on the structure of water in the liquid phase, fractal geometry, and the role of hydrogen-bonded water molecules in determining the tertiary structure of proteins has led to speculations about solvent-mediated influence and stimulatory effects of higher homeopathic potencies (Fisher, 1986). A definite dividing line between effective and lowest noneffective concentrations is requisite in biological research (Horovitz, 1993).

The dominant opinion of the scientific community regards most elements of the periodic table as being biologically nonessential. Several definitions of essentiality do not seem convincingly scientific or experimentally well founded. Some TE are considered as unconvincingly essential in humans but of medical significance in various ways (Davies, 1977). The definition of essentiality is simple in theory and stringent in practice (Klevay, 1987). Rules for proving essentiality are still controversial, and several exceptions and limitations have to be resolved. One criterion employed in the assessment of TE in plant nutrition considers that aluminum and some other TE may be "beneficial" or of "restricted essentiality" for some plant species (Farago, 1985). Because the absence of evidence is not the same as evidence of absence, a nonessential element may be defined as "an element whose essentiality has not yet been proven" (Hatway, 1984). A given element may be essential, toxic, carcinogenic, or beneficial in various ways (growth stimulation, therapeutic effects, diagnostic agents, etc). These properties may depend upon its concentration, chemical form, tissue specificity, biochemical individuality, and several other factors. Essentiality seems to manifest as a dynamic process that may change under the influence of known and, at present, still unknown factors and interactions. It may be manifested more evidently in one or some species and may be absent (or for the time being unknown) in other species (Horovitz, 1988). At least for some TE, essentiality may be considered an adaptive phenomenon (Mertz, 1988). Several of the TE that are still considered nonessential may figure as "supporting" in relation to the central roles played by well-established essential macro- and trace elements. The transition from a beneficial effect through a stimulatory effect and up to an essential role may be a realm in biological processes.

The controversy regarding the essentiality of trace elements represents one of the unsolved issues in biology; it demonstrates the need to be cautious about generalizing on the basis of insufficient data (Sagan, 1994). Different and contradictory opinions about the biological significance of scandium and yttrium (Table 8-40) reflect the challenge faced by inorganic biochemistry in defining "essentiality." It may be speculated that as a result of future progress in analytical methods, chemical speciation, bioinorganic chemistry, biochemistry, physiology, and molecular biology, the relevance of the distinction between "essential" and "nonessential" elements may diminish.

Table 8-40. Biological Significance of Scandium and Yttrium, as Defined by Various Authors

Author(s)	Scandium	Yttrium
Bersin (1963)		Presumed to have some biological activities
Hutchinson (1964)	Approximately 10^7 reactive atoms/cell in mammals	Approximately 10^3 reactive atoms/cell in mammals
Needham (1965)	The complete biological nonentity of the 3*d* period	
Schwarz (1972)	Possible potential importance to mammals	Possible potential importance to mammals
Frieden (1972)	Probably unused	
Schroeder and Darrow (1973)	Nonessential; slightly toxic to some life processes	Nonessential; slightly toxic to some life processes
Luckey and Venugopal (1977)	Potentially stimulatory to vertebrates	Stimulatory to vertebrates
Luckey (1977)	Suspected carcinogen	Suspected carcinogen
Nieboer and Richardson (1980)	Class A binding preference; oxygen-seeking element	Class A binding preference; oxygen-seeking element
Georgievsky *et al.* (1981)	Uncertain biological function in mammals	
Discher *et al.* (1985)	Protein precipitant; does not fit into the broad classes of essential trace metals	Protein precipitant; does not fit into the broad classes of essential trace elements
Pais (1985)		Biological role conceivable
Nielsen (1986)	Apparently required by some forms of life; might have physiological action in higher forms of life	
Ochiai (1987)	Not yet known to be essential to organisms	
Hamilton (1987)	Similar chemistry to Al	Similar chemistry to Ln
Lima-de-Fasia (1988)	Could conceivably enter into the composition of living matter	Could conceivably enter into the composition of living matter
Valcovic (1990)	Nonessential element	Nonessential element
Evans (1990)		No known metabolic role
Frieden (1990)	Strong likelihood as essential for some biological systems	
Loewy *et al.* (1991)	Ultratrace element, not required	Ultratrace element, not universally required
Frausto da Silva and Williams (1991)	Unusual element	Unusual element
Christie and Williams (1995)	Neutral element	Neutral element
Horovitz (1995)	Obviously beneficial element	Obviously beneficial element

8.17 Summary

Several multielement investigations have found a wide distribution of scandium and yttrium in living matter, and highly positive or negative correlations between these elements and numerous others. Because of their relatively low toxicity at low concentrations, scandium and yttrium induce no visible morphological and cytological changes in lower organisms (bacteria and yeast). Several investigations have found specific stimulatory effects of scandium on some fungi (*Aspergillus*, *Rhizoctonia*, *Cercospora*) and of yttrium on other fungal species (*Tetrahymena*, *Chlamydomonas*).

Similarly to several Ln^{3+} ions, Y^{3+} and Sc^{3+} bind in large quantities to the external surfaces of microorganisms. The data concerning the transport of scandium and yttrium into yeast, bacteria, and algae are controversial. A very low or a limited absorption and translocation of these elements in plants has been found in several experiments. Interspecies variability and pH dependence are prominent. However, some data indicate that a slight translocation of scandium and yttrium to aerial organs may occur and that these elements are incorporated in cytoplasm of stem and leaf. A higher tolerance to scandium than to yttrium was observed in aluminum-tolerant plant cultivars.

It is speculated that $Al(OH)_3$ as well as the trihydroxides of other trivalent metal ions, are able to diffuse through cell membranes without a definite transport system. A specific interaction between aluminum, scandium, and other trivalent metal ions and voltage-dependent closure of the mitochondrial channel at very low concentrations supposedly indicates a mechanism for cell–metal interactions. Despite intensive research on transport mechanisms, no specific pathway for any of the polyvalent metal ions is recognized. The binding of Sc^{3+} or another metal to transferrin does not imply that this carrier necessarily has a role in the transport or metabolism of the element in question; however, supposedly transferrin may be one of the vehicles through which these elements are distributed throughout the body.

Although Sc^{3+} and Y^{3+} are similar in size, bonding coordination geometry, and donor atom preference, there are several differences in their relative body distribution. Similarly to the lighter lanthanides, scandium tends to accumulate predominantly in the liver and kidney, whereas yttrium and the heavier lanthanides prefer bone; spleen, lung, endocrine glands, and several other organs may accumulate these and other TE. One important characteristic of Sc^{3+} and Y^{3+} is their ability to replace Ca^{2+} in many biochemical events. As a result of the high affinity of Sc^{3+} and Y^{3+} for proteins, several investigations found various degrees of *in vitro*, as well as *in vivo*, inhibition or activation of various enzymes by these metal ions; these reactions are concentration- and pH-dependent.

The manner in which the elemental abundances and associations in the natural environment are reflected in human nutrition and health is of major importance in modern society. The TE present in the environment in naturally occurring chemicals and metabolic products of plants, animals, and microorganisms and ingested in the diet find their way into metabolic processes. Like many essential TE, scandium and yttrium are presumably metabolized by different reactions simultaneously and may undergo metabolism by several different pathways. Several investigations established a poor intestinal absorption of these elements in mammals, apparently because they form insoluble precipitates and bind to proteins, phospholipids, and carbohydrates. However, a relatively significant percentage of absorbed radioactive ^{90}Y in mammals may pass across the placenta and deposit in the fetus; moreover, when ingested by chickens, scandium and yttrium may be transferred to the egg yolk.

Increased, decreased, or unchanged concentrations of scandium in various tissues have been linked with various disease and pathological conditions. Few data are available for yttrium. The usefulness of TE analysis for diagnosis of cancer, Alzheimer's disease, and other pathological conditions is questionable, because sex and age differences or medication may complicate the interpretation of the results obtained. It is still unknown whether changes in TE concentrations, representing mineral imbalances, are a cause or an effect of several diseases. Only a few studies have examined the effects of scandium and yttrium on genetic and immunologic processes. Reports of Chinese investigators claimed that administration of REE to laboratory animals did not cause major genetic effects (chromosomal aberrations, sister chromatid exchange). Yttrium causes an enhanced release of superoxide anions in alveolar macrophages and macrophage-like differentiated cells.

Some data concerning the participation of scandium and yttrium in physiological processes (heart and muscle activities) indicate that inhibitory effects may be reversible with changes in pH and at low concentrations. Species, sex, and age differences are manifest in some events, but no mechanisms are known for some physiological effects, such as increase of life span. New experimental and theoretical approaches are needed to verify the belief that a basic distinction exists between essential and nonessential chemical elements in the biological world. Further studies of physical and chemical properties of scandium and yttrium are required, and questions regarding their existence in lower oxidation states, with coordination numbers other than six, transition-metal shielding, functioning in isomer complexes, catalysis, supramolecular structures of metal–ligand complexes, chemical speciation, and others, await resolution.

The beneficial and stimulatory effects of scandium and yttrium, which are well proven in several biological processes and medical applications justify further intensive research of their biochemical, physiological, and toxicological properties. It may be speculated that discovery of unknown biochemical properties of many chemical elements will yield new insights into the mysteries of life.

References

Achor, L. B., Geoffroy, R. F., and Chapin, L. R., 1962. Studies on the physiological disposition of ^{46}Sc in the rat, *Fed. Proc.* 21:311.

Ahmad, S., Chaudhary, M. S., Mannan, A., and Qureshi, I. H., 1983. Determination of toxic elements in tea leaves by INAA, *J. Radioanal. Chem.* 78:375–383.

Aitio, A., Antero, A., Jaervisalo, J., and Vainio, H., 1990. *Trace Elements in Health and Disease*, Royal Society of Chemistry, London.

Amellal, M., and Landry, Y., 1983. Lanthanides are transported by ionophore A23187 and mimic calcium in the histamine secretion process, *Br. J. Phamacol.* 80:365–370.

Anbar, M., and Ernst, N., 1962. A distribution study of ^{18}F-labelled cationic fluorocomplexes in rats, *Int. J. Appl. Radiat. Isot.* 13:47–51.

Anderson, J. W., and Beardall, J., 1991. *Molecular Activities of Plant Cells*, Blackwell, London.

Ando, A., and Ando, I., 1990. Accumulation of each element in the lysosome of tumor and liver cells as systematized according to its location in the periodic system, *Appl. Radiat. Isot.* 41: 327–331.

Ando, A., Ando, I., Yamada, N., Hiraki, T., and Hisada, K., 1987a. Distribution of ^{46}Sc and ^{51}Cr in tumor-bearing animals and the mechanism for accumulation in tumor and liver, *Nucl. Med. Biol.* 14:143–151.

Ando, A., Ando, I., Hiraki, T., Yamada, N., and Hisada, K., 1989b. Determination of the lysosomal role in tumor accumulation of ^{67}Ga by dual-tracer studies, *Appl. Radiat. Isot.* 40:521–524.

Andrews, G. A., 1956. The effect of dose on the mobilization of rare earth radioisotopes. Preliminary clinical studies, in: *Rare Earths in Biochemistry and Medical Research* (G. C. Kyker and E. B. Anderson, eds.), U.S. Atomic Energy Commission Report, Oak Ridge, Tennessee, pp. 214–262.

Andronikashvili, E. L., and Mosulishvili, L. M., 1980. Human leukemia and trace elements, in: *Metal Ions and Biological Systems* (H. Sigel, ed.), Marcel Dekker, New York, Vol. 10, pp. 167–206.

Anke, M., 1991. The essentiality of ultra trace elements for reproduction and pre- and post-natal development, in: *Trace Elements in Nutrition of Children* (R. K. Chandra, ed.), Raven Press, New York, pp. 119–144.

Appanna, V. D., and Huang, J., 1992a. Bioaccumulation of yttrium in *Pseudomonas fluorescens*, *Bull. Environ. Contam. Toxicol.* 49:620–625.

Appanna, V. D., and Huang, J., 1992b. Microbial precipitation of yttrium, *Microbios* 772:129–136.

Ashmarin, L. P., Lelekova, T. V., and Sanzhieva, L. T., 1992. On the efficiency of ultra-low doses and concentrations of biological compounds, *Izv. Akad. Nauk, Ser. Biol.* 4:531–536 (Russian).

Babenko, H. A., Mazepa, I. V., Mosulishvili, M., Belokobylskii, A. I., and Shavliashvili, N. S., 1980. Metal content in nucleic acids in experimental leukosis, *Ukr. Biokim. Zh.* 52:551–555 (Russian).

Bachem, C., 1907. Pharmacological experiments on some rare earths, *Arch. Int. Pharm. Ther.* 17:363–386 (German).

Bai, B. Z., Kastori, R., and Petrovic, N., 1990. Effect of elements from scandium and lanthanide groups on growth and morphological characters of young sunflower plants, *Helia* 13:1–9.

Balasiewicz, W., and Kierylowicz, H., 1961. Metabolism of ^{46}Sc in rats, *Nowotwory* 11:231–235 (Polish).

Ballatori, N., 1991. Mechanisms of metal transport across liver cell plasma membranes, *Drug Metab. Rev.* 23:83–132.

Baur, P. S., Clark, R. S., Walkinshaw, C. H., and Scholes, V. E., 1974. Uptake and translocation from Apollo 11 lunar material by lettuce seedlings, *Phyton* 32:133–142.

Beisel, W. R., 1976. Trace elements in infectious processes, *Med. Clin. North Am.* 60:831–849.

Bender, D. A., 1993. *An Introduction to Nutrition and Metabolism*, University College Press, London, pp. 276–281.

Bender, M., Hillman, K., Tirri, L., Graben, M., Fenster, D., and Rosensaft, M., 1980. The influence of the composition of the bathing solutions on frog skin membrane potential, in: *Bioelectrochemistry, Ions, Surfaces, Membranes* (M. Blank, ed.), American Chemical Society, Washington, D.C., pp. 409–443.

Bergethon, P. R., and Simons, E. R., 1990. *Biophysical Chemistry. Molecules to Membranes*, Springer-Verlag, New York, p. 153–170.

Bernhard, U., Kasperek, K., Höck, A., Vyska, K., Freundlieb, C., and Feinendegen, L. Z., 1977. Changes in trace elements concentrations in human serum after parenteral application of amino acids, *J. Radioanal. Chem.* 37:383–391.

Bers, D. M., and Langer, G. A., 1979. Uncoupling cation effects on cardiac contractility and sarcolemmal calcium binding, *Am. J. Physiol.* 237:H332–H341.

Bers, D. M., Philipson, K. D., and Nishimoto, A. Y., 1980. Na–Ca exchange and sidedness of isolated cardiac sarcolemmal vesicles, *Biochim. Biophys. Acta* 601:358–371.

Bersin, T., 1963. *Biochemie der Mineral- und Spurenelemente*, Akademische Verlagsgesellschaft, Frankfurt, pp. 464–486.

Beveridge, T. J., 1989. Metal ions in bacteria, in: *Metal Ions and Bacteria* (T. J. Beveridge and R. J. Doyle, eds.), John Wiley and Sons, New York, pp. 1–30.

Bhandari, H. P. S., Lal, G., Sidhu, N. P. S., Mittal, V. K., and Sahota, H. S., 1987. TE analysis of hair of mentally retarded children, *J. Radioanal. Nucl. Chem. Lett.* 119:379–385.

Bibov, K., Riis, G., and Salbu, B., 1984. *Naringsf. Arg.* 28:84. Cited by T. Mumcu *et al.*, 1988.

Bittar, E. E., and Huang, Y.-P., 1990. The behavior of the ouabain-insensitive sodium efflux in single barnacle muscle fibers toward the microinjection of aluminum, *Toxicol. Appl. Pharmacol.* 106:71–79.

Bogomolov, A. D., Agzamova, S. S., Mikholskaya, I. N., and Zhuk, L. I., 1990. Changes in blood trace elements in lymphedema. *Sov. Med.* 11:21–23 (Russian).

Bohinski, R. C., 1987. *Modern Concepts in Biochemistry*, Allyn & Bacon, Boston, pp. 1–5.

Bollard, E. G., 1983. Involvement of unusual elements in plant growth and nutrition, in: *Encyclopedia of Plant Physiology* (A. Läuchli and R. L. Bieleski, eds.), Springer-Verlag, Berlin, Vol. 15B, pp. 695–744.

Boroughs, H., Townsley, S. J., and Hiatt, R. W., 1956. The uptake, accumulation and loss of ^{91}Y by marine fish, and the importance of short-lived radionuclides in the sea, *Biol. Bull.* 111:352–357.

Boroughs, H., Townsley, S. J., and Ego, W., 1958. The accumulation of ^{90}Y from an equilibrium mixture of ^{90}Sr–^{90}Y by *Artemia salina*, in: Progress Report 1957–1958, Hawaii Marine Laboratory, U.S. Atomic Energy Commission Report, pp. 20–31.

Boström, H., and Wester, P. O., 1969. Pre- and postoperative excretion of trace elements in primary hyperparathyroidism, *Acta Endocrinol.* 60:380–388.

Bowen, J. M., 1972. Effects of REE and yttrium on striated muscle and the neuromuscular junction, *Can. J. Physiol. Pharmacol.* 50:603–611.

Braunstein, H. M., Pack, D. J., and Oakes, T. W., 1982. Intercomparison of stable element content of foods by statistical methods, in: *Trace Substances in Environment and Health* (D. D. Hemphill, ed.), University of Missouri, Columbia, Vol. 16, pp. 377–390.

Brodzinski, R. L., 1972. Measurement of radiation exposure of astronauts by radiochemical technique, NASA Technical Memo X-2440, pp. 162–167. CA 79:123174y.

Brown, P. H., Rathjen, A. H., Graham, R. D., and Tribe, D. E., 1990. Rare earth elements in biological systems, in: *Handbook on the Physics and Chemistry of Rare Earths* (K. A. Gschneidner, Jr., and L. Eyring, eds.), North-Holland, Amsterdam, Vol. 13, pp. 423–452.

Burgess, J., 1996. Man and the elements of groups 3 and 13, *Chem. Soc. Res.* 1996:85–92.

Byrd, B. L., Watson, E. E., Cloutier, R. J., and Hayes, R. L., 1975. Effect of stable scandium on the long-term whole body retention of intravenously administered ^{46}Sc citrate in the rat, *Health Phys.* 29:375–379.

Cai, R., Huang, H., Wang, G., and Zeng, Y., 1994. Study and analytical application of rare earth inhibition of laccase, *Talanta* 41:735–738.

Cairns-Smith, A. G., 1982. *Genetic Takeover and the Mineral Origin of Life*, Cambridge University Press, Cambridge, pp. 93–118, 160–189, 317–319.

Calabrese, E. J., 1994. Primer on BELLE, in: *Biological Effects of Low Level Exposures. Dose–Response Relationships* (E. J. Calabrese, ed.), Lewis Publishers, Boca Raton, Florida, pp. 27–42.

Castillo Mercado, R., and Bibby, B. G., 1973. Trace elements effects on enamel pigmentation, incisor growth and molar morphology in rats, *Arch. Oral Biol.* 18:629–635.

Castillo Mercado, R., and Ludwig, T. G., 1973. Effect of yttrium on dental caries in rats, *Arch. Oral Biol.* 18:637–640.

Chahal, S. S., and Rawla, G. S., 1977. Trace elements requirement for the growth of *Cercospora granati*, *Ind. Phytopathol.* 30:47–50.

Cheng, P. T., 1991. Pathologic mineralization in humans, in: *Mechanisms and Phylogeny of Mineralization in Biological Systems* (S. Suga and H. Nakahara, eds.), Springer-Verlag, Tokyo, pp. 241–245.

Christie, G. L., and Williams, D. R., 1995. Classification of metal ions in living systems, in: *Handbook of Metal–Ligand Interactions in Biological Fluids* (G. Berthon, ed.), Marcel Dekker, New York, pp. 29–37.

Clarkson, D. T., and Sanderson, J., 1969. The uptake of a polyvalent cation and its distribution in the root apices of *Allium cepa*, *Planta* 89:136–154.

Clarkson, D. T., and Sanderson, J., 1971. Inhibition of the uptake and long-distance transport of calcium by aluminum and other polyvalent cations, *J. Exp. Bot.* 22:837–851.

Clayton, R. A., 1959. *In vitro* inhibition of selected enzymes by rare earth chlorides, *Arch. Biochem. Biophys.* 85:559–560.

Clemente, G. F., 1976. Trace elements pathways from environment to man, *J. Radioanal. Chem.* 32:25–41.

Clemente, G. F., Cigna Rossi, L., and Santaroni, G., 1978. Studies on the distribution and relative health effects of the trace elements in Italy, in: *Trace Substances in Environment and Health* (D. D. Hemphill, ed.), University of Missouri, Columbus, Vol. 12, pp. 23–30.

Cochran, K. W., Doull, J., Mazur, M., and DuBois, K. P., 1950. Acute toxicity of zirconium, niobium, strontium, lanthanum, cesium, tantalum, and yttrium, *Arch. Ind. Occup. Hyg.* 1:637–650.

Cohen, B. L., 1985. Transport of elements from soil to human diet. An alternative approach to pathway analysis, *Health Phys.* 49:239–245.

Collins, Y. E., and Stotzky, G., 1989. Factors affecting the toxicity of heavy metals to microbes, in: *Metal Ions and Bacteria* (T. J. Beveridge and R. J. Doyle, eds.), John Wiley and Sons, New York, p. 52.

Colombini, M., 1991. Aluminum and membrane channels, in: *Aluminum in Chemistry, Biology and Medicine* (M. Nicolini *et al.*, eds.), Cortina International, Verona, Italy, pp. 23–43.

Colombini, M., 1994. Personal communication, with permission.

Coomes, R. K., 1968. Effective uptake of ^{90}Y by spermatozoa from ^{90}Sr–^{90}Y equilibrium mixtures, PhD thesis, Cornell University, Diss. Abstr. 29:4819B.

Cooper, J. A. D., Bulklrey, G. J., and O'Conor, V. J., 1956. Intraprostatic injection of radioactive yttrium chloride in the dog, in: *Rare Earths in Biochemical and Medical Research* (G. C. Kyker and E. B. Anderson, eds.), U.S. Atomic Energy Commission Report, Oak Ridge, Tennessee, pp. 323–331.

Corain, B., Nicolini, M., and Zatta, P., 1993. Relevance of metal speciation in directing the biological activity of Al^{3+} *in vitro*, in: *Alzheimer Disease* (B. Corain *et al.*, eds.), John Wiley and Sons, Chichester, England, pp. 273–279.

Corrigan, F. M.. Finlayson, J. D., Muir, M., Ashcroft, G. W., and Ward, N. I., 1987. Multielement analysis of serum in affective disorders, *Trace Elem. Med.* 4:139–143.

Corrigan, F. M., Glen, A. I. M., Glen, E. M. T., Skinner, F. K., Horrobin, D. F., and Ward, N. I., 1989. Multielement and lipid analysis of blood in alcoholic subjects, *Trace Elem. Med.* 6:108–113.

Corrigan, F. M., Reynolds, G. P., and Ward, N. I., 1990a. Multielement analysis of the frontal cortex, temporal cortex and basal ganglia in schizophrenia, *Trace Elem. Med.* 7:1–7.

Corrigan, F. M., Van Rhijn, A. G., Horrobin, D. F., Skinner, E. R., and Ward, N. I., 1990b. Element lipid correlations in schizophrenia, *Trace Elem. Med.* 7:101–105.

Cousins, R. J., 1994. Metal elements and gene expression, *Annu. Rev. Nutr.* 14:449–469.

Cunningham, W. C., and Stroube, W. B., Jr., 1987. Applications of an INAA procedure to analysis of food, *Sci. Total Environ.* 63:29–43.

Cutler, P., and Blair, J. A., 1987. The effect of lead and aluminum on rat dihydropteridine reductase, *Arch. Toxicol.*, Suppl. 11, pp. 227–230.

Daigneault, E. A., 1963. The distribution of intravenously administered yttrium chloride (carrier free) in the rhesus monkey, *Toxicol. Appl. Pharmacol.* 5:331–343.

Dano, K., and Reich, E., 1979. Plasminogen activator from cells transformed by an oncogenic virus, *Biochim. Biophys. Acta* 566:138–151.

Danson, M. J., 1993. Central metabolism in the archaea, in: *The Biochemistry of Archaea* (M. Kates, D. J. Kushner, and A. T. Matheson, eds.), Elsevier, Amsterdam, pp. 1–30.

Darnall, D. W., and Birnbaum, E. R., 1970. Rare earth metal ions as probe of calcium ion binding in proteins, *J. Biol. Chem.* 245:84–88.

Davidson, D. L. W., and Ward, N. I., 1988. Abnormal aluminum, cobalt, manganese, strontium, and zinc concentrations in untreated epilepsy, *Epilepsy Res.* 2:323–330.

Davies, I. J. T., 1977. Medical significance of the essential biological metals, *J. Radioanal. Chem.* 37:39–64.

Degn, H., Holden, A. V., and Olsen, L. F., eds., 1987. *Chaos in Biological Systems*, Plenum Press, New York.

DeNardo, S. J., Kukis, D. L., Kroger, L. A., O'Donnell, R. T., Lamborn, K. R., Miers, L. A., DeNardo, D. G., Meares, C. F., and DeNardo, G. L., 1997. Synergy of Taxol and radioimmunotherapy with ^{90}Y-labeled chimeric L6 antibody: Efficacy and toxicity in breast cancer xenografts, *Proc. Natl. Acad. Sci. USA* 94:4000–4004.

Depledge, M. H., and Rainbow, P. S., 1990. Models of regulation and accumulation of trace metals in marine invertebrates, *Comp. Biochem. Physiol.* 97C:1–7.

Deuber, R., and Heim, T., 1991. Yttrium, in: *Metals and Their Compounds in the Environment* (E. Merian, ed.), VCH Weinheim, pp. 1299–1308.

Diatloff, E., Asher, C. J., and Smith, F. W., 1995. Effects of rare earth elements on the growth and mineral nutrition of plants, in: *Rare Earths in Agricultural Sciences*, Australian Academy of Technological Sciences, Carlton South, pp. 11–23.

Discher, C. A., Medwick, T., and Bailey, L. C., 1985. *Modern Inorganic Pharmaceutical Chemistry*, Waveland Press, Prospect Heights, pp. 379–401.

Dobroliubskyi, O. K., 1962. The role of scandium in redox processes occurring in plants, *Dokl. Akad. Nauk SSSR* 144:1174–1177 (Russian).

Dobrovolsky, V. V., 1994. *Biogeochemistry of the World's Land*, Mir Publishers, Moscow, pp. 36–40.

Doutremepuich, C,, ed., 1991. *Ultra Low Doses*. Taylor & Francis, London.

Draskovic, R. J., Bozanic, V., and Bozanic, M., 1987. Statistical investigations of some element distributions in healthy and pathologically altered human colon mucosa, *J. Radioanal. Nucl. Art.* 116:409–445.

Duffield, J. R., 1991. Low molecular mass aluminum complex speciation in biofluids, *J. Coord. Chem.* 23:277–290.

Duncan, B., 1966. The action of cerium, other rare earth elements and yttrium on yeast metabolism, PhD thesis, University of Arkansas.

Durbin, P. W., 1960. Metabolic characteristics within a chemical family, *Health Phys.* 2:225–238.

Durrant, S. F., and Ward, N. I., 1989. Multi-elemental analysis of human milk by ICP-MS: Implications for infant health, *J. Micronutr. Anal.* 5:111–126.

Ehmann, W. D., Markesbery, W. R., Alauddin, M., Hossain, T. I. M., and Brubaker, E. H., 1986. Brain trace elements in Alzheimer's disease, *NeuroToxicology* 7:197–206.

Ellis, K. J., 1997. Lanthanide elements in biochemistry, biology and medicine, *Inorg. Perspect. Biol. Med.* 1:101–135.

Enomoto, S., Liu, B., Amono, R., Ambe, S., and Ambe, F., 1995. *In vivo* behavior of various REE in lipogenous diabetes model mice using the multitracer technique, *J. Inorg. Biochem.* 59:353.

Erametsa, D., and Sihvonen, M. L., 1971. Rare earths in the human body. 2. Yttrium and lanthanides in the spleen, *Ann. Med. Exp. Fenn.* 49:35–37.

Erdos, J. J., and Maguire, M. E., 1983. Hormone-sensitive magnesium transport in murine S49 lymphoma cells. Characterization and specificity for magnesium, *J. Physiol.* 337: 351–371.

Essington, E., Nishita, H., and Wallace, A., 1963. Effect of the chelating agents on the uptake of ^{91}Y, ^{106}Ru, ^{144}Ce, and ^{147}Pm by beans grown in a calcareous soil, *Soil Sci.* 95:331–337.

Etherington, G., Bailey, M. R., Dorrian, M. D., and Hodgson, A., 1989. A study of the biokinetics of i.v. administered ^{88}Y in man, in: *Radiation Protection. Theory and Practice* (E. P. Goldfinch, ed.), 4th Int. Symp., Malvern, pp. 445–448. National Radiological Protection Board, Chilton, England.

Evans, C. H., 1990. The occurrence and metabolism of lanthanides, in: *Biochemistry of the Lanthanides*, Plenum Press, New York, pp. 285–337.

Evans, W. H., 1914. The influence of the carbonates of the rare earth elements (Ce, La, Y) on growth and cell division in hyacinths, *Biochem. J.* 7:349–355.

Fähnrich, P., 1964. Investigations on the influence of boron on pollen germination and on growth of pollen micella, in: *Pollen Physiology and Fertilization* (H. F. Linskens, ed.), North-Holland, Amsterdam, pp. 120–127 (German).

Farago, M. E., 1985. Metal ions and plants, in: *Frontiers in Bioinorganic Chemistry* (A. A. Xavier, ed.), VCH Weinheim, Germany.

Farrar, G., 1988. The intestinal absorption and tissue distribution of gallium and scandium in the rat. Prospective models for aluminum, PhD thesis, Ashton University, Birmingham, U.K.

Farrar, G., Morton, A. P., and Blair, J. A., 1987. The intestinal absorption and tissue distribution of aluminum, gallium and scandium. A comparative study, *Biochem. Soc. Trans.* 15:1164–1165.

Fekete, F. A., and Barton, L. L., 1991. Effects of Fe^{3+} analogues on growth and pseudobactin synthesis in a chromium-tolerant *Pseudomonas* isolate, *Biol. Metals* 4:211–216.

Felix, C. C., Hyde, J. S., Sarna, T., and Sealy, R. C., 1978. Interactions of melanin with metal ions. Electron spin resonance evidence for chelate complexes of metal ions with free radicals, *J. Am. Chem. Soc.* 100:1922–1926.

Fisher, P., 1986. Homeopathy. 200 years on non-animal methods, *Altern. Lab. Anim.* 14:24–32.

Foreman, H., and Finnegan, C., 1956. Chelation in physiological systems, in: *Rare Earths in Biochemical and Medical Research* (G. C. Kyker and E. B. Anderson, eds.), U.S. Atomic Energy Commission Report, Oak Ridge, Tennessee, pp. 136–142.

Frausto da Silva, J. J. R., and Williams, R. J. P., 1991. *The Biological Chemistry of the Elements. The Inorganic Chemistry of Life*, Clarendon Press, Oxford, pp. 100–103, 245–248, 545–546.

Freed, B. R., 1977. The tissue distribution of ^{47}Sc generated by decay of ^{47}Ca *in vivo*, *Health Phys.* 32:165–178.

Freed, B. R., 1996, personal communication, with permission.

Frey, C. M., and Stuehr, J., 1974. Kinetics of metal ion interactions with nucleotides and base-free phosphates, in: *Metal Ions in Biological Systems* (H. Sigel, ed.), Marcel Dekker, New York, Vol. 1, pp. 51–115.

Frieden, E., 1972. The chemical elements of life, *Sci. Am.* 227:52–60.

Frieden, E., 1990. Personal communication, with permission.

Ganrot, P. O., 1986. Metabolism and possible health effects of aluminum, *Environ. Health Perspect.* 65:363–441.

Gardner, R. A., and Willard, J., 1986. A new description of catalysis applied to carbonic anhydrase and other reactions, *J. Mol. Catal.* 34:103–121.

Garg, A. N., Vivek Singh, Weginwar, R. G., and Sagdeo, V. N., 1994. An elemental correlation study in cancerous and normal breast tissues with successive clinical stages by NAA, *Biol. Trace Elem. Res.* 46:185–202.

Gensicke, F., Nitschke, H. W., and Hölzer, F., 1966. Comparative investigation of the metabolism of ^{91}Y, ^{144}Ce, ^{143}Pr and ^{147}Pm, *Stud. Biophys.* 1:347–352 (German).

George, A., Draganac, P., and Farkas, W. R., 1985. Depolymerization of the interferon-induced 2′5′-oligoadenylate by metal ions, *Inorg. Chem.* 24:3627–3631.

Georgievsky, V. I., Annenkov, B. N., and Samokhin, V. T., 1981. *Mineral Nutrition of Animals*, Butterworths, London, pp. 11–17.

Gibson, R. S., and Scythes, C. A., 1984. Chromium, selenium and other trace elements intakes of a selected sample of Canadian premenopausal women, *Biol. Trace Elem. Res.* 6:105–116.

Goldberger, A. L., and West, B. J., 1987. Chaos in physiology: Health or disease, in: *Chaos in Biological Systems* (H. Degn, A. V. Holden, and L. F. Olsen, eds.), Plenum Press, New York, pp. 1–3.

Graul, E. H., and Hundeshagen, H., 1959. Investigations of ^{90}Y metabolism, *Int. J. Appl. Radiat. Isot.* 5:243–252.

Grubbs, R. D., Wetherill, C. A., Kutschke, K., and Maguire, M. E., 1985. Magnesium transport in murine S49 lymphoma cells: Pharmacology and divalent cation selectivity, *Am. J. Physiol.* 248:C51–C57.

Gulati, N., and Mangal, P. C., 1985. Trace elements profile in skin-tumor and tissues of tumor-bearing mice, *Ind. J. Cancer* 22:308–314.

Gutknecht, J., 1961. Absorption, accumulation and loss of radioactive scandium by marine macroalgae, *Biol. Bull. Mar. Biol. Lab. Woods Hole* 121:374–375.

Hagan, P. L., Chauncey, D. M., Jr., Halpern, S. E., and McKegney, M. L., 1977. Comparison of viable and nonviable tumor uptake of ^{46}Sc, ^{54}Mn, ^{65}Zn, ^{111}In and ^{195}Au with ^{67}Ga citrate in a hepatoma model, *Eur. J. Nucl. Med.* 2:225–230.

Haley, T. J., Komesu, N., Mavis, L., Cawthorne, J., and Uphan, H. C., 1962. Pharmacology and toxicology of scandium chloride, *J. Pharmacol. Sci.* 51:1043–1045.

Hamilton, E. I., 1987. The periodic table of the elements. Geochemical and biochemical associations, in: *Pollutant Transport and Fate in Ecosystems* (P. J. Cougtrey *et al.*, eds.), Blackwell, Oxford, pp. 5–33.

Hart, H. E., 1956. Modification of distribution and excretion of REE by chelating elements, in: *Rare Earths in Biochemical and Medical Research* (G. C. Kyker and E. D. Anderson, eds.), U.S. Atomic Energy Commission Report, Oak Ridge, Tennessee, pp. 118–135.

Hart, H. E., Greenberg, J., Lewin, R., Spencer, H., Stern, K. G., and Laszlo, D., 1955. Metabolism of lanthanides and yttrium chelates, *J. Lab. Clin. Med.* 46:182–192.

Hatway, D. E., 1984. *Molecular Aspects of Toxicology*, Royal Society of Chemistry, London.

Hayes, R. L., and Edwards, C. L., 1973. The effect of stable scandium on red blood cells and on the retention and excretion of ^{67}Ga in humans, *South Med. J.* 66:1339–1340.

Hayes, R. L., Byrd, B. L., Rafter, J. J., and Carlton, J. E., 1980. The effect of scandium on the tissue distribution of ^{67}Ga in normal and tumor-bearing rodents, *J. Nucl. Med.* 21:361–365.

Hayes, R. L., Rafter, J. J., Byrd, B. L., and Carlton, J. E., 1981. Studies of the *in vivo* entry of ^{67}Ga into normal and malignant tissue, *J. Nucl. Med.* 22:325–331.

Hayes, R. L., Rafter, J. J., Carlton, J. E., and Byrd, B. L., 1982. Studies of the *in vivo* uptake of ^{67}Ga by an experimental abscess, *J. Nucl. Med.* 23:8–14.

Hird, V., Stewaet, J., Snook, D., Dhokia, B., Coulter, C., Lambert, H., Mason, W., Soutter, W., and Epenetos, A., 1990. I.p. administration of ^{90}Y-labeled MAb as a third line of treatment in ovarian cancer, *Br. J. Cancer* Suppl. 10:48–51.

Höber, and Spaeth, R. A., 1914. The influence of rare earth elements on the contractibility of the muscle, *Pfügers Arch. Gesamte Physiol.* 159:433–456 (German).

Hoekstra, J. A., and Van Ewijk, P. H., 1993. Alternatives for the no-observed-effect level, *Environ. Toxicol. Chem.* 12:187–194.

Holten, V. Z., Kyker, G. C., and Pulliam, M., 1966. Effects of lanthanide chlorides on selected enzymes, *Proc. Soc. Exp. Biol. Med.* 123:913–919.

Hood, S. L., and Comar, C. L., 1956. Tissue distribution and placental transfer of ^{91}Y in rats and cattle, in: *Rare Earths in Biochemical and Medical Research* (G. C. Kyker and E. B. Anderson, eds.), U.S. Atomic Energy Commission Report, Oak Ridge, Tennessee, pp. 280–300.

Horovitz, C. T., 1965. Influence of scandium on higher plants, *Naturwissenschaften* 52:623–624 (German).

Horovitz, C. T., 1968. A comparative study of the action of the elements of the fourth period in plant physiology and biochemistry, *Prog. Stiintei* 4:102–110 (Romanian).

Horovitz, C. T., 1975. Biological significance of scandium, in: *Scandium, Its Occurrence, Chemistry, Physics, Metallurgy, Biology and Technology* (C. T. Horovitz, ed.), Academic Press, London, pp. 513–534.

Horovitz, C. T., 1988. Is the major part of the periodic system really inessential for life? *J. Trace Elem. Electrolytes Health Dis.* 2:135–144.

Horovitz, C. T., 1993. Could scandium and yttrium be required for life? in: *Trace Elements in Man and Animals* (M. Anke *et al.*, eds.), Media Touristik, Gersdorf, Germany, pp. 747–749.

Horovitz, C. T., 1995. Two hundred years of research and development of yttrium, *J. Trace Elem. Electrolytes Health Dis.* 12:153–160.

Horovitz, C. T., and Bondy, S. C., 1994. The effect of metal ions on free radical formation in lung and brain of rats, *J. Trace Elem. Electrolytes Health Dis.* 8:167–171.

Horovitz, C. T., and Wegmann, K., 1975. Response of *Chlorella* to scandium and potassium at various pH, *Physiol. Plant.* 33:113–117.

Hoshishima, K., 1983. "Play" behavior and trace dose of metals in mice, *Dev. Toxicol. Environ. Sci.* 11:525–528.

Hoshishima, K., and Shimai, S., 1985. The trace amount of metals prenatally administered and the circadian drinking rhythm in mice, in: *Trace Substances in Environment and Health* (D. D. Hemphill, ed.), University of Missouri, Columbia, Vol. 19, pp. 417–424.

Hoshishima, K., Tsujii, H., and Aota, S., 1978. The combined effects of two kinds of metals administered to mice upon their bitter tasting and their spontaneous activity, in: *Trace Elements Metabolism in Man and Animals* (M. Kirchgessner, ed.), Technische Universität München, Freising, pp. 199–202.

Hoshishima, K., Shimai, S., and Kano, K., 1983. The combined administration of certain metals in trace dose upon the postnatal development of behavior in mice, *Dev. Toxicol. Environ. Sci.* 11:529–532.

Howells, D. J., and Coult, D. B., 1971. Ions of the rare earths as possible reactivators of acetylcholinesterase inhibited by some organophosphorus compounds, *Biochim. Biophys. Acta* 244:427–431.

Huber, C. T., and Frieden, E., 1970. The inhibition of ferroxidase by trivalent and other metal ions, *J. Biol. Chem.* 245:3979–3984.

Huljev, D., Graf, D., Draskovic, J., and Rajkovic-Huljev, Z., 1989. Comparison between concentration of trace elements in tumorous and normal human breast tissue, *Radiol. Iugosl.* 23: 165–168.

Hutcheson, D. P., Gray, D. H., Venugopal, B., and Luckey, T. D., 1975. Studies of nutritional safety of some heavy metals in mice, *J. Nutr.* 105:670–675.

Hutcheson, D. P., Venugopal, B., Gray, D. H., and Luckey, T. D., 1979. Lanthanide markers in a single sample for nutrient studies in humans, *J. Nutr.* 109:702–707.

Hutcheson, D. P., Kay, M., Enderle, M., Francis, R., and Luckey, T. D., 1983. The kinetics of lanthanide oxide markers in the alimentary tract of the pig, *Nutr. Res.* 3:913–921.

Hutchinson, G. E., 1964. The influence of the environment, *Proc. Natl. Acad. Sci U.S.A.* 51:930–934.

ICRP (International Commission on Radiological Protection), 1993. *Report of the Task Group on Reference Man*, ICRP Publication 23, Pergamon Press, Oxford.

Ila, P., and Jagam, P., 1980. Multi-element analysis of food spices by INAA, *J. Radioanal. Chem.* 57:205–210.

Il'enko, A. I., 1973. Radioecology of wild animals, in: *Radioecology* (V. M. Klechkovsky *et al.*, eds.), John Wiley and Sons, New York, pp. 241–274.

Inarida, M., Kaji, T., Hoyano, M., and Nozaki, T., 1987. Investigations of trace elements content in meat and edible viscera of Japanese cattle, swine and chicken by NAA, *J. Jpn. Soc. Nutr. Food Sci.* 40:328–332 (Japanese).

Iskander, F. Y., and Morad, M. M., 1986. Multi-element determination in wheat and bran, *J. Radioanal. Nucl. Chem. Lett.* 105:151–156.

Iyengar, C. V., 1989. Trends in the daily dietary intake of minor and trace elements by human subjects. An analytical appraisal, in: *Nutrition Availability, Chemical and Biological Aspects* (D. A. T. Southgate *et al.*, eds.), Royal Chemical Society, London, pp. 69–73.

Izumi, K., 1974. Threshold of bitter tasting and phosphorylcholine-cytidyl transferase activity of tongue epithelium in mice treated with metals, *Fukushima J. Med. Sci.* 20:1–10.

Jacobson, L., and Overstreet, R., 1948. The uptake by plants of plutonium and some products of nuclear fission adsorbed on soil colloids, *Soil Sci.* 65:129–134.

Jacobson, S., and Wester, P.-O., 1977. Balance study of twenty trace elements during total parenteral nutrition in man, *Br. J. Nutr.* 37:107–126.

Jones, F. A., 1994. New concepts in human nutrition in the twentieth century. The special role of micro-nutrients, *J. Nutr. Med.* 4:99–113.

Kalashnikov, V. M., Zaitchik, V. Ye., and Bizzer, V. A., 1983. Trace elements in bone tumors, *Cancer Lett.* 29:48–52 (Russian).

Kamaric, L., 1975. Allosteric reactions of horse serum cholinesterase, *Croat. Chem. Acta* 47:287–297.

Kanapilly, G. M., 1980. *In vitro* precipitation behavior of trivalent lanthanides, *Health Phys.* 39:343–346.

Kanias, G. D., 1991. Nutrient and other trace elements in instant soups, *J. Radioanal. Nucl. Chem. Art.* 151:245–254.

Kanias, G. D., Kouri, E., Arvaniti, H., Karaiosifidi, H., and Kounelis, S., 1994. Trace elements content in breasts with fibrocystic disease, *Biol. Trace Elem. Res.* 43–45:363–370.

Kastori, R., Bai, B. Z., and Petrovic, N., 1990. Effect of elements from scandium and lanthanum groups on some morphological and physiological characters of young soybean plants, *Agrochimica* 34:467–474.

Kauffman, S., 1995. *At Home in the Universe*, Oxford University Press, New York.

Kawin, B., 1953. Metabolism of radioyttrium, *Arch. Biochem. Biophys.* 45:230–231.

Kawin, B., Copp, D. H., and Hamilton, J. G., 1950. Studies of the metabolism of certain fission products and plutonium, U.S. Atomic Energy Commission, UCRL-812.

Khare, S. S., Ehmann, W. D., Kasarkis, E. J., and Markesbery, W. R., 1990. Trace elements imbalance in amyotrophic lateral sclerosis, *NeuroToxicology* 11:521–532.

Kidman, B., Tutt, M. L., and Vaughan, J. M., 1950. The retention and excretion of radiant strontium and yttrium (^{89}Sr, ^{90}Sr and ^{90}Y) in the healthy rabbit, *J. Pathol. Bacteriol.* 62:209–227.

Kidman, B., Tutt, M. L., and Vaughan, J. M., 1951. The retention of radioactive strontium and yttrium (^{89}Sr, ^{90}Sr and ^{90}Y) in pregnant and lactating rabbits and their offsprings, *J. Pathol. Bacteriol.* 63:253–268.

Klechkovsky, V. M., and Tselishcheva, G. N., 1957. Some specific features of the behavior of $^{90}Sr/^{90}Y$ on absorption by plants, in: *On the Behavior of Radioactive Fission Products in Soil, Their Absorption by Plants and Their Accumulation in Crops* (V. M. Klechkovsky, ed.), U.S. Atomic Energy Commission, Washington, D.C., pp. 194–221.

Kleinkauf, H., von Dohren, H., Dornauer, H., and Nesenmann, G., eds., 1986. *Regulation of Secondary Metabolite Formation*, VCH, Weinheim, pp. 381–388.

Klevay, L. M., 1987. Dietary requirements of trace elements in humans, in: *Trace Element Analytical Chemistry in Medicine and Biology* (P. Brätter and P. Schrammel, eds.), W. de Gruyter, Berlin, Vol. 4, pp. 43–60.

Koelling, J. J., 1971. Relationship of trace elements to algae growth, PhD thesis, Washington State University, Pullman.

Kroll, H., Siegel, E., Hart, H. E., Rosoff, B., Spencer, H., and Laszlo, D., 1957. Excretion of yttrium and lanthanum chelates of cyclohexane-1,2-transdiaminetetraacetic acid and diethylenetriamine-pentaacetic acid in man, *Nature* 180:919–920.

Krumbein, W. E., 1986. Biotransfer of minerals by microbes and microbial mats, in: *Biomineralization in Lower Plants and Animals* (B. S. C. Leadbeater, and R. Riding, eds.), Clarendon Press, Oxford, pp. 55–72.

Kuby, A., 1991. *A Study of Enzymes*, CRC Press, Boca Raton, Florida, pp. 191–244.

Kuchel, P. W., Kirk, K., and King, G. R., 1994. NMR methods for measuring membrane transport, in: *Subcellular Biochemistry* (H. J. Hiderson and G. B. Ralston, eds.), Plenum Press, New York, Vol. 23, pp. 247–327.

Kutzner, J., Alfter, G., and Grimm, W., 1986. Effect of carrier on the organ distribution of radioactive yttrium in the rat, *Nucl. Med.* 25:78–80.

Kvicala, J., Havelka, J., Zamrazil, V., and Vondra, K., 1991. Trace elements in serum during oral glucose tolerance test, in: *Trace Elements in Man and Animals* (B. Momcilovic, ed.), IMI, Zagreb, pp. 3.6–3.7.

Kyker,, G. C., 1962. Rare earths, in: *Mineral Metabolism* (C. L. Comar and F. Bronner, eds.), Academic Press, London, Vol. 2B, pp. 499–541.

Laszlo, D., 1956. Distribution of lanthanum and yttrium in the mammalian organism, in: *Rare Earths in Biochemical and Medical Research* (G. C. Kyker and E. B. Anderson, eds.), U.S. Atomic Energy Commission Report, Oak Ridge, Tennessee, pp. 193–221.

Levan, A., 1945. Cytological reactions induced by inorganic salt solutions, *Nature* 156:751–753.

Li, Y.-H., 1984. Why are the chemical compositions of living organisms so similar? *Schweiz. Z. Hydrol.* 46:177–184.

Lima-de Faria, A., 1988. *Evolution without Selection*, Elsevier, Amsterdam, pp. 65–68, 311–329.

Lippard, S. J., and Berg, J. M., 1994. *Principles of Bioinorganic Chemistry*, University Science Books, Mill Valley, California.

Liu, H.-H., Chen, W.-L., and Wu, J.-T., 1990. Sorption of ^{90}Sr and ^{90}Y by *Anacystis* cells, *Sci. Total Environ.* 91:275–282.

Liu, Y.-F., Tang, R.-H., Zhang, Q.-X., Shi, J.-Y., Li, X.-M., Liu, Z.-Q., and Zhao, W., 1986. Stimulation of cell growth of *Tetrahymena pyriformis* and *Chlamydomonas reinhardtii* by trace elements, *Biol. Trace Elem. Res.* 9:89–93.

Loewy, A. G., Siekevitz, P., Menninger, J. R., and Gallant, J. A. N., 1991. *Cell Structure and Function*, Saunders College Publishing, Philadelphia, pp. 81–95.

Lowenstam, H. A., and Weiner, S., 1989. *On Biomineralization*, Oxford University Press, New York, pp. 7–25.

Luckey, 1977. Comparative nutrition, in: *Handbook Series in Nutrition and Food* (M. Rechcigl, Jr., ed.), CRC Press, Cleveland, Vol. 1D, pp. 3–18.

Luckey, T. D., 1980. *Hormesis with Ionizing Radiation*, CRC Press, Boca Raton, Florida.

Luckey, T. D., and Venugopal, B., 1977. *Metal Toxicity in Mammals*, Vol. 1: *Physiology and Chemical Basis for Metal Toxicity*, Plenum Press, New York.

Luckey, T. D., Hartman, R., Knox, T., Palmer, S., Kay, M., and Terry, B., 1979. Lanthanide marker transit times, and rates of flow and passage for three meals in humans, *Nutr. Rep. Int.* 19:561–571.

MacCordick, H. J., Paulus, J. M., and Bontems, M. G., 1990. Preferential mycobacterial adsorption of radionuclides in filiation, *J. Radioanal. Nucl. Chem, Lett.* 145:231–238.

MacSweeney, J. E. Baxter, M. A., and Joplin, G. F., 1990. Heel pad thickness is an insensitive index of biochemical remission in acromegaly, *Clin. Radiol.* 42:348–350.

Maeno, T., and Tanizaki, Y., 1996. Influence of pH of acid irrigation water on the transfer of elements into rice plants from soils, *Radioisotopes* 45:166–175 (Japanese).

Magnusson, G., 1963. The behavior of certain lanthanides in rats, *Acta Pharmacol.* 20(Suppl.):3:1–95.

Maguire, M. E., 1982. Magnesium regulation of the beta-receptor–adenylate cyclase complex. 2. Sc^{3+} as a Mg^{2+} antagonist, *Mol. Pharmacol.* 22:274–280.

Maihara, V. A., and Vasconcellos, M. B. A, 1989. Determination of trace elements in Brazilian rice grains and in biological reference materials by NAA, *J. Radioanal. Nucl. Chem. Art.* 132:329–337.

Mangal, P. C., and Sharma, P., 1981. Effect of leukemia on the concentration of some trace elements in human whole blood, *Ind. J. Med. Res.* 74:559–564.

Mann, H., Fyfe, W. S., and Kerrich, R., 1988. The chemical content of algae and waters: Bio concentration, *Toxic. Assess.* 3:1–16.

Markesbery, W. R., Ehmann, W. D., Alauddin, M., and Hossain, T. I. M., 1984. Brain trace elements concentrations in aging, *Neurobiol. Aging* 5:19–28.

Marquis, J. K., and Black, E. E., 1985. Activation and inactivation of bovine caudate acetylcholinesterase by trivalent cations, *Biochem. Pharmacol.* 34:533–538.

Marquis, J. K., and Lerrick, A. J., 1982. Noncompetitive inhibition by aluminum, scandium and yttrium of acetylcholinesterase from *Electrophorus electricus*, *Biochem. Pharmacol.* 31:1437–1440.

Martin, R. B., and Richardson, F. S., 1979. Lanthanides as probes for calcium in biological systems, *Q. Rev. Biophys.* 12:181–209.

Martoja, M., and Martoja, R., 1984. The bioaccumulation of metals, normal physiological processes and consequences of pollution, *Courr. CNRS* 1:32–37 (French).

Matsumura, K., and Komiyama, M., 1994. Hydrolysis of phosphatidylinositol by rare earths metal ions as a phospholipase C mimic, *J. Inorg. Biochem.* 55:153–156.

May, P. M., 1990. Agents acting on passive ion transport, in: *Comprehensive Medicinal Chemistry* (C. Hansch *et al.*, eds.), Pergamon Press, Oxford, pp. 175–192.

Mayer, S. W., and Morton, M. E., 1956. Preparation and distribution of ^{90}Y fluoride, in: *Rare Earths in Biochemical and Medical Research* (G. C. Kyker and E. B. Anderson, eds.), U.S. Atomic Energy Commission Report, Oak Ridge, Tennessee, pp. 263–279.

Mela, L., 1969. Inhibition and activation of Ca transport in mitochondria. Effect of lanthanides and local anesthetic drugs, *Biochemistry* 8:2481–2486.

Mela, L., 1970. Effect of lanthanides and heavy metals on mitochondrial cation transport, in: *Effects of Metals on Cells, Subcellular Elements, and Macromolecules* (J. Maniloff *et al.*, eds.), Thomas Publishers, Springfield, Illinois, pp. 235–253.

Menczel, J., Rosoff, B., and Spencer, H., 1982. Tissue distribution of orally administered ^{91}Y and ^{46}Sc in mice, *Health Phys.* 42:727–730.

Mertz, W., 1988. Personal communication, with permission.

Meyers, P. R., Rawlings, D. E., Woods, D. R., and Lindsey, G. G., 1993. Isolation and characterization of a cyanide dehydratase from *Bacillus pumilus*, *J. Bacteriol.* 175:6105–6112.
Miller, J. K., and Byrne, W. F., 1970. Comparison of ^{46}Sc and ^{144}Ce as non-absorbed reference materials in studies with cattle, *J. Nutr.* 100:1287–1292.
Mines, G. R., 1910. The action of beryllium, lanthanum, yttrium and cerium on the frog's heart, *J. Physiol.* 40:327–346.
Mines, G. R., 1911. The action of trivalent ions on living cells and on colloidal systems. 2. Simple and complex cations, *J. Physiol.* 42:309–331.
Mlinar, B., and Enyeart, J. J., 1993. Block of current through T-type calcium channels by trivalent metal cations and nickel in neural rat and human cells, *J. Physiol.* 469:639–652.
Moeller, T., Martin, D. F., Thomson, L. C., Ferrus, R., Feistel, G., and Randall, W. J., 1965. The coordination chemistry of yttrium and the rare earth metal ions, *Chem. Rev.* 65:1–50.
Moo, S. P., and Pillay, K. K. S., 1983. Trace elements profiles in the hair of cancer patients, *J. Radioanal. Chem.* 77:141–147.
Moskalev, Yu. I., 1963. Problems of distribution and kinetics of excretion of radioactive isotopes, in: *Diagnosis and Treatment of Radioactive Poisoning*, IAEA, Vienna, pp. 287–306 (Russian).
Mraz, F. R., Wright, P. L., Ferguson, T. M., and Anderson, D. L., 1964. Fission product metabolism in hens and transference to eggs, *Health Phys.* 10:77–782.
Mu, C. J., and Shi, W. L., 1985. The investigation of mutagenicity and teratogenicity from rare earth compounds, in: *New Frontiers in Rare Earth Science and Applications* (G. X. Xu and J. M. Xiao, eds.), Science Press, Beijing, Vol. 2, pp. 1530–1534.
Mumcu, T., Gokmen, I., Gökmen, A., Parr, R. M., and Aras, N. K., 1988. Determination of minor and trace elements in Turkish diet by duplicate portion technique. *J. Radioanal. Nucl. Chem. Art.* 124:289–299.
Muroma, A., 1958. Studies in the bactericidal action of salts of certain rare earth metals, *Ann. Med. Exp. Biol. Fenn.* 36(Suppl. 6):1–54.
Myers, S. L., Slowman, S. D., and Brandt, K. D., 1989. Radiation synovectomy stimulates glycosaminoglycan synthesis by normal articular cartilage, *J. Lab. Clin. Med.* 114:27–35.
Nachsen, D. A., 1984. Selectivity of the calcium binding site in synaptosome calcium channels. Inhibition of calcium influx by multivalent metal cations, *J. Gen. Physiol.* 83:941–967.
Nakamura, Y., Tsumura-Hasegawa, Y., Tonogai, Y., Kanamoto, M., Tsuboi, N., Murakami, K., and Ito, Y., 1991a. Excretion of dysprosium, europium, ytterbium and yttrium in the rat after oral administration, *Jpn. J. Toxicol. Environ. Health* 37:418–425 (Japanese).
Nakamura, Y., Tsumura-Hasegawa, Y., Tonogai, Y., Kanamoto, M., Tsuboi, N., Murakami, K., and Ito, Y., 1991b. Distribution and the histological effects of dysprosium, europium, ytterbium and yttrium in the rat after i.v. administration, *Jpn. J. Toxicol. Environ. Health* 37:489–496; 497–506 (Japanese).
Nakamura, Y., Tsumura, Y., Tonogai, Y., and Ito, Y., 1993. Studies on the biological effects of REE, *Jpn. J. Toxicol. Environ. Health* 39:44–55, 121–131 (Japanese).
Needham, A. E., 1965. *The Uniqueness of Biological Materials*, Pergamon Press, Oxford, pp. 416–455.
Nieboer, E., and Richardson, H. S., 1980. The replacement of the nondescript term "heavy metals" by a biologically and chemically significant classification of metal ions, *Environ. Pollut.* 1B:3–26.
Nieboer, E., and Sanford, W. E., 1984. Essential, toxic and therapeutic functions of metals, in: *Environmental Geochemistry* (M. E. Fleet, ed.), Mineralogical Association, London, Vol. 10, pp. 149–168.
Nielsen, F. H., 1986. Other elements: Sb, Ba, B, Br, Cs, Ge, Rb, Ag, Sr, Sn, Ti, Zr, Be, Bi, Ga, Au, In, Nb, Sc, Te, Tl and W, in: *Trace Elements in Human and Animal Nutrition* (W. Mertz, ed.), Academic Press, Orlando, Vol. 2, pp. 415–463.
Nishita, H., Haug, R. M., and Steen, A. J., 1966. Solubility and plant availability of radionuclides in reactor activated soils, *Health Phys.* 12:615–620.

Ochiai, E.-I., 1987. *General Principles of the Biochemistry of the Elements*, Plenum Press, New York.

Ogawa, Y., Suzuki, S., Naito, K., Saito, M., Kamata, E., Hirose, A., Kaneko, T., Chiba, M., Inaba, Y., and Kurokawa, Y., 1994. Studies on the biological effects of yttrium. Effects of repeated oral administration in rats, *Jpn. J. Toxicol. Environ. Health* 40:374–382 (Japanese).

Ogurusu, T., Wakabayashi, S., and Shigekawa, M., 1991. Functional characterization of lanthanide binding sites in the sarcoplasmic reticulum Ca^{2+}-ATPase. Do lanthanide ions bind to the calcium transport site? *Biochemistry* 30:9966–9973.

Ohsawa, M., 1993. Nutritional and toxicological implications of trace elements in the immune response, *Prog. Clin. Biol. Res.* 380:283–298.

Ohtawa, M., Seko, M., and Takayama, F., 1983. Effect of aluminum on lipid peroxidation in rats, *Chem. Pharm. Bull.* 31:1415–1418.

Ordögh, M., Fazekas, S., and Szabo, E., 1985. The regional distribution of copper and other trace elements in the human brain with special reference to Wilson disease, in: *Metal Ions in Neurology and Psychiatry* (S. Gabay *et al.*, eds.), Alan Liss, New York, pp. 129–137.

Ottoboni, M. A., 1991. *The Dose Makes the Poison. A Plain-Language Guide to Toxicology*, Van Nostrand Reinhold, New York.

Pais, I., 1985. Some aspects of the development in the research of the hardly known trace elements and the importance of the interdisciplinary trace elements research work, in: *New Results in the Research of Hardly Known Trace Elements* (I. Pais, ed.), University of Horticulture, Budapest, pp. 3–14.

Pan, T. C., Lin, T. H., Tseng, C. C., Yang, M. H., and Huang, C. W., 1993. Trace elements in hair of blackfoot disease, *Biol. Trace Elem. Res.* 39:117–127.

Papp, S., Rutzke, M., and Martonosi, A., 1985. The effect of chelating agents on the elemental composition of sarcoplasmic reticulum, *Arch. Biochem. Biophys.* 243:254–263.

Pehrsson, S. K., and Lins, L. E., 1983. The role of trace elements in uremic heart failure, *Nephron* 34:93–98.

Phillips, D. J., and Rainbow, P. S., 1989. Strategies of trace metal sequestration in aquatic organisms, *Mar. Environ. Res.* 28:207–210.

Pietra, R., Sabbioni, E., Ubertalli, L., Orvini, E., Vocaturo, G., Colombo, F., and Zanoni, M., 1985. Trace elements in tissues of a worker affected by rare earths pneumoconiosis, *J. Radioanal. Nucl. Chem. Art.* 92:247–259.

Pitman, M. G., 1982. Ion transport across the root, in: *Membranes and Transport* (A. N. Martonosi, ed.), Plenum Press, New York, pp. 657–662.

Plaha, D. S., and Rogers, H. J., 1983. Antibacterial effect of the scandium complex of enterochelin. Studies of the mechanisms of action, *Biochim. Biophys. Acta* 760:246–255.

Plaha, D. S., Rogers, H. J., and Williams, G. W., 1984. Studies of the antibacterial effect of the scandium complex of enterochelin, *J. Antibiot.* 37:588–595.

Polikarpov, G. G., 1966. *Radioecology of Aquatic Organisms*, Reinhold, New York, pp. 110–117.

Polzer, G., Matejka, M., Lind, I., Weiner-Schmuck, M., and Seidel, A., 1993. *In vivo* effects of $CeCl_3$ and $LaCl_3$ on macrophages, in: *Endomyocardial Fibrosis* (M. S. Valiathan *et al.*, eds.), Oxford University Press, Bombay, pp. 253–265.

Press, O. W., Shan, D., Howell-Clark, J., Eary, J., Appelbaum, F. R., Matthews, D., King, D. J., Haines, A. M., Hamann, P., Hinman, L., Shochat, D., and Bernstein, I. D., 1996. Comparative metabolism and retention of ^{125}I, ^{90}Y ^{111}In radioimmunoconjugates by cancer cells, *Cancer Res.* 56:2123–2129.

Pullin, J. G., Coller, M. A., Das, P., Smith, R. L., Debault, L. E., Johnson, L. L., and Wals, R. C., 1996. Effects of Ho:YAG laser energy on cartilage metabolism, healing, and biochemical properties of lesional and perilesional tissue in a weight-bearing model, *Arthroscopy* 12:15–25.

Qi, T., Liu, P., Song, C., and Wang, X., 1984. Studies on the absorption, transfer and distribution of REE in plants, *J. Chin. Rare Earth Soc.* 2:94–98.

Quadri, S. M., Yi, S., Blum, J. E., Leichner, P. K., Williams, J. R., and Vriesendorp, H. M., 1993. Preclinical evaluation of i.v. administered ^{111}In and ^{90}Y labeled B72.3 immunoconjugates in beagle dogs, *Nucl. Med.* 20:559–570.

Qureshi, I. H., Waheed, S., Rehman, A., and Ahmad, S., 1993. Intake of trace elements through major food articles, in: *Trace and Toxic Elements in Nutrition and Health* (M. Abdulla *et al.*, eds.), Wiley Eastern, Ltd., New Delhi, pp. 318–323.

Rades-Rohkohl, E., Hirsch, P., and Fränzle, O., 1979. NAA for the demonstration of amphibolite rock-weathering activity of a yeast, *Appl. Environ. Microbiol.* 38:1061–1068.

Ramsden, E. N., 1961. A review of experimental work on radio-yttrium comprising: the tissue distribution, the mechanism of deposition in bone, and the state in the blood, *Ind. J. Rad. Biol.* 3:399–410.

Raven, P. H., and Wilson, E. O., 1992. A fifty-year plan for biodiversity surveys, *Science* 258:1099–1100.

RDA, 1989. *Recommended Dietary Allowances*, 10th ed., National Academy Press, Washington, D.C.

Rediske, J. H., and Selders, A. A., 1954. The uptake and translocation of yttrium by higher plants, *Am. J. Bot.* 41:233–242.

Reece, W. O., 1965. Effects of exercise and inhaled yttrium oxide on blood lactic acid, erythrocyte volume and lung histology in dogs, PhD thesis, Iowa State University. Diss. Abstr. 26:3457.

Rengel, Z., 1994. Effects of aluminum, REE and other metals on net $^{45}Ca^{2+}$ uptake by *Amaranthus* protoplasts, *J. Plant Physiol.* 143:47–51.

Rice, T. R., 1956. The accumulation and exchange of strontium by marine planktonic algae, *Limnol. Oceanogr.* 1:123–138.

Richards, M. P., and Steele, N. C., 1987. Trace elements in the development of avian embryo, *J. Exp. Zool* 1987(Suppl. 1):39–51.

Rizzato, G., Locicero, S., Barberis, M., Torre, M., Pietra, R., and Sabbioni, E., 1986. Trace of metal exposure in hard metal lung disease, *Chest* 89:101–106.

Rogers, H. J., 1987. Bacterial iron transport as a target for antibacterial agents, in: *Iron Transport in Microbes, Plants and Animals* (G. Winkelmann *et al.*, eds.), VCH, Weinheim, pp. 223–233.

Rogers, H. J., Synge, C., and Woods, V. E., 1980. Antibacterial effect of scandium and indium complexes of enterochelin on *Klebsiella pneumoniae*, *Antimicrob. Agents Chemother.* 18:63–68.

Rogers, H. J., Woods, V. E., and Synge, C., 1982. Antibacterial effect of the scandium and indium complexes of enterochelin on *Escherichia coli*, *J. Gen. Microbiol.* 128:2389–2394.

Rogers, H. J., Plaha, D. S., and Woods, V. E., 1984. Antibacterial effect of the scandium complex of enterochelin on *Pseudomonas aeruginosa*, *FEMS Microbiol. Lett.* 24:5–8.

Romney, E. M., and Rhoads, W. A., 1966. Neutron activation products from Project Sedan in plants and soils, *Soil Sci. Am. Proc.* 3:770–773.

Rosenblum, M. G., Kavanagh, J. J., Burke, T. W., Wharton, J. T., Cunningham, J. E., Shanken, L. J., Silva, E. G., Thompson, L., Cheung, L., Lamki, L., and Murray, J. L., 1991. Clinical pharmacology, metabolism and tissue distribution of ^{90}Y-labeled MAb B72.3 after i.p. administration, *J. Natl. Cancer Inst.* 83:1629–1636.

Rosoff, B., Feldstein, A., Stevens, E., and Spencer, H., 1961. Excretion and distribution of ^{46}Sc in man and its removal by chelating agents, *Fed. Proc.* 20:403.

Rosoff, B., Siegel, E., Williams, G. L., and Spencer, H., 1963. Distribution and excretion of radioactive rare earths compounds in mice, *Int. J. Appl. Radiat. Isot.* 14:129–135.

Rosoff, B., Spencer, H., Cohn, S. H., and Gusmano, E. A., 1965. Metabolism of ^{46}Sc in man, *Int. J. Appl. Radiat. Isot.* 16:479–485.

Sabbioni, E., Pietra, R., Mousty, F., Colombo, F., Rizzato, G., and Scansetti, G., 1987. Trace metals in human lung, as determined by NAA of bronchoalveolar lavage, *J. Radiol. Nucl. Chem. Art.* 110:595–601.

Sagan, C., 1994. The search for extraterrestrial life, *Sci. Am.* 271:92–99.

Sahm, H., 1993. Membrane transport, in: *Biotechnology* (H. Sahm, ed.), VCH, Weinheim, pp. 70–78.

Saito, K., Watanabe, S., Itoh, T., and Saito, T., 1994. Dynamics of trace and ultra-trace elements in the brain, in: *Dynamics of Trace Elements in Human Body and Disease* (K. Saito, ed.), Hokkaido University, Sapporo, pp. 11–17.

Sato, T., 1986. Elemental composition of laboratory animal diets, *Radioisotopes* 35:24–27 (Japanese).

Schaeffer, H. J., and Walton, J. D., 1990. Aluminum ions induce oat protoplasts to produce an extracellular (1 → 3)β-D-glucan, *Plant Physiol.* 94:13–19.

Scheer, H. W., 1989. Interaction between α-latroxin and trivalent cations in rat striatal synaptosomal preparations, *J. Neurochem.* 52:1590–1597.

Schelenz, R., 1977. Dietary intake of 25 elements by man estimated by NAA, *J. Radioanal. Chem.* 37:539–548.

Schelenz, R., 1983. Essential and toxic food constituents in the daily total diet, Report BFE-R-83-02, Bundesforschungsanstalt für Ernährung, Karlsruhe, Germany, pp. 259–261 (German).

Schnabel, C., Herpers, U., Michel, R., Löer, F., Buchhorn, G., and Willert, H. G., 1994. Changes of concentrations of the elements cobalt, chromium, antimony, and scandium in tissues of persons with joint implants, *Biol. Trace Elem. Res.* 59:389–395.

Schroeder, H. A., and Darrow, D. K., 1973. Relation of trace metals to human health effects, in: *Chemical Analysis of the Environment and Other Modern Techniques* (S. Ahuja *et al.*, eds.), Plenum Press, New York, pp. 81–104.

Schroeder, H. A., and Nason, A. P., 1976. Interactions of trace metals in mouse and rat tissues: Zinc chromium, copper and manganese with 13 other elements, *J. Nutr.* 106:198–203.

Schubiger, P. A., Beer, H.-F., Geiger, L., Rösler, H., Zimmermann, A., Triller, J., Mettler, D., and Schilt, W., 1991. ^{90}Y-resin particles. Animal experiments on pigs with regard to the introduction of superselective embolization therapy, *Nucl. Med. Biol.* 18:305–311.

Schwarz, K., 1972. Elements newly identified as essential for animals, in: *Nuclear Activation Techniques in the Life Sciences*, International Atomic Energy Agency, Vienna, pp. 3–22.

Shi, M. T., *et al.*, 1991. Determination of multiple chemical elements in CSF of Parkinson disease after intracerebral autotransplantation of the adrenal medulla, *Chin. J. Surg.* 29:129–132 (Chinese with English summary).

Shinohara, A., and Chiba, M., 1990. Distribution of administered elements and effect on calcium concentrations in organs of the mice injected with lanthanioles and yttrium compounds, *Biomed. Res. Trace Elem.* 1:273–274 (Japanese).

Silver, F., and Perry, R. D., 1982. Bacterial inorganic cation and anion transport systems, in: *Membranes and Transport* (A. N. Martonosi, ed.), Plenum Press, New York, Vol. 2, pp. 115–121.

Slater, J. V., Broughton, C. T., Hood, L., Egge, A. S., and Kallmeyer, J. L., 1961. Yttrium localization in the desert pocket mouse *Perognathus penicillatus*, *Proc. Soc. Exp. Biol. Med.* 108: 198–199.

Smalley, S. J., Hauser, H. D., and Berg, V. S., 1993. Effect of cations on effective permeability of leaf cuticles to sulfuric acid, *Plant Physiol.* 103:251–256.

Smith, Q. R., 1990. Regulation of metal uptake and distribution within brain, in: *Nutrition and the Brain* (R. J. Wurtman and J. J. Wurtman, eds.), Raven Press, New York, pp. 25–74.

Smith, T., Shawe, D. J., Crawley, J. C. W., and Gumpel, J. M., 1988. Use of single photon emission computed tomography (SPECT) to study the distribution of ^{90}Y in patients with Baker's cysts and persistent synovitis of the knee, *Ann. Rheumat. Dis.* 47:553–558.

Smock, L. A., 1983. The influence of feeding habits on whole-body metal concentrations in aquatic insects, *Freshwater Biol.* 13:301–311.

Smolka, G. E., Birnbaum, E. R., and Darnall, D. W., 1971. Rare earth metal ions as substitutes for the calcium ion in *Bacillus subtilis* α-amylase, *Biochemistry* 10:4556–4561.

Sneddon, J., 1987. Action of di- and trivalent cations on calcium-activated potassium-efflux in rat erythrocytes, *Biochem. Pharmacol.* 36:3723–3730.

Snyder, F., 1960. Rare earth fatty liver, *Nature* 185:480–481.

Snyder, F., Cress, E. A., and Kyker, G. C., 1959. Liver lipid response to intravenous rare earths in rats, *J. Lipid Res.* 1:125–131.

Spallholz, J. E., and Stewart, J. R., 1989. Advances in the role of minerals in immunobiology, *Biol. Trace Elem. Res.* 19:129–151.

Spector, W. S., ed., 1974. *Biological Data Book*, Federation of American Societies for Experimental Biology, Bethesda, Maryland, Vol. 3, pp. 1531–1533.

Speer, R. J., Hill, J. M., and Denton, A. D., Jr., 1961. Colloidal $^{90}Y(OH)_3$ and ^{90}Y phosphate preparations and tissue distribution following i.v. administration, *Proc. Soc. Exp. Biol. Med.* 106:444–448.

Spencer, H., 1960. A study of strontium, calcium and rare earth elements metabolism in humans, Annual Report, Montefiore Hospital, New York, pp. 1–121.

Spencer, H., and Rosoff, B., 1963. Metabolism and removal of rare earths in man, in *Diagnosis and Treatment of Radioactive Poisoning*, International Atomic Energy Agency, Vienna, pp. 171–189.

Spencer, H., and Rosoff, B., 1965. Removal of ^{46}Sc in man, *Health Phys.* 11:1181–1185.

Spivak, J. M., Grande, D. A., Ben-Yishay, A., Menche, D. S., and Pitman, M. I., 1992. The effect of low-level Nd:YAG laser energy on adult articular cartilage *in vitro*, *Arthroscopy* 8:36–43.

Spode, E., and Gensicke, F., 1957. On the distribution of ^{91}Y in animal organs, *Naturwissenschaften* 44:118–119.

Spooner, G. M., 1949. Observations on the absorption of radioactive strontium and yttrium by marine algae, *J. Mar. Biol. Assoc. U.K.* 28:587–625.

Stannard, J. N., 1988. *Radioactivity and Health. A History*, Pacific NW Laboratory, San Diego.

Starzak, M. E., 1984. *The Physical Chemistry of Membranes*, Academic Press, Orlando, pp. 317–320.

Stebbing, A. R. D., 1982. Hormesis—the stimulation of growth by low levels of inhibitors, *Sci. Total Environ.* 22:213–234.

Steinberg, R. A., 1939. Relation of carbon nutrition to trace elements and accessory requirements of *Aspergillus niger*, *J. Agric. Res.* 59:749–763.

Steinberg, R. A., 1950. Growth on synthetic nutrient solutions of some fungi pathogenic to tobacco, *Am. J. Bot.* 87:711–714.

Sternweis, P. C., and Gilman, A. G., 1982. Aluminum: A requirement for activation of the regulatory component of adenylate cyclase by fluoride, *Proc. Natl. Acad. Sci. U.S.A.* 79:4888–4891.

Suhayda, C. G., and Haug, A., 1987. Metal induced conformational changes in calmodulin, *Bull. Environ. Contam. Toxicol.* 38:289–294.

Suschny, O., Dybczynski, R., and Veglia, A., 1982. Comparison of NAA with other analytical techniques for the determination of inorganic components in milk powder, *J. Radioanal. Chem.* 69:147–157.

Suzuki, S., Ogawa, Y., Kisi, Y., and Chiba, M., 1991. The effect of orally administered rare earth metals on the essential metals of the rat femur, *Biomed. Res. Trace Elem.* 2:223–224 (Japanese).

Szeifert, G. T., Julow, J., Slowik, F., Balint, K., Lanyi, F., and Pasztor, E., 1990. Pathological changes in cystic craniopharyngiomas following intracavital ^{90}Y treatment, *Acta Neurochir.* 102:14–18.

Talburt, D. E., and Johnson, G. T., 1967. Some effects of rare earth elements and yttrium on microbial growth, *Mycologia* 59:492–503.

Tang, R.-H., Su, M.-M., Cao, T.-G., and Liu, Y.-F., 1985. Stimulation of proliferation of *Tetrahymena pyriformis* by trace rare earths, *Biol. Trace Elem. Res.* 7:95–102.

Taylor, D. M., 1966. The metabolism of ^{47}Sc produced by the decay of ^{47}Ca *in vivo*, *Br. J. Radiol.* 39:620–622.

Taylor, D. M., and Williams, D. R., 1995. *Trace Elements in Medicine and Chelation Therapy*, Royal Society of Chemistry, London.

Taylor, G. J., 1987. Exclusion of metals from the symplasm: A possible mechanism of metal tolerance in higher plants, *J. Plant Nutr.* 10:1213–1222.

Teherani, D. K., and Badawi, N., 1988. Studies on the scandium distribution in organs of fattened chicken by NAA, *J. Radioanal. Nucl. Chem. Lett.* 127:395–400.

Terai, M., Yasuda, M., Akabane, A., Matsumoto, M., and Hashimoto, K., 1980. Distribution pattern of small and trace elements in human cancerous tissues and normal human tissues, in: *Nuclear Methods in Environmental and Energy Research*, Proc Conf. University Missouri, Columbia, pp. 427–436.

Thomassen, P. R., and Leicester, H. M., 1964. Uptake of radioactive beryllium, vanadium, selenium, cerium and yttrium in the tissues and teeth of rats, *J. Dent. Res.* 43:346–352.

Thompson, C. M., Markesbery, W. R. Ehmann, W. D., Mao, Y.-X., and Vance, D. E., 1988. Regional brain trace elements studies in Alzheimer's disease, *NeuroToxicology* 9:1–8.

Tomita, H., ed., 1990. *Trace Elements in Clinical Medicine*, Springer-Verlag, Tokyo.

Tosteson, D. C., 1989. Introduction. Membrane transport in physiology, in: *Membrane Transport* (D. C. Tosteson, ed.), American Physiological Society, Bethesda, Maryland, pp. 1–14.

Trosper, T. L., and Philipson, K. D., 1983. Effects of divalent and trivalent cations on Na^+–Ca^{2+} exchange in cardiac sarcolemmal vesicles, *Biochim. Biophys. Acta* 731:63–68.

Tso, T. C., Sorokin, T. P., and Engelhaupt, M. E., 1973. Effects of some rare elements on nicotine content of the tobacco plant, *Plant Physiol.* 51:805–806.

Tsujii, H., and Hoshishima, K., 1979. The effect of the administration of trace amounts of metals to pregnant mice upon the behavior and learning of their offsprings, *Shinshu Univ. J. Fac. Agric.* 16:13–28.

Tu, Q., Wang, X.-R., Tian, L. Q., and Dai, L.-M., 1994. Bioaccumulation of the REE lanthanum, gallium, and yttrium in carp (*Cyprinus carpio*), *Environ. Pollut.* 85:343–350.

Tyler, D. D., 1992. *The Mitochondrion in Health and Disease*, VCH, Weinheim, pp. 182–183.

Tyler, D. D., and Sutton, C. M., 1984. Mitochondrion transporting systems, in: *Membrane Structure and Function* (E. E. Bittar, ed.), John Wiley and Sons, New York, Vol. 5, pp. 225–236.

Vaidhyanathan, V. S., 1993. *Regulation and Control Mechanisms in Biological Systems*, PTR Prentice Hall, Englewood Cliffs, New Jersey, pp. 270–285.

Valcovic, V., 1990. Origin of trace elements requirements by living matter, in: *Symmetries in Science* (B. Gruber and J. H. Jopp, eds.), Plenum Press, New York, pp. 213–242.

Vance, D. E., Ehmann, W. D., and Markesbery, W. R., 1988. Trace elements imbalances in hair and nails of Alzheimer's disease patients, *NeuroToxicology* 9:197–208.

Van Toller, S., 1992. Biochemical individual differences and psychonutrition, in: *Handbook of Individual Differences. Biological Perspectives*, John Wiley and Sons, Chichester, pp. 19–44.

Verstraeten, S. V., and Oteiza, P. I., 1995. Sc^{3+}, Ga^{3+}, In^{3+} and Be^{2+} promote changes in membrane physical properties and facilitate Fe^{2+}-initiated lipid peroxidation, *Arch. Biochem. Biophys.* 322:284–290.

Villafranca, J. J., and Nowak, T., 1992. Metal ions at enzyme active sites, in: *The Enzymes* (D. S. Sigman, ed.), Academic Press, San Diego, Vol. 20, pp. 63–94.

Waheed, S., Fatima, I., Mannan, A., Chaudhary, M. S., and Qureshi, I. H., 1985. Trace elements concentration in egg-yolk and egg-white of farm and domestic chicken eggs, *Int. J. Environ. Anal. Chem.* 21:333–344.

Wallach, S., and Chausmer, A., 1990. Metabolism of trace metals in animals and man, in: *Trace Metals and Fluoride in Bones and Teeth*, CRC Press, Boca Raton, Florida, pp. 242–248.

Ward, N. I., 1986. Multi-element analysis of urine by NAA: Application to starvation and anorexia nervosa, *J. Micronutr. Anal.* 2:211–231.

Ward, N. I., and Mason, J. A., 1987. NAA techniques for identifying elemental status in Alzheimer disease, *J. Radioanal. Nucl. Chem. Art.* 113:515–526.

Ward, N. I., and Pim, B., 1984. Trace element concentrations in blood plasma from diabetic patients and normal individuals, *Biol. Trace Elem. Res.* 6:469–487.

Ward, N. I., Bryce-Smith, D., Minski, M., and Matthews, W. B., 1985. Multiple sclerosis, *Biol. Trace Elem. Res.* 7:153–159.

Webb, E. C., 1992. *Enzyme Nomenclature*, Academic Press, San Diego.

Weinberg, E. D., 1986. Regulation of secondary metabolism by trace elements, in: *Cell Metabolism, Growth and Environment* (T. A. V. Subramanian, ed.), CRC Press, Boca Raton, Florida, Vol. 1, pp. 141–149.

Welch, G. R., 1984. Biochemical dynamics in organized states: A holistic approach, in: *Dynamics of Biochemical Systems* (J. Ricard and A. Cornish-Bowden, eds.), Plenum Press, New York, pp. 85–101.

Wen, H. R., Guo, G. R., Zhu, W. H., and Wang, J. X., 1995. Effect of rare earth compounds on alveolar macrophage *in vitro*, *Wuji Huaxue Xuebao* 11:389–393 (Chinese) CA 124:1381136j.

Wenstrup, D., Ehmann, W. D., and Markesbery, W. R., 1990. Trace elements imbalances in isolated subcellular fractions of Alzheimer's disease brains, *Brain Res.* 533:125–131.

Wester, P. O., 1971. Trace element balances in two cases of pancreatic insufficiency, *Acta Med. Scand.* 190:155–161.

Wester, P. O., 1974. Trace element balances in relation to variation in calcium intake, *Atherosclerosis* 20:207–215.

Wheeler, D. M., and Power, I. L., 1994. Comparison of plant uptake and plant toxicity of various ions in wheat, *Plant Soil* 172:167–173.

Wheeler, D. M., Power, I. L., and Edmeades, D. C., 1993. Effect of various metal ions on growth of two wheat lines known to differ in aluminum tolerance, *Plant Soil* 155–156; 489–492.

WHO (World Health Organization), 1989. *Minor and Trace Elements in Breast Milk*, World Health Organization and International Atomic Energy Agency, Geneva, pp. 14–98.

WHO (World Health Organization), 1996. *Trace Elements in Human Nutrition and Health*, Macmillan, London.

Weiner-Schmuck, M., Lind, I., Polzer, G., and Seidel, A., 1990. *In vivo* and *in vitro* effects of rare earth compounds, *J. Aerosol Sci.* 22(Suppl. 1):5505–5508.

Wilkinson, R. E., and Duncan, R. R., 1992. Non-EDTA extractable scandium in roots of three sorghum cultivars, *J. Plant Nutr.* 15:2559–2565.

Williams, M., 1985. Effects of selected heavy metals and trace elements on birds, Report G4365 M5-22, U.S. Fish and Wildlife, San Francisco State University.

Williams, R. J. P., 1991. The chemical elements of life, *J. Chem. Soc. Dalton Trans.* pp.539–546.

Williams, R. J. P., and Frausto da Silva, J. J. R., 1996. *The Natural Selection of the Chemical Elements*, Clarendon Press, Oxford, pp. 81–116, 598–634.

Wyttenbach, A., Bajo, S., Tobler, L., and Zimmerli, B., 1987. The concentration of 19 trace elements in the Swiss diet, in: *Trace Element-Analytical Chemistry in Medicine and Biology* (P. Brätter and P. Schrammel, eds.), Walter de Gruyter, Berlin, Vol. 4, pp. 171–179.

Xiong, B. K., 1995. Application of rare earths in Chinese agriculture and their perspectives of development, in: *Rare Earths in Agricultural Sciences*, Australian Academy of Technology and Engineering, Carlton South, pp. 3–9.

Xu, G. X., and Xiao, J. M., eds., 1985. *New Frontiers in Rare Earth Science and Applications*, Science Press, Beijing.

Yang, C. S., Schaedle, M., and Tepper, H. B., 1989. Phytotoxicity of scandium in solution culture of loblolly pine (*Pinus taeda L.*) and honey locust (*Gleditsia triacanthos L.*), *Environ. Exp. Bot.* 29:155–164.

Yeagle, P. L., 1993. *The Membranes of Cells*, Academic Press, Orlando, pp. 1–5, 190–215.

Young, R. S., 1935. Certain rarer elements in soils and fertilizers and their role in plant growth, *Mem. Cornell Agric. Exp. Stan.*, No 174, pp. 1–70.

Yuan, T. L., 1955. The accumulation and distribution of scandium, chromium, iodine, cesium and thallium in corn plant, *J. Agric. Assoc. China* 9:67–72.

Zaichick, V., Ye, P., and Bagirov, S., 1993. Human saliva electrolytes and trace elements with standard and paradentium diseases, in: *Trace Elements in Man and Animals*, TEMA-8 (M. Anke *et al.*, eds.), Verlag Media Touristik, Gersdorf, Germany, pp. 902–903.

Zaidi, J. H., Qureshi, I. H., Arif, M., and Fatima, I., 1992. Trace elements analysis of food spices by INAA, *Int. J. Environ. Anal. Chem.* 48:33–40.

Zenina, T. A., 1986. Biological uptake coefficients for dispersed elements in land plants, *Geochem. Int.* 23:8–11.

Zhang, D. W., and Colombini, M., 1989. Inhibition by aluminum hydroxide of the voltage-dependent closure of the mitochondrial channel, VDAC, *Biochim. Biophys. Acta* 991:68–78.

Zhang, G., and Taylor, G. J., 1989. Kinetics of aluminum uptake by excised roots of aluminum-tolerant and aluminum-sensitive cultivars of *Triticum aestivum*, *Plant Physiol.* 91:1094–1099.

Zhang, W. D., Li, F. Q., Zhang, X. Y., and Chen, Y., 1988. Effects of traces of REE on the protozoan *Blepherisma* and on mice, *Biol. Trace Elem. Res.* 17:81–85.

Zhuang, G., Wang, Y., Tan, M., Wang, Y., and Zhang, F., 1991. Preliminary study of trace elements in human brain tumor tissues by NAA, *J. Radioanal. Nucl. Chem.* 151:327–335.

Zimakov, Yu. A., and Zimakova, I. E., 1971. Effect of rare earth element salts on antibody formation during primary immunological response, *Tr. 2nd Mosk. Med. Inst.* 3:93–99. CA 79:51706y.

Zollner, H., 1993. *Handbook of Enzyme Inhibitors*, Part A, VCH, Weinheim.

Toxicology, Carcinogenicity, and Pharmacology of Scandium and Yttrium

9

What is food to one may be a fierce poison to another

Lucretius, 95–35 B.C.

Why should we fear to being crushed by savage elements, we who are made of the same elements?

Ralph Waldo Emerson, *The Conduct of Life*, 1860

9.1 Introduction

Scandium and yttrium, like all other ions, complexes, or compounds, are toxic in excess. These and other elements of group IIIA are hard metals that bind weakly to organic ligands and may compete for the binding sites of some other elements. Mechanisms of metal toxicity may include enzyme inhibition, indirect effects (i.e., binding to cofactors, vitamins and substrates), substitution for essential metals, and metal imbalance (Pounds, 1985) and may involve as well such issues as cation and anion antagonism, biphasic response, binding reactions of living cells, and selective toxicity (Albert, 1985). Isolated data exist about the likely effects of scandium, yttrium, and numerous other TE on the immune system (Bendich and Chandra, 1990), alterations of gene expression (Heflich, 1991), and mutagenic or teratogenic effects (Gebhart and Rossman, 1991).

Concerns about toxicity are raised by industrial or other uses of these metals or damage imposed by radioactivity. The latter aspect is related to such issues as the safety of using radioisotopes, the operation of nuclear power plants, and the dangers of nuclear weapons and, as such, is not unique to scandium and yttrium (Kathren, 1984; Warner and Harrison, 1993).

The hazardous effects of scandium and yttrium have found little mention in books dealing with the toxicology, carcinogenicity, and pharmacology of metals (Izraelson, 1967; Venugopal and Luckey, 1978; Carson *et al.*, 1986; Bondy and Prasad, 1988; Goodman and Gilman, 1990; Ennever, 1994) or in handbooks of general and applied toxicology (Hathway, 1984; Ballantyne *et al.*, 1993; Richardson and Gangolli, 1994). Several book chapters, articles, and reviews are dedicated to the toxicology of REE and their radioisotopes (Kyker and Anderson, 1956; Kyker, 1962; Haley, 1965, 1979; Arvela, 1979; Masse, 1982; Goyer, 1986; Haley, 1991; Hirano and Suzuki, 1996). Part of the accumulated data refer to studies in which scandium or yttrium was compared with other chemical elements that have similar chemical-physical properties (Bulman, 1988; Lambotte-Vandeaer and Duverger-van Bogaert, 1988).

The studies of Schulz at the end of the 19th century found that poisonous substances had a stimulatory effect when given in small doses. Subsequent work has indicated that this stimulatory action of a harmful agent when present in subharmful quantities, i.e., hormoligosis or hormology, is a general phenomenon, known today as the Arndt–Schulz law. This effect is noted with chemicals, radiation, heat or cold, and indeed any agent capable of exerting an effect on cells (Luckey, 1980). Many metals may both stimulate and depress, showing beta pattern activity on bone marrow and other organs, tissues, cells, and cellular organelles in various organisms (Luckey *et al.*, 1975).

9.2 Toxic Effects on Microorganisms

Metal–microbe interactions are of importance in theoretical and applied studies, such as investigations of microbiological model systems that have implications for plants, animals, and humans. Patterns of developing resistance of some microorganisms to toxic metals and antimicrobial metal compounds represent a useful approach to solving major industrial and environmental problems associated with the potential toxic effects of metals (Hughes and Poole, 1989).

In vitro assessment of the toxicity of metal salts, by evaluation of their effect on DNA structure and function in intact cells, has been studied by Christie and Costa (1983). The chemical reactivity of metal ions as measured by the parameters of softness and LD_{50} in mice has been shown to correlate well with the LC_{50} in *Drosophila* (Table 9-1). The mode of action of metals on microorganisms in aquatic media is complicated by the antagonistic, synergistic or additive effects that may occur when more than one metal is present. Bioconcentration and bioaccumulation of many TE in aquatic, land, and airborne micro- or macroorganisms have been intensively investigated in recent years (Connell and Miller, 1984).

An exponential correlation was established between the logarithm of the solubility product of metal hydroxides, metal concentration factors in phytoplankton, and sublethal metal toxicity to marine diatoms (defined as the depression of log-linear cell division rates by 50%), as well as in phytoplankton (Fisher, 1986). High log metal concentration factors in phytoplankton of 5.0–5.3

Table 9-1. 14-Day LD_{50} Values in Mice and 4-Day LC_{50} Values in *Drosophila* for Selected Metal Ions[a]

Metal ion	Salt	σ_P[b]	LD_{50}[c]	LD_{50}[d]	LC_{50}[e]
Mg^{2+}	$MgCl_2 \cdot 6H_2O$	0.167	1.04	4.1	430
Hg^{2+}	$HgCl_2$	0.064	0.192	0.024	5.7
Y^{3+}	$YCl_3 \cdot 6H_2O$	0.147	0.451	0.52	25
Cr^{3+}	$CrCl_3 \cdot 6H_2O$	0.107	0.884	0.800	23

[a] Data reprinted from Christie and Costa (1983), with permission of The Humana Press.
[b] Values of the softness parameter σ_P, obtained from Ahrland (1965).
[c] Values of LD_{50} from the literature.
[d] LD_{50} = Lethal dose of an agent which is fatal to 50% of a population 14-day; LD_{50} values, expressed in mmol/kg.
[e] LC_{50} = Lethal concentration of a gas which will kill 50% of the test animals within a given time of experiment 4-day; LC_{50} values, expressed in mM.

for aluminum, cerium, americium, and plutonium were calculated, as compared to a low value of 0.3 for magnesium.

The *in vitro* fungistatic activity of 24 metal ions against *Alternaria tenuis* and *Botrytis fabae* was investigated by Somers (1961). The toxicity index of the median effective doses, ED_{50}, of the metal salts ranged from 1 to 10^{-7} mol/l. No evidence was found that either of the two fungi is consistently more susceptible than the other to the range of metals tested. However, yttrium, aluminum, beryllium, copper, and silver were more toxic to *A. tenuis* than to *B. fabae* (Table 9-2). These and other experimental results suggested that the fungistatic action of metal ions may be related to their strength of covalent binding to surface ionogenic groups of the cell, i.e., to such groups as imidazole, carboxyl, phosphate, or sulfhydryl. Although the primary fungistatic action of the cations was regarded as nonspecific reaction outside the cell protoplast, ultimate fungicidal action may well be caused by secondary reactions following the formation of un-ionized complexes (Somers, 1961).

Table 9-2. Fungitoxicity of Metal Salts to Alternaria tenuis and Botrytis fabae[a]

	ED_{50} (mol metal/1)		pH at ED_{50}	
Salt	*A. tenuis*	*B. fabae*	*A. tenuis*	*B. fabae*
$NaNO_3$	4.6×10^{-1}	1.6×10^{-1}	6.5	6.5
$Al_2(SO_4)_3$	1.3×10^{-5}	1.4×10^{-4}	5.0	4.4
$Y(NO_3)_3 \cdot 6H_2O$	6.3×10^{-6}	1.0×10^{-5}	6.9	6.9
OsO_4	1.7×10^{-7}	2.4×10^{-7}	6.8	6.2

[a] Data reprinted from Somers (1961), with permission of the Association of Applied Biologists.

Table 9-3. Generation Times and Reduction in Growth Yield of Pseudomonas LC-11 When Grown in the Presence of Metal Analog of Fe^{3+} [a]

Metal[a]	Ionic radius (pm)	Generation time (min) +Fe	Generation time (min) −Fe	Percent decrease in growth yield[c] +Fe	Percent decrease in growth yield[c] −Fe
Fe^{3+}	64.5	58	90	0	0
Al^{3+}	53.5	63	78	0	0
Ga^{3+}	62.0	193	—[d]	91	100
Y^{3+}	90.0	84	84	55	20
Sc^{3+}	74.5	110	226	60	50

[a] Data reprinted from Fekete and Barton (1991), with permission of the authors and Springer-Verlag.
[b] Concentration of the metal in the culture medium was 1 mM.
[c] Decrease in growth yield was determined by comparison to control cultures that had not been supplemented with metal analog of Fe^{3+}.
[d] With Ga^{3+}, no growth occurred.

A study with *Escherichia coli*, *Staphylococcus aureus*, and five fungi yielded the following order of decreasing toxicity of 11 mM solutions of the nitrates of yttrium and other Ln: Lu > Y > Er > Ho > Dy > Eu > Nd > Ce (Talburt and Johnson, 1967). Yttrium showed a high toxicity to *Saccharomyces cerevisiae*. Fekete and Barton (1991) found that the growth of *Pseudomonas* sp. was inhibited by analogs of Fe^{3+} in the following order of decreasing toxicity: $Ga^{3+} > In^{3+} > Sc^{3+} > Cr^{3+} > Y^{3+} > Al^{3+}$ (Table 9-3). The inhibition of growth by 1 mM In^{3+}, Sc^{3+}, and Ga^{3+} was greater during iron-limited growth than in media containing 10 μM Fe^{3+} (Fekete and Barton, 1991). Microbial nitrogen fixation was apparently inhibited by yttrium and cerium. Yttrium chloride did not show any genotoxic effect on various strains of *Salmonella typhimurium*, even when administered in combination with 9-aminoacridine (lyehara *et al.*, 1987).

9.3 Toxic Effects on Plants

General aspects of metal toxicity to plants refer to particular interactions with carrier systems in the plasmalemma, metal-bonding proteins of the metallothionein type, or accumulation of metal in a soluble complex of low molecular weight in one species or in an insoluble product in another one. These mechanisms may cause some metals may be excluded in one species but accumulated in another.

Short-term experiments with onion (*Allium cepa*) showed that scandium had similar phytotoxic effects to those caused by aluminum; both of these metal ions showed inhibitory effects on mitosis and root elongation (Clarkson and Sanderson, 1969). Sc^{3+}, as well as Fe^{3+} and Al^{3+}, inhibited the uptake of calcium from acid culture solutions by week-old barley plants (Table 9-4). The inhibition by scandium could be detected at a Sc^{3+}: Ca^{2+} ratio of 1 : 1000. The onset of this inhibition could be quite rapid, and the inhibition persisted for at least 24 h in the absence of polyvalent metals. A relatively nonspecific action of scandium, iron, and aluminum in controlling calcium uptake, dependent primarily on their polyvalency rather than on their precise chemical nature, was presumed (Clarkson and Sanderson, 1971). Reid *et al.*, (1996) measured the membrane fluxes and toxicities of aluminum, gallium, and scandium in internodal cells of *Chara corallina*. AT 20 $mmol/m^3$ and pH 4.4, the inhibition of growth was in the order $Sc^{3+} > Al^{3+} > Ga^{3+}$. Scandium and aluminum were also toxic to mature, nongrowing cells. Influx of ^{46}Sc increased with increasing pH, consistent with membrane permeation by hydroxyscandium species rather than Sc^{3+}. These results do not support the conclusion reached by Clarkson and Sanderson (1971) that scandium may act intracellularly. Apparently, scandium toxicity is not correlated with either the intracellular scandium concentration or the external concentrations of $ScOH^{2+}$ and $Sc(OH)_2^+$, since the relative concentrations of these species are higher at higher pH. The hypothesis was advanced that toxicity is correlated with the concentration of Sc^{3+} in the external solution. This may be due to the high charge of Sc^{3+} being attracted to negatively charged sites not only in the cell wall, but on proteins and glycoproteins embedded in or on the surface of the cell membrane (Reid *et al.*, 1996).

Scandium was found to be 10-fold more toxic to wheat (*Triticum aestivum*) roots than mononuclear aluminum on a concentration basis. (Kinraide, 1991). At 0.2 μM Sc^{3+}, the relative root length was 42% of that of controls, and at 0.4 μM inhibition was complete. The speciation of scandium was unknown, but it was presumed that if polynuclear hydroxyscandium species were present, they would be about as toxic as aluminum (Kinraide, 1991).

Table 9-4. Effects of Fe^{3+}, Al^{3+}, and Sc^{3+} on the Relative Uptake of Calcium by Barley from Solutions Containing 100 μM $CaCl_2$[a]

	Uptake (g/root) as percentage of control		
Concentration of trivalent metal (μM)	Fe^{3+}	Al^{3+}	Sc^{3+}
0.1	100 ± 7	100 ± 7	75 ± 7
10	73 ± 4	67 ± 3	36 ± 4
100	24 ± 2	21 ± 2	16 ± 2

[a] Data reprinted from Clarkson and Sanderson (1971), with permission of Oxford University Press.

Table 9-5. Toxic Effect of Increasing Concentrations of Scandium on Dry Weight (mg/Plant) and Chlorophyll Content (mg/g Fresh Weight) of Loblolly Pine and Honey Locust[a]

	Dry weight (mg/plant)					
	Stem		Root		Chlorophyll (mg/g fresh weight)	
Sc concentration (μM)	Honey locust	Loblolly pine	Honey locust	Loblolly pine	Honey locust	Loblolly pine
0	178.7a[b]	115.9a	70.9a	30.3a	1.94	1.5
10	116.7b	109.7a	57.0b	28.5a	0.91	0.71
25	103.7b	88.2a	61.3ab	24.5a	0.75	0.68
50	55.8c	67.5b	16.8c	17.8b	—	0.58
100	—	52.6b	—	10.7c	—	0.50

[a] Data reprinted from Yang *et al.*. (1989), with permission of the authors and Elsevier Science.
[b] Values in columns with different letters are significantly different at $P < 0.01$.

Scandium induced chlorosis and substantially reduced the chlorophyll concentration in young leaves of honey locust (*Gleditsia triacanthos* L.) and loblolly pine (*Pinus taeda* L.) at all the investigated concentrations (Yang *et al.*, 1989; Table 9-5). Figure 9-1 shows that at a concentration of 0.1 mM Sc^{3+} the loblolly pine roots become bifurcated, thickened, and were hairless.

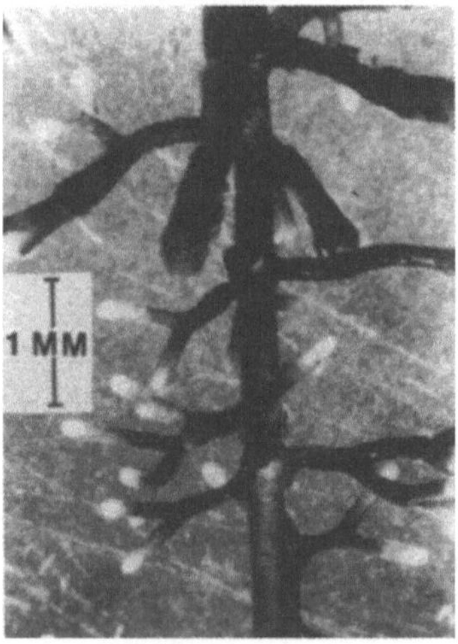

Figure 9-1. Mycorrhiza-like short roots of loblolly pine grown in the presence of 0.1 mM Sc^{3+}. Notice short roots with two- and threefold bifurcation and without root hairs. Reprinted from Yang *et al.* (1989), with permission of the authors and Elsevier Science.

Cytological investigations of the effect of 0.5–5 μM yttrium sulfate solutions, as well as 40 other metal salt solutions, on root meristem of *Allium cepa* showed a common effect of colchicine type mitosis. An induction of pronounced stickiness of chromosomes and a differential staining of the heterochromatin at metaphase–anaphase condition were also found (Levan, 1945).

9.4 Toxic Effects on Animals

Organs affected by metal intoxication are the skin, g.i. tract, liver, kidney, and hematologic organs, as well as the central and peripheral nervous systems (Bryson, 1989). The critical concentration in a cell, critical organ concentration, critical organ, and critical effect are known to some extent for a few elements (Ewers and Schlipkoter, 1991), but scarce information is available for scandium, yttrium, and numerous other TE. Symptoms of REE toxicity in animals include writhing, ataxia, labored respiration, walking on toes with arched back, and sedation. There is a delayed lethality, with the death rate peaking after 48–96 h. A sex difference is apparent, with males being less susceptible than females (Haley, 1979). A comprehensive compilation of metal toxicity in mammals was made by Luckey and Venugopal (1977), in which the chemical toxicity of a considerable number of the elements, including scandium and yttrium, were compared on the basis of vertical groups and horizontal periods of the periodic table. The following toxicity sequence of lethal doses (LD_{50}) of metal ions in mammals (rat, mouse, and rabbit), administered intravenously, intraperitonically, or orally, was proposed: Ag^{+}, Hg^{2+}, Tl^{+}, Cd^{2+} > Cu^{2+}, Pb^{2+}, Co^{2+}, Sn^{2+}, Be^{2+}, In^{3+}, Ba^{2+}, Mn^{2+}, Zn^{2+}, Ni^{2+}, Fe^{2+}, Cr^{3+}, Y^{3+}, La^{3+} > Sr^{2+}, Sc^{3+}, Cs^{+}, Li^{+}, Al^{3+}. In this sequence, the metal concentrations resulting in toxicity were expressed on a molar basis (Venugopal and Luckey, 1978).

Acute, chronic, latent, and recondite toxicity symptoms may be dependent on the susceptibility of the animal, the dose and mode of administration, and the duration of exposure. The complex toxicity of 27 TE to mice was studied by exposing them to low doses of metals, given in drinking water, for a lifetime (Schroeder and Mitchener, 1971; Table 9-6). A 2000-mg/kg dose of technical-grade REE nitrates fed to rats caused a significant reduction of liver weight in female animals, but not in males (Ji *et al.*, 1985). The kidney weight significantly increased, but other organs were not affected.

Intraperitoneal administration of a 4.32-mg/kg dose of yttrium nitrate to rats inhibited gastric excretion in a dose-dependent manner. Intragastric administration at the same dose caused serious gastric mucosal damage and decreased the secretion of Alcian blue-binding mucus in fasted animals (Luo, 1991).

Table 9-6. Weights of Mice Fed with Scandium or Yttrium in Drinking Water (5 mg/l)[a]

Days	Weight (g)			
	Control	Sc fed	Control	Y fed
		Males		
30	28.0	26.8	24.2	17.6[b]
90	45.1	40.5[b]	41.6	37.7[b]
180	49.3	48.1	49.8	47.5[b]
360	48.0	46.2	47.0	48.1
		Females		
30	22.4	21.5	20.5	15.5[b]
90	36.4	31.8[b]	32.2	31.3
180	42.1	39.5[b]	39.6	34.0[b]
360	49.5	44.4[b]	47.8	41.0[b]

[a] Data reprinted from Schroeder and Mitchener (1971), with permission of the American Society for Nutritional Sciences.
[b] Differs from control, $P < 0.001$ or more.

In an investigation of the deposition patterns and fate of large particles inhaled by small laboratory animals via nose breathing, rats and guinea pigs inhaled polystyrene latex microspheres of various diameters labeled with ^{46}Sc (Snipes *et al.*, 1988). Long-term biological half-times for alveolar retention of the 3-μm microspheres were 63 days for rats and 83 days for guinea pigs (Fig. 9-2). The total ^{46}Sc activity on the head skin immediately after exposure of rats was 3.7–6.4% of the initial body burden, whereas for guinea pigs the range was 41–53%. However, these results did not allow that the deposition and retention patterns to be projected for similar particles inhaled by humans (Snipes *et al.*, 1988).

Extensive investigations were made at the Argonne National Laboratory and other American institutions on the accumulation and cumulative excretion of fission products in plants, animals, and humans. Data obtained for rats and mice showed a high affinity of plutonium, strontium, and yttrium for bone (Van Middlesworth, 1947). Following parenteral administration of these metals to rats, excretion occurred in the first few days, whereas not more than 2% was excreted in the next 3–50-day period. Urinary excretion declined to zero after 5–8 days (Fig. 9-3). In rats yttrium has a high affinity for teeth and bones, where accumulation occurs and metabolism is relatively slow (Zhang *et al.*, 1985).

Correlations of several physical and chemical parameters of metal ions and their complexes with a variety of amino acids and other physiological and chemical compounds of interest with the metal ion toxicities to both aquatic and

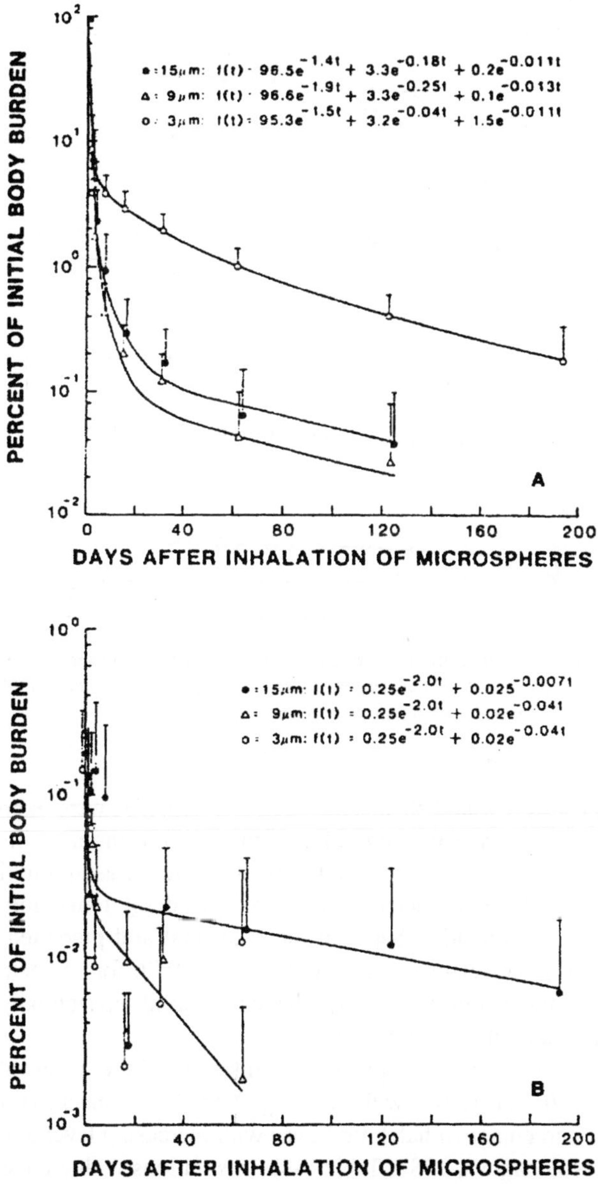

Figure 9-2. Whole-body retention of ^{46}Sc for guinea pigs (A) and rats (B) after inhalation of ^{46}Sc-labeled polystyrene latex microspheres. Reprinted from Snipes *et al.* (1988), with permission of the authors and Taylor and Francis, Inc.

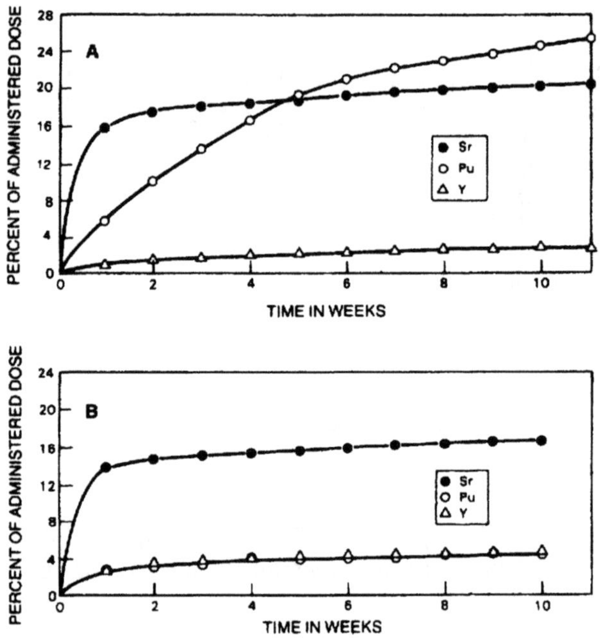

Figure 9-3. Cumulative excretion through urine (A) and feces (B) of radioactive yttrium, strontium, and plutonium after parenteral administraiton to rats. Redrawn by Stannard (1988) from Van Middlesworth (1947); reprinted with permission of Dr. J. N. Stannard.

terrestrial species were established. The metal ion toxicities were expressed as pT values, where pT is the negative logarithm of the ion concentration in moles/liter, associated with the 14-day LD_{50} value (Table 9-7) The measured toxicities of 20 metals were plotted versus calculated values based on ionization and redox potentials. A good correlation between the measured and predicted values was found. (Table 9-7). Correlation coefficients of $r^2 = 0.76$ for 15 metal ions were found; however, the toxicity of many elements varied substantially with their oxidation numbers (Kaiser, 1980, 1985).

Experiments on the development of embryos of the freshwater mollusk *Lymnaea stagnalis* in water with varying $^{90}Sr/^{90}Y$ concentration found a decrease in the percentage of hatching larvae with increasing level of radioactivity (Kulikov and Molchanova, 1982; Table 9-8). In solutions with a concentration of 10^{-3} Ci/l, there was, along with a higher rate of death of embryos, a noticeable delay in embryonic development, as well as weaker pigmentation and diminution of embryo size. Irradiation of the female amphipode *Grandidierella japonica* grown in brackish water with $^{90}Sr/^{90}Y$ caused delayed maturation and hatchability (Kuwabara, 1994).

Table 9-7. Observed and Calculated Toxicities of Elements with Similar Electron Configurations in the Outer Orbitals to Aquatic Biota, in Particular *Daphnia magna*[a]

Metal	Electrochemical potential	Ionization potential differentials	Observed toxicity	Calculated toxicity
Ca^{2+}	2.76	5.76	2.54	2.42
Sc^{3+}	2.08	11.95	—	3.90
Y^{3+}	2.37	8.27	—	4.74
Al^{3+}	1.71	9.62	4.92	4.93
La^{3+}	2.37	7.74	—	5.69

[a] Data reprinted from Kaiser (1980), with permission of the National Research Council of Canada. The toxicities of these chemical elements to *Daphnia magma* were measured and published by Biesinger and Christensen, in 1972. Cited from Kaiser, 1997, personal communicatiuon with permission.

Table 9-8. Effect of $^{90}Sr/^{90}Y$ on Embryonic Development of the Mollusk *Lymnaea stagnalis*[a]

Water radioactivity (Ci/l)	8-day radiation dose (rad)	Percent of larvae hatched	Number of abnormal embryos
Control	—	98.0	1
1×10^{-9}	0.0035	98.0	0
1×10^{-5}	35	96.0	5
1×10^{-3}	3500	61.5	19
1×10^{-2}	35000	0	—

[a] Data reprinted from Kulikov and Molchanova (1982), with permission of Plenum Press.

9.5 Acute Toxicity

Acute metal toxicity is caused by a relatively large dosage during a short period of time (24 h or less), and its symptoms are a result of changes in the molecular structure of proteins, breaking of hydrogen bonds, inhibition of enzymes, changes in potential, and so forth. A subacute toxicity may be the organism's response to a less severe condition than present in an acute effect and usually after a longer term (Connell and Miller, 1984). Exposure of organs to chemicals can be expressed in terms of the incidence, intensity, and severity of response. A 9-ranking scale of empirical severity rating values was proposed, which increases from enzyme induction or other reversible biochemical changes up to pronounced pathological changes and death or pronounced life shortening (Hartung, 1986). Typical toxic endpoints or signs observed in acute animal toxicity are lethality, irritation, necrosis, change in normal homeostatis parameters, and neurologic effects (Brusick, 1987).

Studies of the correlation between the physical and chemical properties of yttrium and several other TE and their acute toxicity in mice and *Drosophila* suggested that the relative covalent/ionic character of the bonds formed may be important in determining the toxicity of the metals. Electronegativity does not yield improved correlations to toxicity (Turner *et al.*, 1985).

As a group, the REE are considered to have a relatively low acute toxicity. The oral acute lethal doses (LD_{50}) may range from >2000 to $>10{,}000$ mg/kg. However, numerous interspecies differences are known (Haley, 1979). The symptoms of acute toxicity caused by $ScCl_3$ in mice were immediate defecation, abdominal stretching, depressed respiration, tremor of the hind legs, and sedation, and the first deaths occurred within 24 h. The LD_{50} was 755 mg/kg and 4 g/kg by i.p. or oral administration, respectively (Haley *et al.*, 1962). The acute toxicity of scandium and yttrium compounds in several animal species by different modes of administration is presented in Table 9-9.

Evaluation of the acute toxicity of yttrium nitrate administered orally to rats showed that a limit dose of 5 g/kg caused 100% mortality by day 3. Toxic signs prior to death included lethargy, hunched posture, and a red nasal discharge. All animals lost weight prior to death. The single-dose acute oral LD_{50} of yttrium oxide and bastnasite mineral concentrate (which contains up to 0.3% yttrium) when administered orally to rats as a 50% solution was greater than 5 g/kg (Lambert *et al.*, 1993).

Calculation of $\log LD_{50}$ for the acute toxicity of scandium and yttrium, expressed in units of mmol metal/kg mice, gave values of -0.211 and 0.346,

Table 9-9. Acute Lethal Toxicity Doses (LD_{50}) of Scandium and Yttrium Salts in Various Species

Salt	Species	Route[a]	LD_{50} (mg/kg)	Reference
$ScCl_3$	Mouse	Oral	3980	Haley *et al.* (1962)
	Mouse	i.p.	755	Haley *et al.* (1962)
	Mouse	i.v.	93	Haley *et al.* (1962)
YCl_3	Mouse	i.p.	58	Floersheim (1995)
	Guinea pig	i.p.	85	Haley *et al.* (1962)
	Rat	i.p.	45	Kyker and Cress (1957)
$Y(NO_3)_3$	Frog	s.c.	350	Steidle and Ding (1929)
	Mouse	i.p.	175	Floersheim (1995)
	Mouse	s.c.	1660	Steidle and Dign (1929)
	Rat	i.v.	20–30	Maxwell and Bischoff (1931)
	Rat	i.p.	350	Cochran *et al.* (1950)
	Rabbit	i.v.	515	Steidle and Ding (1929)
Y_2O_3	Mouse	i.p.	>3000	Floersheim (1995)
	Rat	i.p.	500	Cochran *et al.* (1950)

[a] Abbreviations: i.p., intraperitoneal; i.v., intravenous; s.c., subcutaneous.

respectively (Apak and Tor, 1988). These values were compared with the value of 1.649 mmol Na/kg mice. The LD_{50} may well be a function of total ionization potential, electron affinity, electronegativity, or ionic radius, as well as of Pearson's softness parameter, or a linear combination of these (Apak and Tor, 1988; Table 9-10).

The approximate LD_{50} values for the acute toxicity of a mixture of REE nitrates given intraperitoneally to mice ranged from 225 to 480 mg/kg. For rats, the LD_{50} ranged from 210 to 335 mg/kg. Most of the mice that were acutely poisoned became depressed within an hour after injection (Bruce *et al.*, 1963).

A comparison of the acute LD_{50} toxicity of metallic salts with that of other compounds and natural toxins shows that metal toxicity ranges from negligible (pT 0–1) to moderate (pT 4–5). The pT values for scandium and yttrium are close to those of other group III elements (Venugopal and Luckey, 1978).

Because some metals persist in the body for a long time, a major therapeutic intervention in poisoning is the administration of drugs that enhance their excretion. This may be done with the aid of chelating agents, but these are generally nonspecific with regard to their affinity for metals. A summary of the generalized treatment of toxicosis in animals and humans, including the use of therapeutic and chelating agents (i.e., EDTA and some newer, more elaborate complexing agents), was published by Bryson (1989). The acute toxicity, differential distribution in various tissues, and elimination $ScCl_3$, $^{46}ScCl_3$, Sc-EDTA, and ^{46}SC-EDTA in mice were investigated in detail by Lachine *et al.* (1976) (Table 9-11). $ScCl_3$ was extensively deposited in the liver and spleen, whereas Sc-EDTA was rapidly taken up by the kidney, with subsequent elimination via the urine. Whole-body desaturation kinetics for ^{46}Sc-EDTA were found to fit a three-compartment model. The LD_{50} doses for $ScCl_3$ were 440 and

Table 9-10. Correlations between Acute Toxicities of Metals to Mice and Physical and Chemical Parameters[a]

Metal	σ_P[b]	I_t/r_i[c] (eV/Å)	Log LD_{50} (mmol metal/kg mice)
Na^+	0.209	6.01	1.6490
Al^{3+}	0.135	71.03	0.1981
Fe^{3+}	0.095	44.98	−0.0168
Sc^{3+}	0.140	41.88	−0.2111
Y^{3+}	0.148	32.37	−0.3458

[a] Data reprinted from Apak and Tor (1988), with permission of the authors.
[b] Pearson's softness parameters.
[c] Radius/total ionization potential.

Table 9-11. Distribution of ^{46}Sc in the Tissues of Mice, Following i.v. Injection of $^{46}ScCl_3$ or ^{46}Sc-EDTA[a]

	5 min after injection of:		120 min after injection of:	
Tissue	$ScCl_3$	Sc-EDTA	$ScCl_3$	Sc-EDTA
Liver	0.58	0.23	1.22	1.63
Spleen	0.56	0.25	1.14	1.25
Lung	1.17	0.84	0.84	0.75
Heart	0.31	0.35	0.26	0.38
Kidney	0.23	2.42	0.24	12.0
Bone	0.13	0.42	0.24	1.88
Muscle	0.03	0.23	0.03	0.63
Stomach	0.03	0.05	0.05	200.4
Intestine	0.04	0.19	0.09	15.8
Gallbladder	0.37	0.87	0.36	9.63
Brain	0.05	0.05	0.02	0.25
Urinary bladder	0.03	0.63	0.03	10.5
Pancreas	—	0.24	—	0.63

[a] Data are expressed as the ratio of the percent dose per gram of tissue to the percent dose per milliliter of whole blood. Tissues from 6 or more mice were used at each data point. Data reprinted from Lachine *et al.* (1976), with permission of Elsevier Science.

24 mg/kg, respectively, after i.p. and i.v. injection. The fast elimination phase ($t_{1/2} = 12.75$ min) accounted for 74.6% of the dose; the intermediate phase ($t_{1/2} = 40.2$ min) for 21.8%, and the slow ($t_{1/2} = 5351$ min), for 3.6% (Lachine *et al.*, 1976).

9.6 Chronic Toxicity

Chronic toxicity of metals is caused by prolonged or repeated exposure to relatively small doses and is manifested primarily in the nervous systems, followed by the kidney. Typical toxic endpoints in chronic animal toxicology are nonreversible tissue degeneration, carcinogenicity, and change in life expectancy; the effects of chronic toxicity may also include those which are typical of acute toxicity (Ji *et al.*, 1985; Brusick, 1987).

For male white mice injected with various concentrations of $ScCl_3$ up to 45 mg, the death rate after 28 days ranged from 25% to 90%. The observation of two periods of high mortality, on the 7th and 16th days (Fig. 9-4), was attributed to two different modes of metal action in the mouse organism (Dolezal *et al.*, 1966).

Figure 9-4. Daily death rate frequency during a period of 28 days following i.p. injection of $ScCl_3$ into mice, showing high mortality on the 7th and 16th days. Reprinted from Dolezal *et al.* (1966). We made efforts without success to trace the copyright holders.

Depression of growth rate was recorded in mice when $ScCl_3$ was added to drinking water at a dose of 5 mg Sc^{3+}/kg (Schroeder and Mitchener, 1971). However, this salt was scarcely toxic to rats when fed for two weeks at a concentration of 1% of their diet (Hutcheson *et al.*, 1975).

The relative chronic toxicity of 42 metal ions (mainly as chlorides), following i.p. injection to mice, was investigated by Bienvenu *et al.* (1963). The LD_{50} at 30 days was evaluated on a semilogarithmic scale, relative to a value of unity for NaCl. The scandium and yttrium chlorides showed a relatively low toxicity, similar to that of aluminum (Table 9-12). A correlation of the toxicity with the electronic structure of the elements was assumed (Bienvenu *et al.*, 1963). Intestinal adhesions, but no effect on growth, were observed when YCl_3 was administered intraperitoneally to rats for 5 months (Bruce *et al.*, 1963).

Table 9-12. Relative Chronic Toxicities of Several Metal Ions Administered to Mice by i.p. Injection as Chloride Solution[a]

Metal	LD_{50}[b]	Relative toxicity[c]
Sodium	44.5 ± 0.720	1.0
Aluminum	0.80 ± 0.060	55.6
Yttrium	0.66 ± 0.012	67.4
Scandium	0.62 ± 0.010	72.1
Lanthanum	0.493 ± 0.017	90.3
Indium	0.022 ± 0.0009	1824.5
Mercury	0.019 ± 0.0006	2283.0

[a] Data reprinted from Bienvenu *et al.* (1963), with permission of Publications Elsevier Paris.
[b] Lethal dose LD_{50} expressed in milliatoms of metal ion/kg weight of mice at 30 days of exposure.
[c] Relative to sodium, considered as unity.

9.7 Toxic Ocular and Skin Effects

Toxic effects of chemicals, and in particular scandium and yttrium, involving the eyes or vision and produced either by direct contact or by systemic routes were described in detail by Grant and Schuman (1993). The immediate or delayed effects and systemic actions of numerous agents on the cornea, conjunctiva, lens, iris, anterior chamber, and posterior eye segment were outlined.

Application of $ScCl_3$ to rabbit eyes with an intact corneal epithelium produced a permanent opacification of the corneal stroma with a removed epithelial layer. Direct application of 0.1 ml of 1 : 1 $ScCl_3$ solution to the eyes of rabbits caused slight translucency involving less than half of the cornea, with loss of reaction to light by the iris (Grant and Kern, 1956a); this corneal reaction completely cleared within 2 days. In addition, considerable redness and edema of the conjunctiva with small ulcerations were observed but cleared within 4 days (Haley *et al.*, 1962). Administration of 0.1 M solutions of REE chlorides as eyedrops on the scraped cornea of rabbit eyes for 10 min induced opacities after 17 days for scandium and yttrium, as compared to 6–12 days for several REE (Grant and Kern, 1956b). Scandium and yttrium caused total and permanent opacity of the cornea, as well as a characteristic sequence of changes. The cornea generally remained clear during the actual exposure to neutral solutions of $ScCl_3$, but after a latent period of several hours, diffuse gray opacities gradually developed. Subsequently, over a period of up to 3 weeks, opacities changed to dense yellowish-white plaques accompanied by extensive vascularization. The dense whitish opacities developed during the first and second weeks following exposure to scandium, yttrium, and other group IIIA metals are expected to be associated with secondary calcification (Grant and Schuman, 1993).

Selye *et al.* (1966b) conducted an experiment for the production of cerebral and ocular changes in rats, which simulated certain naturally occurring morbid lesions in humans. Intravenous injection of $ScCl_3$ to rats caused a sensitization with extensive thrombosis of the ophthalmic and intracranial veins. The lesions thus induced were prevented by prophylatic administration of dibenzyline, an adrenolytic agent (Selye *et al.*, 1996b). A year long study of the effects of the toxic rabbit eye irritation of REE found that yttrium nitrate was severely irritating to both unwashed and washed eyes, whereas yttrium oxide and bastnasite showed mild, minimally irritating effects (Lambert *et al.*, 1993; Table 9-13). Instillation of 100 mg of yttrium chloride into rabbit eyes caused severe irritation (Richardson and Gangolli, 1994).

The exposure of the rabbit's lens to Nd : YAG laser irradiation induced complex damage, which was probably due to heat-related changes in the lens fibers. Swelling of the fibers on the lens equator may result from damage to the cell membranes in the anterior part of the lens (Lazo and Makarenkova, 1995). Whether these effects were due to the laser irradiation alone, or due to the

Table 9-13. Scores of Rabbit Eye Irritation after Application of Yttrium Nitrate, Yttrium Oxide, and Bastnasite[a]

	2h h		48 h		72 h	
Compound	Unwashed	Washed	Unwashed	Washed	Unwashed	Washed
Yttrium nitrate	34.0	27.7	37.7	29.0	30.7	29.0
Yttrium oxide	13	9	3.7	2	0.7	0
Bastnasite	3.3	3.3	2.0	1.3	0.7	0

[a] A 0.1-g dose was applied to the left eye of six rabbits. Three of the treated eyes were rinsed with saline 30 s post instillation. Eyes were examined at 24, 48, and 72 h by the method of Draize *et al.* (1944). Data reprinted from Lambert *et al.* (1993), with permission of the authors and the Executive Director of the *Journal of the American College of Toxicology.*

combined effect of irradiation and of the chemical compounds of the laser, e.g., aluminum, yttrium, and neodymium, was not further studied. YAG laser posterior capsulotomy in human eyes may cause cystoid macular edema (diffusion of prostaglandins), retinal detachment (loss of hyaluronic acid from the vitreous), endophthalamitis (spread of bacteria), or rubeosis iridis (angiogenic factors). Intracapsular posterior chamber lens implantation more effectively maintains the protective aqueous vitreous barrier and blood aqueous barrier (Schalnus *et al.*, 1995).

Direct application of 0.5 g $ScCl_3$ to intact rabbit skin produced no reaction throughout the entire 14-day observation period (Haley *et al.*, 1962). There was a very severe reaction on abraded skin, with a maximum irritation index of 8 at 24 h. No changes in irritation index occurred up to 14 days. Nevertheless, healing with scar formation was observed at 45 days. Intradermal injection of $ScCl_3$ to rabbits at concentrations of $1:10^4$ to $1:10^6$ produced necrosis within 1 h (on a relative scale of 1 to 10) and an irritation index of 8, which persisted for 7 days. Healing occurred at 14 days, with epilation and scar formation in the injected area. Irritating effects of yttrium chloride to the skin, eyes, mucous membranes, and upper respiratory tract in unspecified mammalian species were reported. Application of 500 mg of yttrium nitrate to the skin of rabbits caused severe dermal irritancy after 24 h (Richardson and Gangolli, 1994).

Intravenous injection of 4 mg $ScCl_3$ to rats did not cause skin calcification at the site of injection, a process known as calcergy, whereas several REE (dysprosium, erbium, holmium, and thulium) did induce it (Tuchweber and Savoie, 1968). The intensity of splenic calcification was proportional to the amount of adminstered REE (Tuchweber and Savoie, 1968). Other investigations (Garrett and McClure, 1981) obtained evidence that subcutaneous or intradermal injection of salts of yttrium and all 14 REE caused calcergy in mice. Examination of the calcification site revealed close association of metals with the collagen fibers of the dorsal fascia, with a peripheral accumulation of multinucleated giant cells.

In a study by Lambert *et al.* (1993), employing a similar method to that used to study eye irritation, 0.5 g of yttrium nitrate was applied to one intact and one abraded skin site of rabbits and occluded for 24 h. Results from evaluation of the sites at 24 and 72 h after exposure are given in Table 9-14. Slight erythema and edema were noted at most sites by 24 h post dose. No signs of dermal irritation were observed with yttrium oxide and bastnasite concentrate.

Shvidko *et al.* (1972) reported that the radioisotopes of cerium, neodymium, praseodymium, and yttrium were distributed in the skin mainly in insoluble protein (82–94%), with less in lipids (6–12%), and water extracts (2–4%). Dose conversion factors for a point source at different skin depths were calculated by Pöllänen and Toivonen (1995) for ^{90}Y, ^{91}Y, and other radioisotopes that are detected in nuclear fuel particles. The nuclide specific beta dose rate per unit activity was estimated at different skin depths (0.07, 0.4, and 3 mm), as a function of particle size. Following exposusre to a range of single doses of $^{90}Sr/^{90}Y$ β-rays, a reduction in relative pig skin thickness developed in two distinct phases (Rezvani *et al.*, 1994). The first phase was between 12 and 20 weeks post irradiation. No further changes were then seen until 52 weeks post irradiatioin, when a second phase of skin thinning was observed. This second phase was completed after 76 weeks, and no further changes in relative skin thickness were seen up to 129 weeks. The timings of these phases of damage were independent of the radiation dose; the severity of both phases of radiation-induced skin thinning was dose-related. The estimated half-timing of the short and long repair processes in pig skin irradiated with $^{90}Sr/^{90}Y$ were 0.20 and 6.6 hours, respectively; these estimates were independent of the assumed distribution of target cells in the skin (Millar *et al.*, 1996).

Table 9-14. Primary Irritation Index of Rabbit Skin by Yttrium Nitrate, Yttrium Oxide, and Bastnasite[a]

	24 h		72 h	
Compound	Abraded skin	Intact skin	Abraded skin	Intact skin
Yttrium nitrate	5.3	4.9	2.6	1.5
Yttrium oxide	0	0	0	0
Bastnasite	0	0	0	0

[a] A 0.5-g dose was applied to one intact and one abraded skin site of 6 rabbits and occluded for 24 h. The patches were then removed, and the test sites wiped. The sites were evaluated at 24 and 72 h after initial exposure, according to the method of Draize *et al.* (1944). Data reprinted from Lambert *et al.* (1993), with permission of the authors and the Executive Director of the *Journal of the American College of Toxicology*.

Table 9-15. Effective Dose (ED) Associated with Selected Probabilities of Acute Ulceration for Discrete $^{90}Sr/^{90}Y$ Beta Particles on Pig Skin, at 1.1 mm² Dose Gy at 16 μm Depth.[a]

Distance from source	ED_{50}	ED_{10}	ED_5
1 mm	253	104	63
2 mm	119	82	72

[a] The 1-mm²-diameter $^{90}Sr/^{90}Y$ values need to be adjusted upward by a factor of 2.3, based on dose film studies. Data reprinted from Baum and Kaurin, (1991), with permission of the authors and Nuclear Technology Publishing.

Baum and Kaurin (1991) worked out a model for the effect of "hot particles" on the skin of pigs, monkeys, and humans. A threshold of 6×10^9 β-particles from a point source on skin of mixed fission product β particles, or about 10^{10} β particles from $^{90}Sr/^{90}Y$, was calculated. Table 9-15 presents doses which may produce acute ulceration with 50% and 10% probabilities, respectively. Other results concerning toxic ocular and skin effects of yttrium and scandium radioisotopes are presented in Section 9.16.

9.8 Toxic Effects on Heart, Liver, Lung, and Other Organs

In animals and humans, the liver removes a major part of the scandium, yttrium and other metals that enter the body. Hepatotoxicity caused by lighter REE, such as cerium, is manifested by increased levels of numerous enzymes and the formation of a fatty liver (Evans, 1990). Injection of YCl_3 into rats did not increase the lipid content of their livers, whereas some REE caused a considerable increase of lipids (Snyder, 1960).

Intratracheal administration of dusts of Y_2O_3 to rats caused a marked emphysema of the lung, an enlargement of the bronchial lymph nodes, formation of characteristic granuloma nodules, and sometimes diffuse sclerosis of the pulmonary tissue (Mogilevskaya and Raikhlin, 1967). Brief inhalation exposure of dogs to coal dust aerosols labeled with ^{46}Sc to predict coal retention following chronic exposure indicated a mean biological half-time for pulmonary ^{46}Sc retention of not shorter than 990 days, and probably much longer (Morrow *et al.*, 1979). However, the balance of the data for this radioisotope appeared ambiguous.

In life-span studies, a single exposure of beagle dogs to ^{90}Y or ^{91}Y by inhalation of an aerosol of the radioisotope had late-occurring biological effects (Hahn *et al.*, 1986; Table 9-16). Tracheobronchial lymph node tumors were observed in dogs that inhaled relatively insoluble forms of beta emitters (Table 9-17). The importance of dose-effect modifying factors with respect to the resulting risks was emphasized by Hahn *et al.*, (1986). Inhalation exposure of laboratory animals to environmental air pollutants is used to simulate the mode of human exposure to such agents. To determine pulmonary deposition, translocation, and clearance of inhaled fly ash, hamsters (*Mesocricetus auratus*) received a single 95-min nose-only exposure to radioactive fly ash (Wehner *et al.*, 1980). Estimates with ^{46}Sc found that an average of 63 μg of fly ash, or 2–3% of the inhaled fly ash, was initially retained in the respiratory tract. The estimated biological half-times of the fly ash were 2.6 and 34.5 days, probably for the airways and for the deep lung, respectively. Acute toxicity tests showed no appreciable decline in lung burden between 7 days and 30 days post exposure. After 99 days, the mean lung burden had decreased to about 10% of the initial value (Table 9-18). Close agreement between the ^{46}Sc and ^{59}Fe measurements indicated fly-ash particulate levels in the lungs, carcass, head, pelt, g.i. tract, and feces. By extrapolating from the clearance data, it can estimated that near-complete clearance of fly ash from the lung would have been achieved in about 200 days post exposure.

Intratracheal instillation of YCl_3 into rats caused a very slow pulmonary clearance of yttrium; the half-life was estimated to be 168 days. (Hirano *et al.*, 1990). The accumulation of yttrium in the supernatant of bronchoalveolar lavage fluid (BALF) did not exceed 5 μg Y^{3+}/lung, even when a dose of 200 μg Y^{3+}/rat was administered. These results indicated a long pulmonary half-life of yttrium. Of the six biochemical or elemental indices measured in the BALF supernatant, β-glucuronidase was the most sensitive index for the detection of lung injury. The activity of this enzyme, as well as calcium and phosphorus concentrations in the BALF supernatant, increased significantly even at the lowest dose of 10 μg Y^{3+}/rat.

Transmission electron microscopy and x-ray microanalysis suggested that yttrium was localized in lysosomes of alveolar and interstitial macrophages and

Table 9-16. Initial Lung Burdens and Observation Period for Studies with Beagle Dogs Exposed by Inhalation to Aerosols of Relatively Insoluble Forms of Beta-Emitting Radioisotopes and Plutonium Oxide[a]

Radioisotope	Effective half-time in lung (days)	Range of initial lung burdens (μCi/kg body weight)	Range of observation times (years after exposure)
^{90}Y-FaP[b]	2.5	80–5200	11.5–13.0
^{91}Y-FAP	50	11–360	11.0–12.5
^{91}Sr-FAP	600	0.12–96	7.8–12.8
$^{239}PuO_2$	750–1800	0.00067–2.0	3.6–5.7

[a] Data reprinted from Hahn *et al.* (1986), with permission of the authors.
[b] FAP Fused aluminosilicate particles.

Table 9-17. Comparison of Initial Lung Burdens and Cumulative Risk Factors in Dogs of ^{91}Y, ^{14}Ce, and ^{90}Sr for the Induction of Pulmonary and Tracheobronchial Lymph Node (TBLN) Tumors[a]

Radioisotope	Mean initial lung burdens (μCi/kg)	Mean survival time (days after exposure)	Risk factor[b] (lung tumors/ 10^6 rads)	Lung dose (rads)	Number of tumors per lung	Risk factor[b] (lung + TBLN tumors/10^6 rads)
^{91}Y	105	1856	42 ± 5	11,000–25,000	21	53 ± 5
^{144}Ce	16	2385	5 ± 7	16,000–61,000	15	27 ± 3
^{90}Sr	4.9	2400	12 ± 6	17,000–57,000	30	22 ± 2

[a] Data reprinted from Hahn *et al.* (1986), with permission of the authors.
[b] To 10 years after exposure.

basement membranes (Hirano *et al.*, 1993; Fig. 9-5). Electron microscopic analyses of rat organs to which yttrium was i.v. injected revealed that the colloidal material was taken up by phagocytic cells in the liver and spleen (Fig. 9-6). Intravenously injected YCl_3 caused acute hepatic injury; surprisingly, the lung concentration of yttrium increased for about 3 months following the treatment. It was postulated that the increase in the lung content of yttrium was due to the migration of yttrium-loaded white cells from bone marrow to the lung (Hirano *et al.*, 1993). Lung granulomas, consisting of monocytes and some multinuclear giant cells (Xiao *et al.*, 1992), cytotoxic effects on alveolar macrophages (Zhou *et*

Table 9-18. Fly-Ash Burdens in Tissue and Excreta of Hamsters as a Function of Time after Exposure, Based on ^{46}Sc Data[a]

	Time of sacrifice after exposure				
	15 min	1 day	7 days	99 days	102 days (control)
Lung	65.8 ± 49.3	48.8 ± 15.2	27.9 ± 11.6	6.4 ± 4.43	$< 0.09 \pm 0.03$
Liver	0.52 ± 0.45	0.41 ± 1.9	0.35 ± 0.17	0.36 ± 0.24	$< 0.05 \pm 0.02$
Kidneys	$< 0.12 \pm 0.07$	0.18 ± 0.09	0.13 ± 0.09	$< 0.21 \pm 0.31$	$< 0.07 \pm 0.02$
Carcass	26.9 ± 13.0	10.0 ± 8.2	1.7 ± 1.7	$< 0.70 \pm 0.37$	$< 0.10 \pm 0.02$
G.I. tract	814 ± 244	259 ± 108	$< 4.3 \pm 1.1$		
Head	102 ± 57.2	20.3 ± 10.8	2.0 ± 1.1		
Pelt	80.6 ± 39.1	23.5 ± 12.6	2.9 ± 2.0		$< 0.08 \pm 0.02$
Feces		873 ± 340	127 ± 168		$< 2.4 \pm 1.3$
Urine		1.0 ± 0.7	0.9 ± 1.6		$< 0.05 \pm 0.02$

[a] All values are in micrograms and represent the means ± SD for six samples. Data reprinted from Wehner *et al.* (1980), with permission of Academic Press.

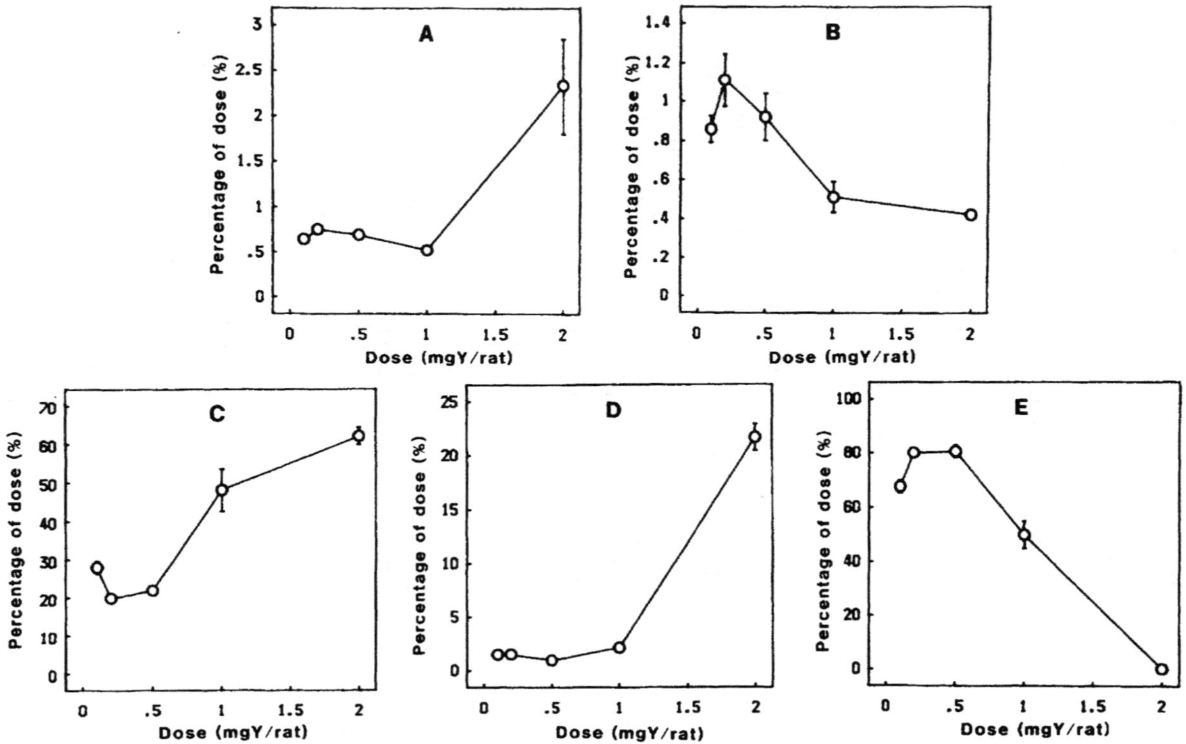

Figure 9-5. Dose-related changes in tissue distribution of yttrium in rats at 1 h after i.v. injection of YCl_3 at doses of 0.1, 0.2, 0.5, 1, and 2 mg Y^{3+}/rat. The ordinate denotes the percentage of the initial dose. (A) Lung; (B) kidney; (C) liver; (D) spleen; (E) whole blood. Reprinted from Hirano *et al.* (1993), with permission of the authors and Academic Press.

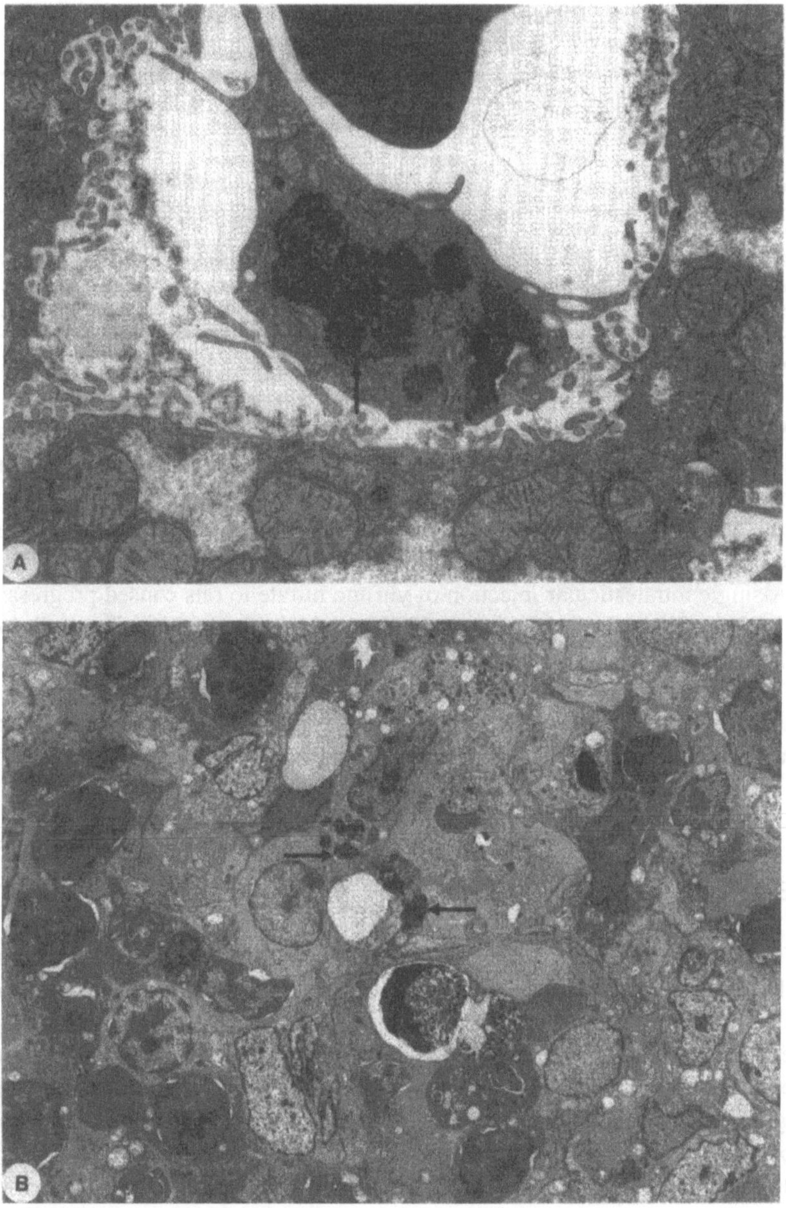

Figure 9-6. Transmission electron micrographs of rat liver (A; ×17,000) and spleen (B; ×3400) at 7 days post i.v. injection of YCl_3 at a dose of 1 mg yttrium/rat. An X-ray microanalysis revealed that yttrium was deposited in lysosomal inclusions of Kupffer cells (A) and macrophages (B) (arrows). Reprinted from Hirano *et al.* (1993), with permission of the authors and Academic Press.

Table 9-19. Effect of Yttrium on Calcium Concentration in Serum and Testis of Rats[a]

	3 days		8 days		15 days	
Treatment	Serum (mg/100 ml)	Testis (mg/g)	Serum (mg/100 ml)	Testis (mg/g)	Serum (mg/100 ml)	Testis (mg/g)
Control	7.8	6.1	7.9	7.1	7.9	6.4
Calciferol	9.9	5.8	8.0	6.4	8.1	6.6
$Y(NO_3)_3$	7.5	7.1	8.0	9.6	7.6	13.1
Calciferol + $Y(NO_3)_3$	9.8	8.1	7.5	10.2	7.3	14.2

[a] Data reprinted from Sharma *et al.* (1972), with permission of the authors and of Gustav Fischer Verlag.

al., 1992), and definite lung fibrosis (Zhao *et al.*, 1992) were induced by yttrium powder and its mineral, when injected tracheally to rats. The activities of several enzymes (succinate dehydrogenase, isocitrate dehydrogenase, malate dehydrogenase, lactate dehydrogenase and glucose-6-phosphate dehydrogenase) were inhibited by yttrium metal ions in a dose-dependence manner.

A single intratesticular injection of yttrium nitrate to rats caused progressive calcification of both the seminiferous tubules and the interstitium (Sharma *et al.*, 1972). After 8 days, there was intense calcification of the tunica propria and the walls of blood vessels; also, a progressive increase in calcium concentration of the testis occurred (Table 9-19). Similar to the effects of lanthanum, thorium, and uranium, yttrium induced on acute sterility in males; none of the females rats that were mated with these animals 3–15 days after injection became pregnant.

9.9 Biochemistry and Molecular Biology of Toxicity

The cell membrane is the first barrier met by a toxic agent upon reaching a cell. The selective permeability of the plasma membrane is vital for the regulation of cell volume and ionic composition, entry of nutrients, and elimination of wastes, initiation and conduction of action potential (of nerve and muscle), homeostatis in epithelial cells of many tissues (kidney, liver, and intestine), and vital processes of subcellular organelles (oxidative phosphorylation in mitochondria or photosynthesis in chloroplasts). The membrane theory of toxicity of heavy metals is gaining increased attention (Kinter and Pritchard, 1977). Toxicosis of most metals affects multiple organ systems, and the targets for toxicity may be specific or nonspecific biochemical processes (enzymes), which act on membranes of cells and organelles. The toxic effect of the metal usually involves an interaction between the free metal ion and the toxicological target (Goyer and Cherian, 1995). However, induction of oxidative enzymes, enhanced biotransformation, and biliary or urinary excretion occur too late to avoid toxic effects of xenobiotics. The development of chelating agents and human monoclonal antibodies are potential approaches in the treatment of acute poisoning (Goodman and Gilman, 1990).

An *in vitro* study showed cytotoxic effects of yttrium compounds on alveolar macrophages. The poisonous effect of yttrium chloride at a concentration of 1 mmol/dm^3 was expressed as an increase of cellular calcium. The greater the cytotoxicity, the greater were the increase of calcium accumulation and the increase of the Ca-ATPase activity of erythrocytic membranes (Wen *et al.*, 1995).

As compared with other REE (yttrium, lanthanum, praseodymium, and ytterbium), $ScCl_3$ exhibited the weakest inhibition of the secretion (enterosorption) induced by the enterotoxins of *Vibrio cholerae* and *Escherichia coli* in the rabbit illeum *in vivo* (Leitch and Amer, 1975). Injection of colloidal solutions of ^{90}Y induced the binding of the isotope to SH groups of blood, liver, kidney, spleen and testes of rats, which initially caused an increase in such groups, followed by a decrease on the 3rd day, with a return to normal on the 15th day (Pavlyushchik *et al.*, 1970). In a very early study (Höber and Spaeth, 1914), an irreversible toxic effect of YCl_3 at a higher concentration was illustrated with the contractility of frog muscles (Fig. 9-7).

The REE are antagonists of thrombokinase (Haley, 1965). Comparative dose–effect profiles of lactate dehydrogenase (LDH) activity in BALF supernatant revealed that 1 mol of YCl_3 is equivalent in potency to about 0.3 mol of cadmium compounds and about 3 mol of ZnO for acute pulmonary toxicity (Hirano *et al.*, 1990; Fig. 9-8). Further investigation of the acute toxicity of i.v. injected YCl_3 in rats showed that yttrium was predominantly distributed to plasma in the blood (Hirano *et al.*, 1993). At doses greater than 0.2 mg Y^{3+}/rat, most plasma yttrium appeared to be in colloidal material, which was composed of proteins and some minerals. At a dose of 1 mg Y^{3+}/rat, the concentration of yttrium in liver was slowly cleared with a half-time of 144 days. Part of the plasma calcium was translocated to the colloidal material, and plasma calcium concentration increased transiently following i.v. injection of YCl_3, apparently because of bone resorption.

The interaction of thrombin with REE chlorides caused a fast inhibition of the enzymatic activity, which was dependent upon the concentration of fibrinogen (Ducastaing *et al.*, 1973; Fig. 9-9). The anticoagulative action of scandium was inferior to that of yttrium and other REE. Investigation of the changes in the coagulation of rabbit blood following i.v. injection of scandium and yttrium acetate found that the decrease of blood coagulation reached a peak in 2 h and continued for up to 24 h (Lakin *et al.*, 1970a; (Fig. 9-10). The introduction of EDTA with a scandium acetate solution scarcely affected the concentration of scandium in blood but grossly augmented its passage through the kidneys, thus

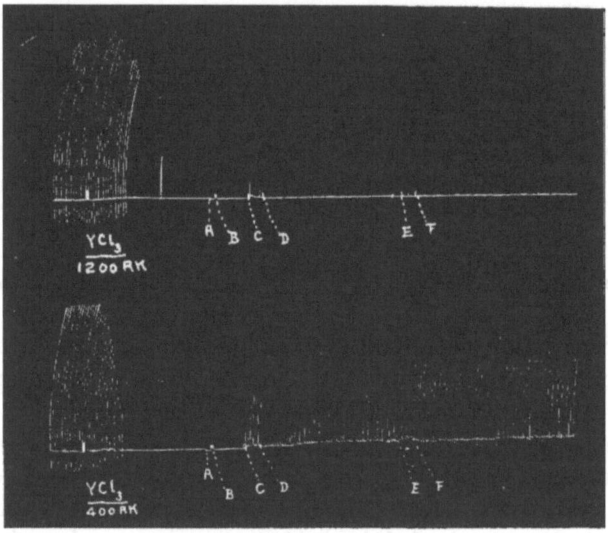

Figure 9-7. Influence of YCl_3 on contraction of frog (*Rana esculenta*) muscle. *Top:* Muscle in Ringer solution with 0.04 M YCl_3 at the beginning of the contraction; *bottom:* same, during 14 hours of interchange of the Ringer solution with $NaHCO_3$. Reprinted from Höber and Spaeth (1914), with permission of Springer-Verlag.

diminishing its acute toxicity and the anticoagulant action. Complexation of scandium and yttrium with EDTA sharply reduced the anticoagulative action and reduced their acute toxicity (Thies *et al.*, 1962; Lakin *et al.*, 1970a,b). Young beagle dogs placed in a chamber whose atmosphere contained yttrium oxide manifested a significant increase in blood lactic acid concentration and total leukocyte counts. A slightly significant decrease was noted in erythrocyte volume. No significant difference was noted in plasma volume and hemoglobin concentration. Hypertrophy, hyperplasia, and desquamation of the alveolar epithelial cells and leukocyte infiltration were noted in the lungs. No connective tissue increase was observed (Reece, 1965).

Accurate *in vivo* LD_{50} doses in rats and mice could be predicted for at least 75% of selected chemicals on the basis of cytotoxicity in hepatocyte cultures (Shrivastava *et al.*, 1992). Cytotoxicity was expressed in terms of CT_{50}, which represents the minimum test concentration inducing morphometric changes in 50% of cells or 50% cell death or a 50–100% increase in hepatocyte LDH release, all values being compared to control. A correlation is supposed to exist between *in vitro* mammalian cytotoxicity and human acute toxicity (Shrivastava *et al.*, 1992). A number of malignancies associated with surgical implants in humans have been reported. An investigation of the carcinogenicity and biochemical

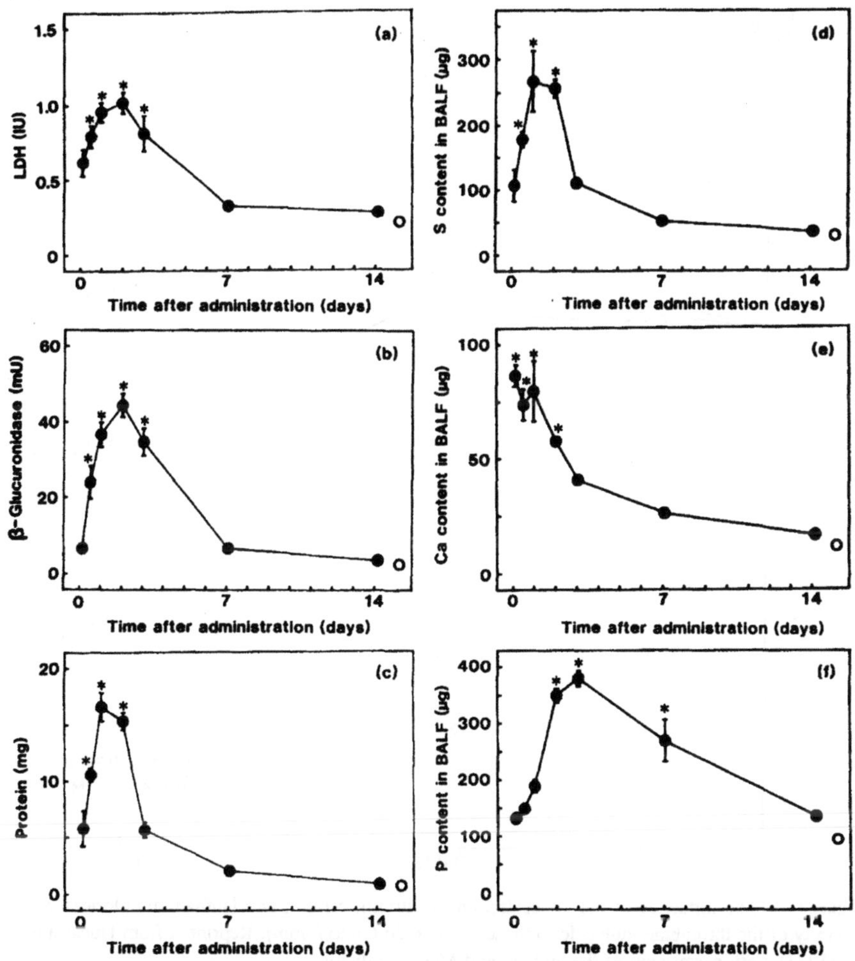

Figure 9-8. Time-course profiles of lactate dehydrogenase (LDH) (a) and β-glucuronidase (b) activities and protein (c), sulfur (d), calcium (e), and phosphorus (f) contents in the supernatant of bronchoalveolar lavage fluid of the rat lung, following intratracheal instillation of YCl_3 at a dose of 100 μg Y^{3+}/rat. Data are presented as means ± SE. Open circules represent control values. Reprinted from Hirano *et al.* (1990), with permission of the authors and Academic Press.

changes associated with several orthopedic implants, including alumina and zirconia (which contain 3% yttrium oxide), was performed in mice, with nickel alloy used as positive control implant (Takamura *et al.*, 1994). Observations of mice with implants during a 25-month period indicated that nickel alloy caused high carcinogenicity, whereas other materials showed no such effects. Also, no

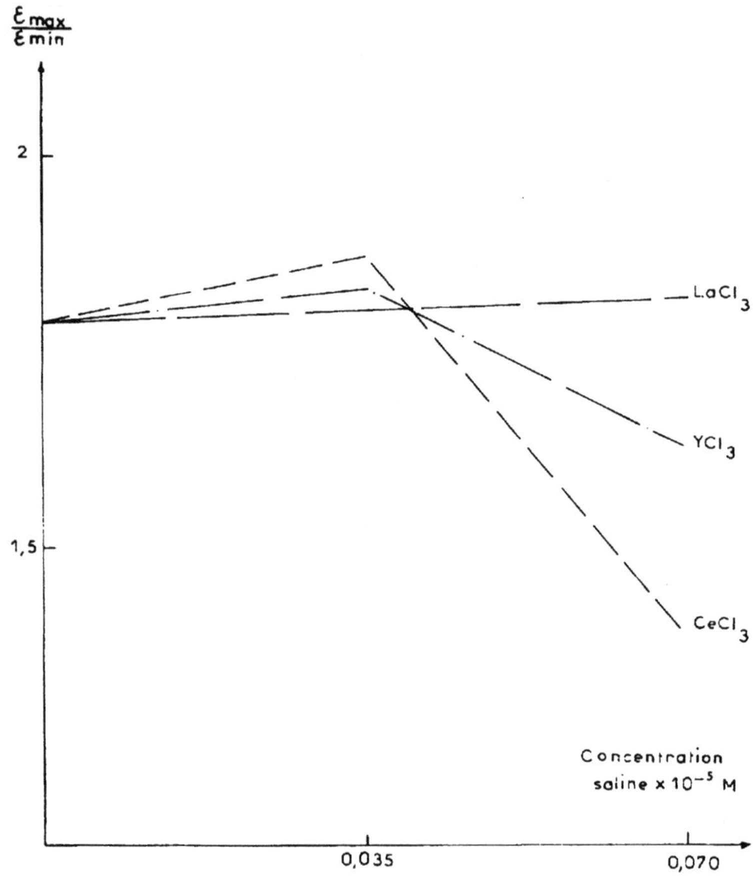

Figure 9-9. The influence of the concentration of yttrium, cerium, and lanthanum chlorides on the integrity of the thrombine molecule to those of the ratio (εmax/εmin). Reprinted from Ducastaing *et al.* (1973), with permission of the authors and Masson Editeur.

significant changes in the biochemical characteristics of blood (red blood corpuscles (RBC), white blood cells (WBC), blood urea nitrogen (BUN), and glutamate oxyacetic transaminase (GOT)) were measured, which indicated that alumina, zirconia, and hydroxyapatite did not cause chronic toxicity to mice (Takamura *et al.*, 1994).

Use of a YAG laser for anterior capsulotomies is complicated by the elevated intraocular pressure. Biochemical investigation indicated an increased prostaglandin synthesis in the eye tissues in response to laser treatment (Lin *et al.*, 1988). A correlation between prostaglandin and protein concentration in aqueous

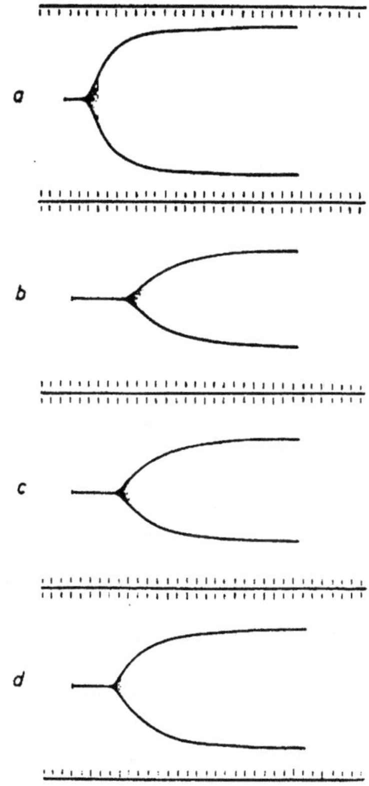

Figure 9-10. Thromboelastograms showing changes in blood coagulation caused by i.v. injection into rabbits of scandium acetate (20 mg/kg body weight). (a) Before scandium administration; (b)–(d) 2, 6, and 24 h after scandium administration. Reprinted from Lakin *et al.* (1970a). We made efforts without success to trace the copyright holders.

humor after YAG laser treatment was found (Fig. 9-11). Application of inhibitors of prostaglandin synthesis prior to laser treatment was suggested.

9.10 Pathologic Effects

Histopathological examination of the heart, lung, liver, kidney, spleen, pancreas, adrenals, and small intestine of rats revealed no substantial changes that could be related to the ingestion of $ScCl_3$ at dietary levels (Haley *et al.*, 1962). Intraperitoneal administration of a 60-mg/ml solution of YCl_3 or an acid solution of a mixture of REE (which included 0.2% yttrium) to rats caused an acute peritoneal reaction, which subsequently developed into granulomatous peritonitis with ascites formation. Occasionally, focal hepatic necrosis was observed. Histologic observations in patients given radioactive yttrium via i.p. or i.v. administration found fatty livers and serosal inflammation at higher concentra-

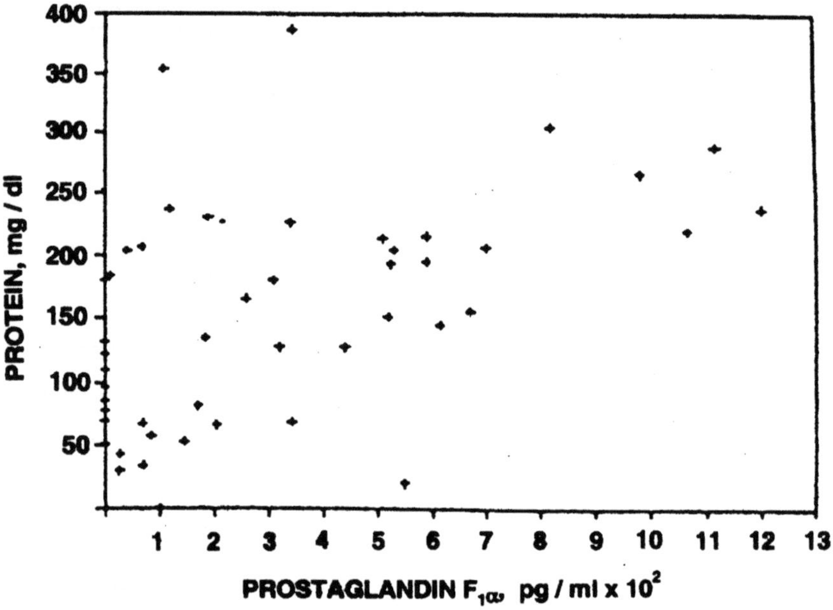

Figure 9-11. Correlation between prostaglandin and protein concentrations in aqueous humor of patients after YAG laser treatment. Reprinted from Lin *et al.* (1988), with permission of the American Society of Contemporary Medicine, Surgery & Ophthalmology.

tions of yttrium chloride (Steffee, 1959). Injection into rats of mica samples of respirable size that contained 28–36 mg Sc^{3+}/kg and several other TE resulted in enhanced dust-induced pulmonary reaction with characteristic abscess formation at later periods, up to 360 days (Garg *et al.*, 1987). Also, the cellular reaction became intense around bronchioles and in the parenchyma (Fig. 9-12). The analysis of industrial and mine mica samples found up to 0.28 mg Sc^{3+}/kg and very high concentrations of barium, zinc, and other heavy metals (Garg *et al.*, 1987).

Chargaff and Green (1978) conducted tests to distinguish between antithromboplastic and antithrombic effects of several metal ions. At concentration of 0.1 μM, Sc^{3+} showed the strongest inhibition of the thromboplastic effect, whereas Y^{3+}, Cu^{2+}, and Nd^{3+} exhibited only a moderate effect. At a concentration of 0.05 mM, 15 of 17 metals tested showed a strong effect. These effects were possibly connected with the reaction of TE with prothrombin and with the displacement of calcium ions that are necessary for the formation of thrombin (Chargaff and Green, 1948). Studies by Solymoss *et al.* (1966) on experimental thrombosis produced in rats showed that $ScCl_3$ delayed the blood coagulation *in vitro*. In its presence, there was a decrease in thromboplastin generation, prothrombin content, and heparin tolerance and an increase in

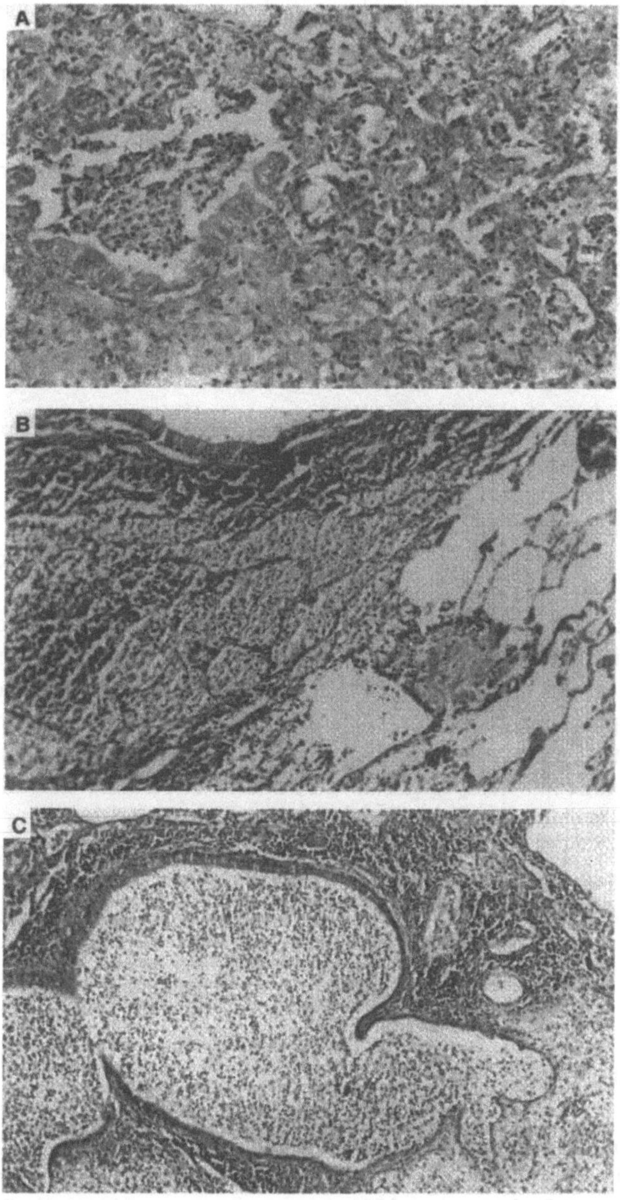

Figure 9-12. Rat lung showing pathological effects of respirable mica. (A) Polymorphonuclear leukocytic infiltration and edema at 48 h after inoculation of a mine sample of mica, ×280; (B) infiltration of polymorphonuclear leukocytes around bronchi, after 180 days of exposure, ×150; (C) abscesses in bronchiole, after 360 days, ×150. Reprinted from Garg *et al.* (1987), with permission of Elsevier Science.

recalcification and partial thromboplastin and thrombin times, but there was no thrombus formation in animals treated with this salt. Adrenaline aggravated the blood coagulation changes produced by $ScCl_3$, without altering their character (Solymoss *et al.*, 1966). The i.v. administration of $ScCl_3$ to rats caused a predominant response of the paw to all doses of adrenaline, characterized by sensitization and inflammatory edema. Strong sensitization with larger $ScCl_3$ doses caused exclusively thrombosis with hemorrhagic reactions in response to adrenaline (Selye et al., 1966a). On the basis of these findings, Selye *et al.* (1966b) concluded that a given dose of adrenaline may produce two qualitatively distinct types of local reactions, depending upon the amount of $ScCl_3$. In another experiment, Selye *et. al.* (1965) found that the blood of rats sensitized with $ScCl_3$ clotted in the blood vessels after death much more rapidly than that of normal controls.

Bland *et al.* (1974) reported that murine tumors induced by irradiation of bone marrow with $^{90}Sr/^{90}Y$ were primitive mesenchymal cell tumors of angiosarcomatous form. These tumors contained tumor giant cells and occasional cells containing acid phosphatase. The latter were interpreted as tissue phagocytes, but they did not manifest osteoclasis. Lymphomas also occurred when the marrow was irradiated.

$ScCl_3$ caused an increasing depression of intestinal tonus and contractility, terminating in complete paralysis of the rabbit ileum, when applied in a dosage range of 10–30 mg (Haley *et al.*, 1962). Such depressant action also occurred with guinea pig preparations, for which the ED_{50} values for blocking the circular and longitudinal muscle contractions were 11.7 and 5.6 mg, respectively. This muscular contraction was postulated to be related to pressure stimulation of the enteric ganglia; thus, scandium appeared to have ganglionic blocking properties. However, experiments with preparations of the superior cervical ganglion of the cat indicated that this is unlikely (Haley *et al.*, 1962). Examples of pathologic changes caused by stable and radioactive isotypes of yttrium are summarized in Table 9-20.

9.11 Genetic Effects

A 15-minute exposure of sprouts of *Crepis capillaris* to a $^{90}Sr/^{90}Y$ solution, containing 1.5×10^{-4}Ci/ml, caused 8% chromosomal rearrangement after 5 h (Shevchenko, 1970). More than 30% rearrangement was observed after 90–180-min exposure. Twenty-four hours after a 180-min exposure, 80% of cells were tetraploid, with arrangements and chromatid aberrations formed during the second cell cycle. Also, somatic cells of *C. capillaris* seeds exposed to internal radiation of 4200 rad from incorporated $^{90}Sr/^{90}Y$ were affected 20 times more than cells exposed to external β irradiation of equal dosage from the same source (Kaplan *et al.*, 1973). The difference was postulated to be due to transmutation.

A positive genotoxic dominant lethal effect of yttrium nitrate on goat sperm was observed *in vivo* (Richardson and Gangolli, 1994). In a study by Brooks and McClellan (1969), Chinese hampsters were injected with various doses of $^{90}Sr/^{90}Y$, and the biological damage was estimated over a prolonged period. In all cases, death appeared to be related to total accumulated radiation dose and cellular depletion of the bone marrow. Changes in bone-marrow cells at each time point included an increase in the number of chromosome breaks per cell as a function of dose accumulated at the time of death and an increase of the bone-marrow mitotic index above control levels at a low total dose but decrease as the dose passed a critical point. The number of chromatid plus isochromatid deletions increased with the elevated injected activity and dose rate. The relative frequency of chromosome exchanges increased with time (Fig. 9-13).

Table 9-20. Pathologic Effects of Stable and Radioactive Yttrium

Compound	Species	Route of administration	Effects	Reference
YCl_3	Rat	Intraperitoneal	Ascites, fibrinopurulent and granulomatous peritonitis, focal hepatic necrosis	Steffee (1959)
YCl_3	Guinea pig	Intravenous	Subcutaneous hemorrhage, renal petechiae, toxic hepatitis	Graca *et al.* (1962)
$^{91}YCl_3$	Rat	Intravenous	Liver focal necrosis giant cells	Magnusson (1963)
Y_2O_3	Rat	Intratracheal	Granulomatous nodules, giant cells, fibroses	Mogilevskaya and Raikhlin (1963)
Y metal	Guinea pig	Subcutaneous	Granulomatous inflammation, multinucleated giant cells, neoplasia	Talbot *et al.* (1968)
^{90}Y-FAP[a]	Beagle dog	Inhalation	Pneumonitas, vasculitis, lung fibrosis, squamous metaplasia, lymphopenia	Hobbs *et al.* (1972)
^{90}Y	Pigs	Catheterization	Renal shrinkage	Zimmerman *et al.* (1995)
^{90}Y	CFW mice	Inhalation	Pneumonitis, adenomas	Lundgren *et al.* (1981)
^{91}Y-FAP	Beagle dog	Inhalation	Pneumonitis, lung fibrosis and carcinoma	Hahn *et al.* (1987)
^{90}Y	Human	Implantation	Bilaterial aneurysm of carotid arteries	McConachie and Jacobson (1994)

[a] FAP, Fused aluminosilicate particles.

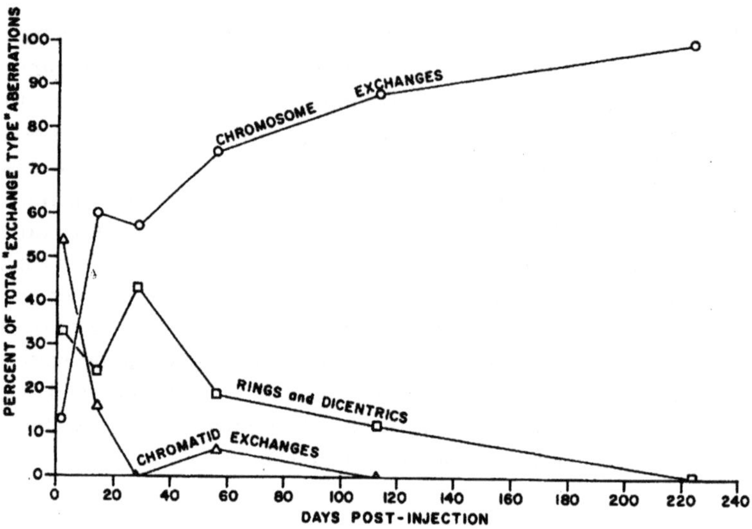

Figure 9-13. Relative frequencies of different types of exchange aberrations in Chinese hamsters given graded doses of ^{90}Y, expressed as percentages of the total number of exchange aberrations, as a function of time after injection. Reprinted from Brooks and McClellan (1969), with permission of the authors and Taylor and Francis, Inc.

An examination of cDNA of the p53 gene in murine tumors of the skin and bone of mice produced by repeated exposure to ^{90}Sr/^{90}Y β-radiation found that 31% of mutations were minute deletions, including aberrant splicing, 5% were insertions of 4–8 base pairs, and 9% were base-pair substitutions, including substitutions at CpG sites (Ootsuyama *et al.*, 1994). All but one of these mutations were in the central region of the p53 gene. Since Ha-ras mutation was absent in these tumors, it was concluded that repeated β-irradiation createed a cell-growth stimulating effect similar to that of ras mutation.

Radioisotope synovectomy of rheumatoid arthritis and of other diseases constitutes an effective alternative to surgical or drug treatment (Gumpel and Stevenson, 1975). The possible chromosomal aberrations (CAs) in patients receiving ^{91}Y β-radiation were evaluated by Cannas *et al.* (1994). A significant increase in CAs was registered after 3 days, 30 days, and 12 months. CAs seemed to decline 1 year after treatment, although not to the control values. Micronuclei (MNi) still showed a small increase in comparison with the control (Table 9-21). The persistently high level of CAs or MNi would indicate the permanent genotoxicity of this kind of therapy. A reasonable suspicion was supported that chromosome damage may be related to development of cancer (Cannas *et al.*, 1994).

McFee *et al.* (1994) quantified cytogenetic damage in marrow cells of mice at various times up to 14 days after injection of ^{90}Y-labeled MAbs. Aberrations,

Table 9-21. Effect of ^{90}Y Radiation of Rheumatoid Arthritis Patients on Chromosomal Aberrations and Micronuclei Formation[a]

Cytogenetic changes	3 days	30 days	12 months
Chromosomal aberrations			
Differences	+2.6	+2.7	+2.2
Percent differences	+147	+192	+200
Micronuclei			
Differences	10.2	+9.2	+15.5
Percent differences	+81	+77	+159
Number of cells with micronuclei			
Differences	+9.0	+7.5	+12.5
Percent differences	+79	+69	+137

[a] Data are compared with average baseline values, $P < 0.001$ and $P < 0.01$. Data reprinted from Cannas *et al.* (1994), with permission of the authors and Chapman & Hall.

predominantly of the chromatid type, were sharply elevated at 24 h post injection and then declined in a curvilinear fashion over 14 days. Micronuclei numbers among polychromatic erythrocytes peaked 3–4 days after treatment and then declined exponentially but remained at higher than expected levels (Fig. 9-14). Sister-chromatid exchanges were roughly double the control rate, with no apparent relation to post-treatment time. Since the evaluation of radiation-induced cytogenetic endpoints is complicated by the dynamic nature of the bone-marrow cell population, reasonable estimates of the dose from internal emitters could be feasible through cytogenetic evaluation (McFee *et al.*, 1994). Further genetic effects of yttrium radioisotopes are presented in Table 9-22.

9.12 Immunologic Effects

Frouin and Ledebt (1913) found that sulfates of yttrium, as well as of lanthanum, neodymium, praseodymium, and thorium, had an agglutinating action *in vivo*. This varied in intensity with the concentration of the metals and the species of corpuscles. The agglutinating action was inhibited by a trace of serum. The salts had an antihemolytic action toward the preparations, but not toward normal sera.

Powdered yttrium oxide and hexaboride administered intratracheally to rats caused a decrease of lysozyme to ~10% of the initial level after 15 days of administration. Immunologic resistance was decreased, and complete recovery did not occur even 6 months after the treatment (Olefir, 1967). Yttrium propionate

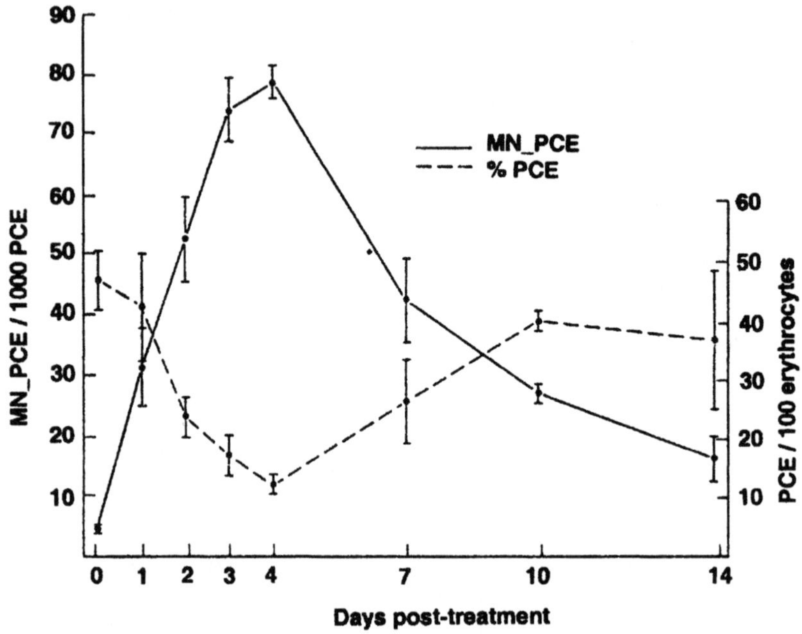

Figure 9-14. Cytogenetic damage in marrow cells of mice after injections of ^{90}Y-labeled MAbs, as indicated by the number of micronuclei and proportions of polychromatic cells among the marrow erythrocytes. Reproduced from McFee *at al.* (1994), with permissin of the authors and Elsevier Science.

Table 9-22. Genetic Effects of Yttrium Radioisotopes

Radioisotope	Species	Effects	Reference
^{90}Sr/^{90}Y	Hamster	Increased number of chromatid and isochromal deletions; frequency of chromosomes exchanges	Brooks and McClellan (1969)
^{90}Y	Salmon	Embryonic chromosome aberrations; multiple mitosis; formation of micronuclei	Migalovskaya (1971)
^{90}Y	*Crepis capillaris*	Increase of double-stroke chromosomes; ablation in somatic cells	Kaplan *et al.* (1973)
^{90}Y	*Crepis capillaris*	Chromosome aberrations in radicle metaphase; Breaks at stage of second mitosis; tetraploid cells	Shevchenko (1970)

increased the number of antibody-producing cells in the spleen and hemolysin titters in the serum of CBA mice immunized with sheep erythrocytes (Zimakov and Zimakova, 1971). The cytotoxicities of dusts of yttrium and cerium oxides as well as several mixed rare earth dusts were compared, and the former induced a higher extent of cell damage and membrane fluidity in guinea pig alveolar macrophages (Zhou *et al.*, 1992).

It was shown that immunotoxin (IT) could be labeled with ^{90}Y by a cyclic anhydride procedure with little overall effect on IT potency, as measured by assays of the inhibition of protein synthesis (Frankel, 1988). Indirect evidence indicated that the galactose binding site of ricin was inhibited 10-fold by its exposure to ^{90}Y. The findings also support the earlier contention that the β-chain-binding and the β-chain-enhancing properties of it may indeed be separable. Radiolabeled IT perhaps may be useful for delivering dual cytotoxic signals to tumors and for the rapid elimination of tumor cells (Frankel, 1988).

The interest in the toxicology of REE, especially after their inhalation, stimulated studies on their effects on excitable cells and on functions of immunocompetent cells. Cerium, lanthanum, and yttrium chlorides changed the morphology and functions of bovine alveolar macrophages (BAM), such as release of tumor necrosis factor (TNF). These effects were demonstrated *in vitro* (Polzer *et al.*, 1993) and *in vivo* (Wiener-Schmuck *et al.*, 1990). The REE inhibited the chemotactic response of BAM in a time- and dose-dependent manner. Whereas the REE alone had no effect, when they were incubated together with zymosan, they impaired release of the superoxide anion by HL-60 macrophages or BAM (Fig. 9-15). During the past few years, evidence has been obtained that particles of high-T_c superconducting (HTS) materials can cause functional and structural disorders in alveolar macrophages, including disturbance of certain immunofunctions. Experiments with BAM and HL-60 cells showed that HTS materials can disturb important immunofunctions of macrophages. The REE components alone can be responsible for the effects on chemotactic mobility and release of chemotactic factors, perhaps by interference with calcium metabolism. Most probably, the induction of interleukin-8 and TNF is due to a combined influence of two or more of the constituents of HTS materials (Polzer *et al.*, 1994). The cytotoxicity of one such material ($YBa_2Cu_3O_{6-7}$) has been shown to be almost the same as that of quartz (Wilczek *et al.*, 1989). Further experiments (Wilczek *et al.*, 1990) found that HTS materials affect the viability of BAM, through release of LDH and *N*-acetylglucosamidase. Also, their effects affect the transfer of fatty acids in phospholipids and triacylglycerols. Apparently, cooper and REE components cause these effects after dissolution of the materials inside the lysosomes. Other experiments found that HTS materials such as $YBaCuO_{3.5}$, (YBa_{111}) and $YBa_2Cu_3O_{6-7}$) (YBa_{123}) are potent inducers of important proinflammatory

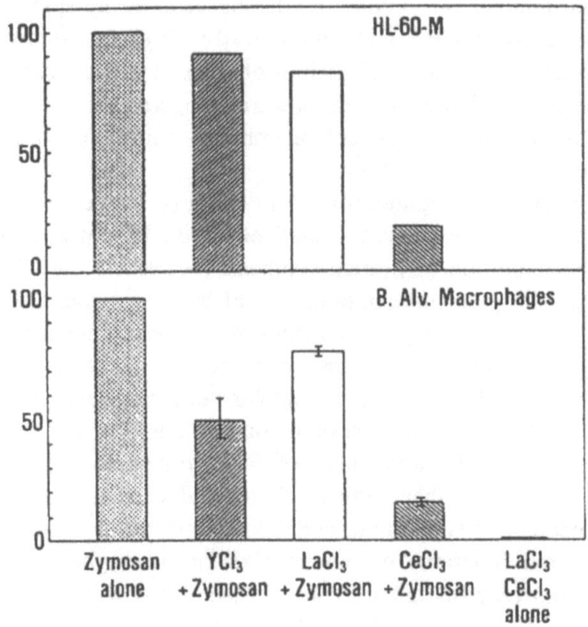

Figure 9-15. Influence of zymosan and yttrium, lanthanum, and cerium chlorides, alone or in combination, on the production of the superoxide anion by macrophage-like differentiated cells Hl-60-M or bovine alveolar macrophages. Reprinted from Polzer *et al.* (1993), with permission of Oxford University Press.

mediators *in vitro*. Particles of these materials should be considered as potentially dangerous when inhaled. In view of their rapid dissolution in biological systems, heavy metal accumulation in liver, kidney, and bone might also occur (Baierl *et al.*, 1992).

An efficient method of treatment of patients with ovarian cancer was found to be the administration of the radioimmunoconjugate ^{90}Y-CITC-DTPA-HMFG1 at a dose of 18.5 mCi/m^2 (Maraveyas *et al.*, 1994). However, a type III, monophasic serum sickness reaction resulted, and the bone-marrow toxicity was the limiting factor. Analysis of sera from most patients found human anti-mouse antibody responses and other reactions, which are believed to be the result of the administration of xenogenic protein to immunocompetent individuals (Maraveyas *et al.*, 1994). The evaluation of the radioimmunoreagent obtained by ^{90}Y labeling of T101, a murine IgG2a MAb that recognizes CD5, indicated a selective induction of cytotoxicity to cells in culture (Naruki *et al.*, 1995).

9.13 Pharmacological Effects

Early pharmacological studies of yttrium in various animal species were performed by Bachem (1907). Steidle and Ding (1929) summarized the results of extensive investigations performed by Ding (1927). Experiments with lower (*Saccharomyces, Spirogyra, Tradescantia, Paramecium*, worms) and higher (mice, frog, rabbit) organisms found that REE nitrates showed toxic, stimulatory, or no effects, depending on concentration. A similar range of effects was obtained with isolated muscle and other tissues.

On the basis of their pharmacological properties and their poor absorption, scandium and its compounds are classified as being slightly toxic. Oral toxicity in mice is perhaps due to a mass effect, i.e., the rupture or blocking of the g.i. tract by an accumulation of inert salts, which are either poorly absorbed or poorly metabolized (Venugopal and Luckey, 1978).

Within the dosage range of 0.5–2 mg/kg $ScCl_3$ produced no observable pharmacological effects in the cat. At a dosage of 5 mg/kg, there was a transient drop in blood pressure, and blood flow decreased slightly (Haley *et al.*, 1962). Complete cardiovascular collapse occurred at 10 mg/kg and was followed by respiratory paralysis. Within the dosage range investigated, $ScCl_3$ did not affect the pharmacological responses to various stimulatory agents. None of its effects could be counteracted or modified by atropine or epinephrine.

In rats, pretreatment with $ScCl_3$ followed by i.m. injection of adrenaline in the neighborhood of the vena cava induced thrombosis in the vena cava (Varga *et al.*, 1967). In some of the treated animals, a deterioration of the venous endothelium was observed and extensive endothelial alterations were noticed after 5 min up to 24 h. No thrombosis and only slight endothelial changes were present in the animals receiving adrenaline alone. A dose of 4 mg of $ScCl_3$ injected into rats increased the severity of endothelial lesions produced by adrenaline alone. The evolution of the lesions of the endothelium produced by adrenaline + $ScCl_3$ is shown in Fig. 9-16; the endothelial changes observed might affect the thrombus formation (Varga *et al.*, 1967).

Intravenous administration of scandium, yttrium, or any of the Ln chlorides plus epinephrine to rats produced topical thrombohemorrhagic lesions (THL) at the sites of epinephrine administration. Depending upon dosage and timing, it was possible to produce such lesions at either the epinephrine injection sites alone or in distant organs also, particularly in the kidney (Gabbiani *et al.*, 1966). The THL were prevented by pretreatment with α-adrenergic blocking agents. $ScCl_3$, and to a lesser extend, YCl_3, produced characteristic highest THL in the kidney and changes limited to the injection site of the dorsal skin (Table 9-23).

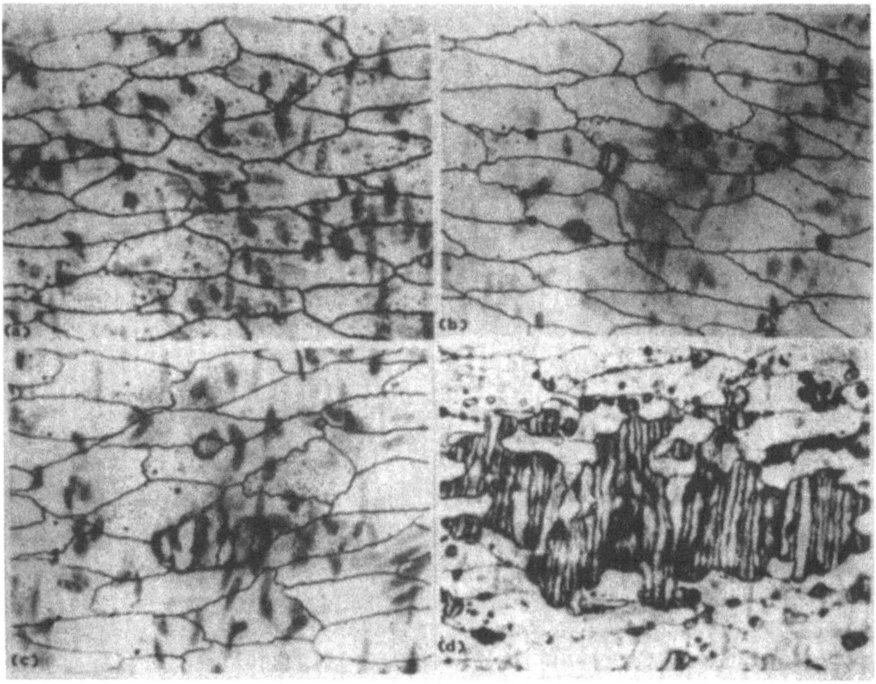

Figure 9-16. Lesions of the endothelium in rats after adrenaline or adrenaline + $ScCl_3$ treatment. (a) Endothelium of vena cava of control; (b) mild lesions, 5 min after adrenaline injection; (c) lesion, 5 min after adrenaline + $ScCl_3$ treatment; (d) serious alteration, 1 h after adrenaline + $ScCl_3$ treatment. Reprinted from Varga *et al.* (1967), with permission of the authors.

Table 9-23. Thrombohemorrhagic Lesions Induced by Combined Treatment with Rare Earth Metal Cations and Epinephrine i.v. Adminstered to Rats[a]

	$ScCl_3$[b]	YCl_3[b]	$LaCl_3$[b]
Thrombohemorrhagic lesions			
Paw	1.8 ± 0.30	1.3 ± 0.26	1.7 ± 0.26
Dorsal skin	2.2 ± 0.29	0.2 ± 0.20	0
Heart	0.1 ± 0.10	0.6 ± 0.22	0
Lung	0.1 ± 0.10	0.6 ± 0.22	0
Kidney	1.9 ± 0.26	0.1 ± 0.13	0
Percent mortality	20	40	0

[a] Thrombohemorrhagic lesions were rated on a scale of 0–3. Data reprinted from Gabbiani *et al.* (1966), with permission of the American Society for Pharmacology and Experimental Therapeutics.

[b] In addition, all animals received epinephrine.

9.14 Toxic Effects on Humans

Metal poisoning is the fourth most frequent type of poison exposure among 10 categories of poisoning due to nonpharmaceutical agents. Several metals cause severe g.i. injuries and other injuries in humans. A review of the usual methods of diagnosis and treatment of metal poisoning includes 29 metals, but not scandium and yttrium (Ellenhorn and Barceloux, 1988). Effects of acute intoxication are classified as g.i. symptoms, damage to the respiratory tract, and cardiovascular and nervous system effects. Among group IIIA metal, elevated toxicity occurs with an increase of electronegativity. These metals affect the blood and blood-producing organs, and their effects are mediated by their high affinity for amino acids and imino and SH groups (Geldmacher-von Mallincrodt, 1991).

$ScCl_3$ is described in the U.S. Environmental Protection Agency Toxic Substances Control Act (TSCA) and considered a poison by the intraperitoneal route but moderately toxic by ingestion (Sax and Lewis, 1992). No ocular injuries caused by REE in humans have been reported so far (Grant and Schuman, 1993). Effective reversal of the toxic effects of $ScCl_3$ on corneal hydration and mucoprotein extractability may be obtained in contaminated human eyes, if early irrigation with NaEDTA is carried out (Grant and Kern, 1956b).

Clinical manifestations of poisoning as a result of yttrium chloride ingestion occurred in two stages (Filov and Khudolei, 1993). The first stage was characterized by dyspnea and pulmonary edema. The second stage was characterized by oedema of the liver, congestion of the portal vein, subpleural hemorrhages, and hypoemia of the lungs.

An investigation of the potential inhalation hazard associated with TE emissions from coal-fired power plants showed that scandium has a low emission and deposition ratio, but a relatively high respirability ratio (Ragaini and Ondow, 1976). This is because scandium, like iron and aluminum, is predominantly associated with large silicate fly-ash particles.

Varous therapeutic trials found that ^{90}Y escapes from radioconjugates and is taken up by the skeleton. The resultant bone-marrow hematopoietic toxicity was dose-limiting. In experiments with mice, ^{90}Y toxicity caused decreases in the platelet and white blood cell counts. Because the bone and marrow of mice are much smaller than those of humans, the results obtained with mice probably underestimate the importance of ^{90}Y escape in the skeleton of patients. The efficacies of two high-affinity chelators, methylbenzyldiethylenetriaminepenta-acetic acid (MX-DTPA) and bromoacetamidobenzyl-1,4,7,10-tetraazocyclododecane-*N*,*N*′,*N*″,*N*‴-tetraacetic acid (BAD), were compared by DeNardo *et al.*, (1994).

Radiolabeled immunoglobulin therapy (RIT), a new method of cancer treatment, is dose-limited owing to side effects of bone-marrow toxicity and hematopoietic stem-cell damage. Also, platelet consumption in irradiated normal liver tissues and apoptosis occur. Investigation of the effects of chelate immunoglobulins labeled with ^{90}Y used for therapy concluded that the bone-seeking properties of free ^{90}Y may not be a major component of the hematologic damage caused by yttrium-labeled immunoglobulins (Vriesendorp *et al.*, 1996). The observed effects in human subjects whose occupations involved radium dial painting were compared with the osteosarcoma in dogs, which is the most common late side effect resulting from radioactive irradiation. After high cumulative activities of bone-seeking metals were given in fractionated schedules, many dogs developed bone-marrow damage in the form of myeloproliferative disorders or bone-marrow aplasia. In patients with Hodgkin's disease receiving relatively low doses of RIT, the hematopoietic recovery (granulocyte and platelet levels) was usually not up to pre-RIT levels at the time of the second RIT cycle (Vriesendorp *et al.*, 1996). When patients were treated with low activities of RIT (20 mCi of ^{90}Y-labeled antiferritin), the ratio between the highest and the lowest granulocyte or platelet level per cycle appeared to be lower for the second RIT cycle (Fig. 9-17).

Estimates of human exposure to soil indicate the potential risks from toxicants released into the environment. Evaluation of average daily soil ingestion in young children gave a mean of 85 mg Y^{3+}/day, as compared to values of 153–459 mg/day for aluminum, silicon, titanium, and vanadium (Calabrese and Theil, 1989). Calculations of soil ingestion over a lifetime were made, based on levels determined in children (Sedman and Mahmood, 1994).

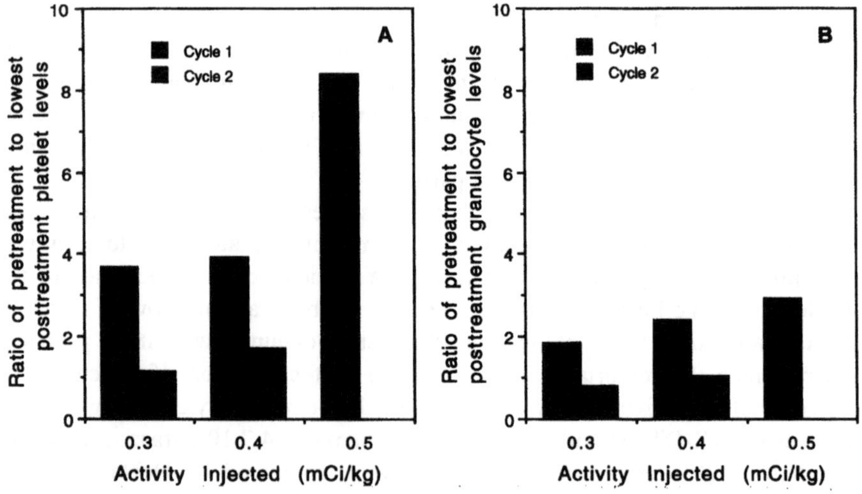

Figure 9-17. Platelet (A) and granulocyte (B) levels after first and second cycle of ^{90}Y-labeled antiferritin therapy for Hodgkin's disease. Bars represent averages of five or more patients. Reprinted from Vriesendrop *et al.* (1996), with permission of the authors and the International Society for Experimental Hematology.

"Lifestyle" factors, such as smoking and alcohol and drug abuse, may have a direct and even indirect influence on toxicity. Cigarette smoke and cigarette ash contain several toxic metals, which may have pulmonary effects and reduce the intake of essential minerals from the diet. The hazardous effects of smoking are well demonstrated and are correlated to the content of numerous toxic TE in tobacco and the inhaled smoke. Cigarette tobacco of some American brands contained up to 2.5 mg Sc^{3+}/kg. This figure was correlated with higher concentrations of aluminum, bromine, titanium, and several other metals of potential health concern. About 60–70% of scandium detected in cigarette tobacco was found to be transferred into the smoke (Iskander *et al.*, 1986). Fewer data exist about the concentration of yttrium in tobacco and in other products consumed by humans and animals.

The pharmacological and toxicological effects of metal salts that are used effectively to inhibit sweating have been investigated (Reller and Luedders, 1977), and the capacity of various metal salts to form polymeric, gelatinous hydroxide precipitates at or below physiological pH has been evaluated. Principal compounds in use are $Al_2(OH)_5Cl{\cdot}2H_2O$ and $ZrO(OH)Cl{\cdot}3H_2O$. Backs of male volunteers were used to screen the antiperspirant effectiveness of various other agents. The metal chlorides investigated fell into two categories: effective (Al, Cr, Fe, Ga, In, Sc, Sn, and Zr) and ineffective (Ca, Co, Cu, La, Mg, Mn, Mo, Ni, Y, and Zn) antiperspirant agents. Several metal salts that did not produce an effect within 1 h did show significant effectiveness the following day, when the salts effective at 1 h were checked for persistance of antiperspirant effect (Table 9-24).

Table 9-24. Immediate (1 h) and Delayed (24 h) Antiperspirant Effectiveness on Humans of Metal Ions in 0.25 M Aqueous Salt Solutions[a]

			Effect	
Salt	pK_1[b]	Reference salt	Immediate	Delayed
$CaCl_2$	10.3	$AlCl_3$	0	0
$CuCl_2$	7.3	$AlCl_3$	0	66
$AlCl_3$	4.4	ACH[c]	88	83
$ScCl_3$	5.0	$AlCl_3$	80	62
YCl_3	9.1	$AlCl_3$	5	6

[a] The antiperspirant effectiveness in aqueous solutions was measured with back tests over a period of <4 h. The concentrations of metal ions varied between 0.25 and 0.95 moles of metal ion per liter of solution. Data reprinted from Reller and Luedders (1977). With permission of Wiley and Sons.

[b] Negative log of dissociation constant.

[c] ACH, $Al_2(OH)_5Cl \cdot H_2O$.

This delayed effect was distinctly different from the aluminum type of antiperspirant effect. No explanation was given for these differences.

A significant exposure to REE occurs in an occupational setting, such as in the production of carbon-arc lamps for various purposes. Also, increased uses of yttrium and other REE in the production of lasers, phosphors, semiconductors, color televisions, and several other industrial devices may represent health hazards (Brown *et al.*, 1990; Haley, 1991; Nasisse *et al.*, 1882; Hrebinko, 1996). More information about toxic effects of scandium and yttrium on humans is presented in the following sections.

9.15 Carcinogenicity, Mutagenicity, and Teratogenicity

The covalent binding of a carcinogen to DNA represents classical mechanism of tumor induction in mammals. It is presumed that this mechanism does not operate with metals, since no covalent binding of these agents to DNA occurs. However, somatic mutations, typical consequence of DNA damage, may be induced by metals (Schlatter, 1985).

Although the concentrations of scandium and yttrium in nature are relatively low, these elements may appear with increasing frequency in urban and rural environments. The burning of fossil fuels, industrial activities, and urbanization are constant sources of pollution of the atmosphere, water, soil, and food by chemical elements (Brakhnova, 1975). One-third to two-thirds of all cancers are considered to be related to the innumerable exogenous chemicals present in our modern environment. Such may be the scandium and yttrium metals, their inorganic and organic compounds, radioisotopes, and REE minerals which contain numerous TE. Many of them, such as asbestos, may be considered potentially carcinogenic inorganic chemicals. Several metals have been reported to be mutagenic or co-mutagenic *in vivo*, or at physiological concentrations *in vitro*. However, some of these micronutrients may possess antimutagenic or co-mutagenic activity *in vivo* and *in vitro* (Chow, 1990). Diverse metals, including ytterbium and thorium, have shown positive or equivocal results in assessments of chromosome damage, mutagenicity, and teratogenicity (Gebhart and Rossman, 1991).

The assessment of risk to humans of established (As, Be, Cr^{6+}, and Ni), suspected (Co, Cu, Pb, Li, and Zn), or possible carcinogens (Gd, Ge, In, Pd, Rh, Te, Ti, V, Yb, and perhaps other TE) refers to the following classes of exposure: iatrogenic, occupational, and general environmental. Few, if any, attempts have been made to assess the carcinogenicity of REE through epidemiological studies, investigations of dose–response relationships, animal studies, or short-term assays, in the presence or absence of carcinogenic cofactors (Al, As, Co, Cu, Mn, Ni, Se, and Zn) (Gilman and Swierenga, 1984). Carcinogenic effects of yttrium radioisotopes have been observed in dogs (Table 9-25). Furst and Radding (1979) included scandium and yttrium in the group of carcinogenic TE but a later review (Furst, 1987) did not maintain this claim. England *et al.*, (1989) made an attempt to extrapolate the existing data on metal toxicities in animal species to humans. Indicators of toxicity for biological systems of increasing levels of complexity were chosen, and numerical values of these indicators were normalized to a scale of 100 for Mg^{2+} (essentially nontoxic) and 0 for Cd^{2+} (very toxic). For yttrium, five physical-chemical properties of the metal indicators were employed, and the predicted normalized toxicity of this metal by extropolation to humans was 48. Comparatively, values of 3 and 112, were calculated for Be^{3+} and Sr^{2} respectively. No explanation was given for this approach.

The analysis of smelter slags of copper and nickel and of feldspar and novaculite, which are used as substitutes for sand in abrasive blasting operations, showed up to 800 mg Sc^{3+}/kg and up to 68 mg Y^{3+}/kg and very high concentrations of other toxic metals. Intratracheal instillation of these materials into male rats caused alveolar wall fibrosis, which may indicate a carcinogenic potential (Stettler *et al.*, 1988).

Teratogenic effects in animals and humans are not well defined. The brain and eyes are the most susceptible organs to teratogenic agents. Chemicals are considered teratogenic only when they disrupt development of the fetus at doses not toxic to the adult. Many TE [As, Bi, Cd, Hg (organomercury species), Ni, Pb, Sb, and Se] are considered teratogenic, but no data are available for other TE (Schardein, 1993). Yttrium-90 showed no teratogenic effects on pregnant and lactating rabbits, whereas ^{90}Sr produced malformations of the long bones in both mice and rats, as well as skeletal, tail, and eye defects in rats (Kidman *et al.*, 1951). A single intratesticular injection of yttrium nitrate to goats, at the lowest tissue dose of 5 mg/kg, reduced spermatogenesis (Richardson and Gangolli, 1994).

Schroeder and Mitchener (1971) evaluated possible inborn toxic effects of several metals, in terms of growth and survival of mice. These indices indicated

Table 9-25. Carcinogenic Effects of Yttrium Radioisotopes, Observed in Dogs

Compound	Effect	Reference
$^{91}YCl_3$	Hemangiosarcoma	Muggenburg *et al.* (1986)
^{90}Y-FAP[a]	Lymphosarcoma	Muggenburg *et al.* (1986)
$^{91}YCl_3$	Squamous cell carcinoma of nasal cavity	Boecker *et al.* (1986)
^{91}Y	Tracheobronchial lymph node tumor	Hahn *et al.* (1986)

[a] FAP, Fused aluminosilicate particles.

the following decreasing order of toxicity: Ga > Y > Sc > In ≈ Cr > Pd > Rh. Yttrium nitrate added to drinking water at a concentration of 5 mg/l was somewhat carcinogenic, whereas scandium, gallium, and indium seemed to be inert in this respect. An examination of the effect of yttrium chloride on *Salmonella typhimurium* strains did not find any mutagenic effects (Hiroaki *et al.*, 1987). Mutagenic effects of nitrates of scandium, yttrium, and other REE on human and other mammalian somatic cells were investigated by Mu and Shi (1985). They reported that when concentrations of 2.5–266 mg/kg of the salts were added to the culture medium of human peripheral blood lymphocytes, no differences from controls were observed. Also, no change in the rate of chromosomal aberrations or the frequency of sister chromatid exchanges and no teratogenic effects were found. Chinese investigators claimed that no teratogenic effects on rats and their offspring were caused by a mixture of REE nitrates (which included scandium and yttrium) at concentrations varying between 16 and 2000 mg/kg (Zhong *et al.*, 1992).

9.16 Radiotoxicity

Ariel and Galey (1964) investigated the effect of intralymphatic administration of cancer-destroying agents [anticancer drugs such as phenylalanine mustard (melphalan) and methotrexate] and radioisotopes (^{46}Sc, ^{90}Y, ^{131}I, and ^{198}Au) as a means of destroying cancerous cells within the lymphatic system and lymph nodes. Administration of radioisotopes resulted in relief of symptoms, and the effectiveness of the treatment was confirmed by histological evidence of radiologic effects. The intralymphatic administration of a small dose of ^{46}Sc in men produced multiple areas of micronecrosis in lymph nodes, which were removed 2 days later. Histological alterations and radioautographs indicated the feasibility of administering certain radioisotopes.

During WWII studies on plutonium, one of the main fission products, Van Middlesworth (1947) and others (Stannard, 1988) extensively investigated the comparative behavior of ^{239}Pu, ^{89}Sr, and ^{88}Y in bone. The fission product series showed considerable accumulation and varying degrees of retention by the skeleton (Fig. 9-18). The rates of elimination after accumulation in the skeleton were less than the rates of radioactive decay (Hamilton, 1949). Most of the research regarding effects of fission products and other components of fallout on aquatic organisms has been performed in laboratory experiments. Figure 9-19 provides a chronological overview of the work done in the late 1950s and the 1960s. CFs of several radioisotopes, and in particular ^{90}Y and ^{91}Y, are higher in fresh water plants, and lower in invertebrates and fish (Table 9-26). Nonetheless, a CF of 10,000 as measured in egg membranes of fish (Polikarpov, 1966). Accumulation of radioactivity by plants from soil is dependent on many factors.

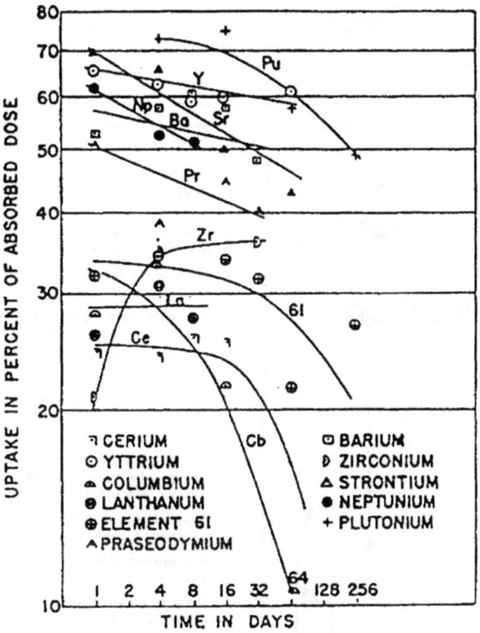

Figure 9-18. Deposition of carrier-free fission products in the skeleton of the rat after parenteral administration. Reprinted from Hamilton (1949), with permission of the Massachusetts Medical Society.

A coefficient (ratio of radiation concentration in plants to that in soil) of ^{91}Y accumulation from soil by subterranean parts of pea and barley plants varied between 0.08 and 0.80, depending upon the type of soil; the value of this coefficient for ^{90}Sr varied between 0.86 and 6.8 (Kulikov and Molchanova, 1982). Fission products (^{91}Y, ^{89,90}Sr), as well as neutron-induced radioisotopes (^{46}Sc, ^{88}Y), were found in plants grown on ejecta during a 3-year period following the detonation in 1962 of a subterranean thermonuclear device in Nevada (Romney *et al.*, 1967). As a result of several French and Chinese nuclear tests in the 1960s, 17 γ-emitting radioisotopes, including ^{46}Sc and ^{88}Y, were measured in different species of Pacific salmon (Jenkins, 1969). During World War II, plutonium was produced at the Hanford site, in Washington State for the U.S. nuclear weapons program. Residues of ^{90}Y, ^{46}Sc, and other nonvolatile radioisotopes were released in the period 1944–1971 into the Columbia River an into the atmosphere (Heeb *et al.*, 1996). Yttrium, like beryllium, cerium, selenium, and vanadium (although in different proportions) was taken up in calcified tissues and teeth of rats; these radioisotopes were retained in the molars in an almost constant amount for prolonged periods (Thomassen and Leicester,

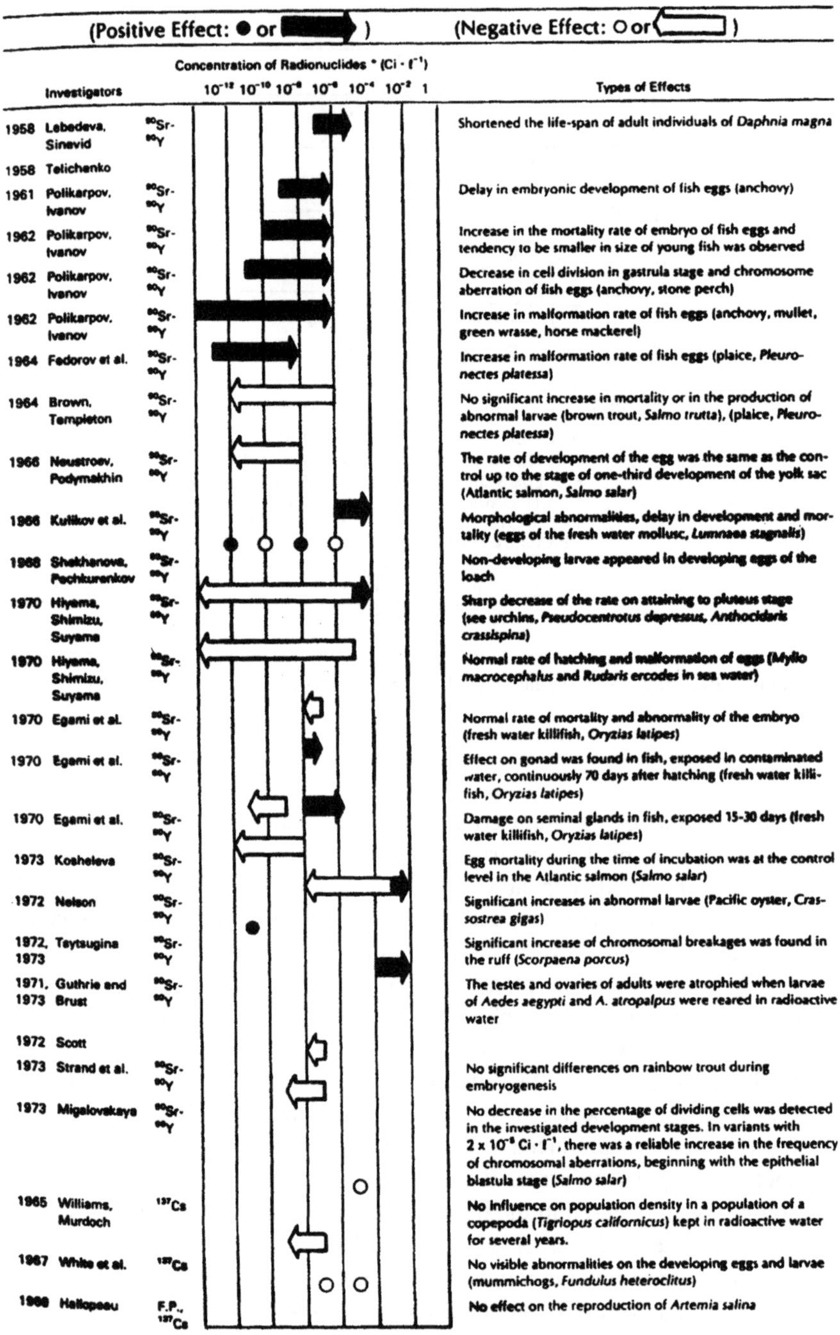

Figure 9-19. Summary of experimental studies on the toxic effects of ^{90}Sr and ^{90}Y on aquatic organisms. Reprinted from Stannard (1988), with permission of Dr. J. N. Stannard.

Table 9-26. Concentration Factors (CFs) of ^{90}Y and ^{91}Y in Various Groups of Organisms

Organisms	Range	Mean	Reference[a]
Freshwater plants	920–40,000	14,440	1, 2
Freshwater animals	123–2020	765	1, 2
Marine algae	160–900		1, 2
Marine animals	40–250		1, 2
Marine fish	10–250	20	1, 2
Freshwater fish	30–210		1, 2
Phytoplankton		100	3
Zooplankton		1	3

[a] References: 1, Timofeeva-Resovskaya, 1963; 2, Polikarpov, 1966; 3, Santschi and Honeyman, 1989.

1964). Yttrium-91, which is initially deposited in liver, may be slowly transferred to bone at a later time (Speer *et al.*, 1961) and deposited in the teeth (Lloyd, 1961; Ramsden, 1961). Chronic ^{90}Y β-radiation of mouse skin tumors caused a weak and positive characteristic of a stage 1 tumor promoter. (Mitchel and Trivedi, 1992). This promoting activity was considered to be the result of stimulation of a natural process. The biological half-times of ^{46}Sc retention in the skeleton of rats is 3410 days, which is scarcely higher than those for other radioisotopes (^{241}Am, ^{45}Ca, ^{244}Cm, ^{22}Na, ^{239}Pu, ^{226}Ra, and ^{85}Sr), with the exception of ^{65}Zn (half-time of 598 days). Whether the rates of loss of bone-seeking metals from the skeleton determined in experimental animals may be similar in humans cannot yet be answered with confidence (Taylor, 1983). Irradiation of rats with sublethal doses of γ-rays caused substantial changes in the dynamics of TE in the 30 days after irradiation (Marinov *et al.*, 1981). The dynamics of scandium and several other TE in spleen and liver of irradiated animals are shown in Fig. 9-20. Chemical and autoradiographic studies showed that the deposition of ^{91}Y in the bones of rabbits of different ages, following a single i.v. injection of a carrier-free dose, was extremely rapid (Rayner *et al.*, 1953). Yttrium-91 was concentrated in the site of active bone growth beneath the epiphysis, but not beneath the endosteum or periosteum. New bones in young animals formed after the original injection contained traces of ^{91}Y, presumably due to the capacity of bone tissues to complex with ^{91}Y present in extremely low levels in blood (Rayner *et al.*, 1953). Gastrointestinal irradiation of lambs with an insoluble ^{90}Y-tagged sand showed after 15 days, at autopsy, severe damage in the ventral compartment of the rumen (Sasser *et al.*, 1972). Typical changes in the rumen were depapillation, denudation, and superficial necrosis of the mucosa. With increasing levels of ^{91}Y sand in the diet, the size and intensity of the necrosis and the severity of the injury increased. Figure 9-21 indicates a decrease of feed consumption with an increase in the amount ingested ^{90}Y, which was reflected also in a significant loss of body weight.

Investigation of the distribution of single radioisotopes injected intraperitoneally into meadow voles (*Microtus pennsylvanicus*) yielded exponential curves consistent with a one-component model of distribution in the case of ^{88}Y and a two-component model in the case of ^{46}Sc and ^{85}Sr (Morimoto *et al.*, 1985; Fig. 9-22; Table 9-27). The first or fast component k_1 was assumed to represent the rapid elimination of extracellular radioisotopes via the urine and feces. The second or slow component k_2 was assumed to represent the loss of radioisotopes incorporated into tissues. However, extremely high variability was observed for ^{46}Sc, which may be attributed to experimental errors or behavioral and physiological variables.

Beagles exposed to aerosols of ^{90}Y in fused clay particles were used to study the radiation dose to the g.i. tract (Hahn *et al.*, 1975). The daily fecal excretion of ^{90}Y indicated a short-term retention of ^{90}Y in the g.i. tract (Fig. 9-23). Focal colonic lesions, g.i. lesions, focal colitis, nasal dermatitis, and radiation pneumonitis were observed at different times after inhalation of ^{90}Y. Inhalation of ^{90}Y, ^{144}Ce, and ^{143}Pr by mice resulted in a relatively equal distribution of the radioactivity in the trachea and the g.i. tract (Gensicke *et al.*, 1966). The lung clearance of ^{91}Y was 19 days, as compared to 60 days in the case ^{144}Ce. The maximal accumulation of ^{91}Y in the liver occurred on the 1st day after radioactivity inhalation and was followed by a steady decrease over the next 14 days; the accumulation in the skeleton reached a maximum on the 14th day. Exposure of beagle dogs to aerosols of ^{91}Y in fused aluminosilicate particles caused initial lung burdens (Hahn *et al.*, 1994; Fig. 9-24). The cumulative dose of 990–55,000 rads of β-radiation from ^{90}Y in beagle dogs caused pneumonitis and deaths, which occurred at 60–163 days. Hobbs *et al.*, (1972) emphasized the importance of dose rate on acute radiation damage to the lung as illustrated in Fig. 9-25.

Mice treated with Sc^{3+} displayed a sensitization to ^{90}Co ray radiation. A reduction of survival time from 12 days (control animals, which were only irradiated) to 9.5 and 8.0 days was observed with 32.5 and 65 mg Sc^{3+}/kg, respectively (Floersheim, 1995; Fig. 9-26). A synergistic type of interaction of several metals and radiation was postulated to be the result of catalytic action of these metals on free-radical-mediated reactions. Also, this type of reaction would be the result of metal oxidation of endogenous SH compounds and variations in redox potentials. An apparent role of beryllium and indium as well as of lanthanum, cerium, and scandium as sensitizers in particular situations of radiation-induced tumor response in human xenografts was suggested, and the need for further study of these effects for therapeutic options was underlined (Floersheim, 1995)

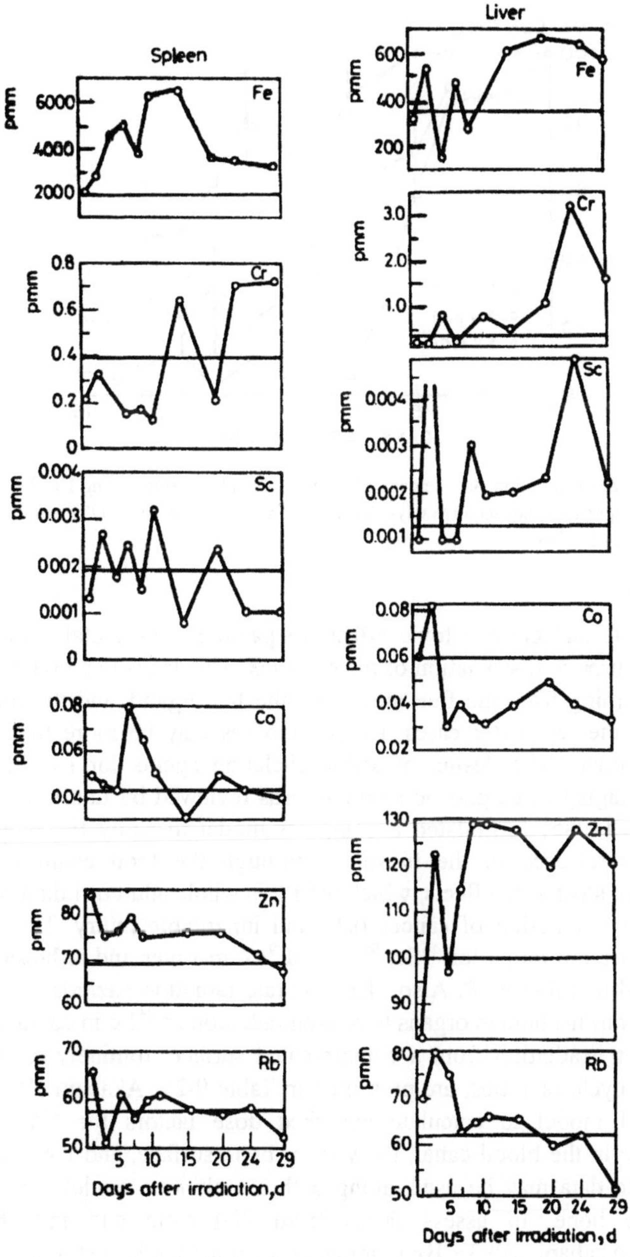

Figure 9-20. Dynamics of iron, chromium, scandium, cobalt, zinc, and rubidium in spleen and liver of rats irradiated with a sublethal dose of γ-rays during the 30 days after irradiation. Results for controls are denoted by the horizontal lines on the graphs. Reprinted from Marinov *et al.* (1981), with permission of the authors and Akademiai Kiado.

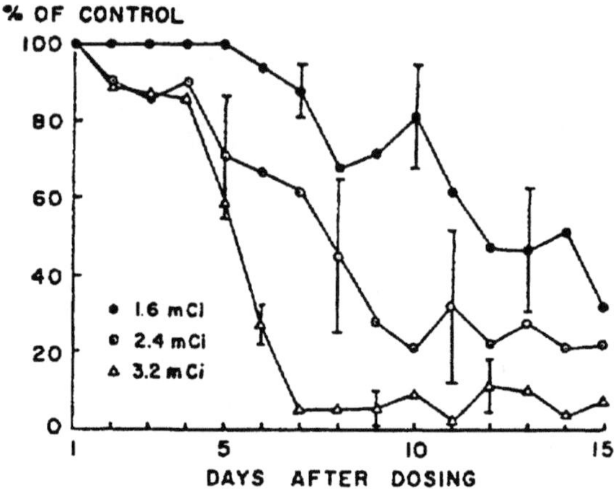

Figure 9-21. Feed consumption, as a percent of control, by lambs after ingesting 1.6, 2.4, and 3.2 mCi ^{90}Y/kg body weight for 3 consecutive days. Reprinted from Sasser *et al.* (1972), with permission of the Health Physics Society.

The acute and chronic toxic effects of paramagnetic metal complexes are accentuated through dissociation of the complex. This leads to a higher degree of toxicity stemming from the free metal ion, the free ligand, and the intact metal complex. In the two latter cases, the metabolites may be more toxic than the parent compound. The design of stable chelating agents for use in magnetic resonance imaging in diagnostic medicine was reviewed by Laufer (1987).

Laurens (1985) formulated a computer model to study the movement of radioisotopes released in the biosphere through the food chain to humans. Calculation of stochastic effects, which refer to a whole adult or infant body dose, permitted the estimation of cancer risk, and inheritable injury. The calculated dose/concentration ratios for ^{46}Sc, ^{90}Y, and ^{91}Y in water and airborne effluents are presented in Table 9-28. Also, the dose-rate radiation hazards from external exposure of various human organs to photon radiation of ^{46}Sc in contaminated air and water, or irradiation from a contaminated surface, routinely released from nuclear fuel cycle facilities, are presented in Table 9-29. Akabani (1993) used a mathematical model to calculate absorbed dose factors for ^{90}Y and other radioisotopes in the blood canal, the wall or bone surface, and the bone matrix. The tabulated data may be used, along with a biokinetic model for radioactive materials in bone, to assess doses from diagnostic and radiotherapeutic procedures (Akabani, 1993). By employing a Monte Carlo method, Cross *et al.* (1992a) calculated β-ray dose distributions at an air–water interface. Distributions were calculated for more than 130 β-emitters, including ^{46}Sc, ^{47}Sc, ^{49}Sc, ^{90}Y, ^{91}Y, and $^{90}Sr/^{90}Y$. The depth–dose distributions were calculated for arbitrarily

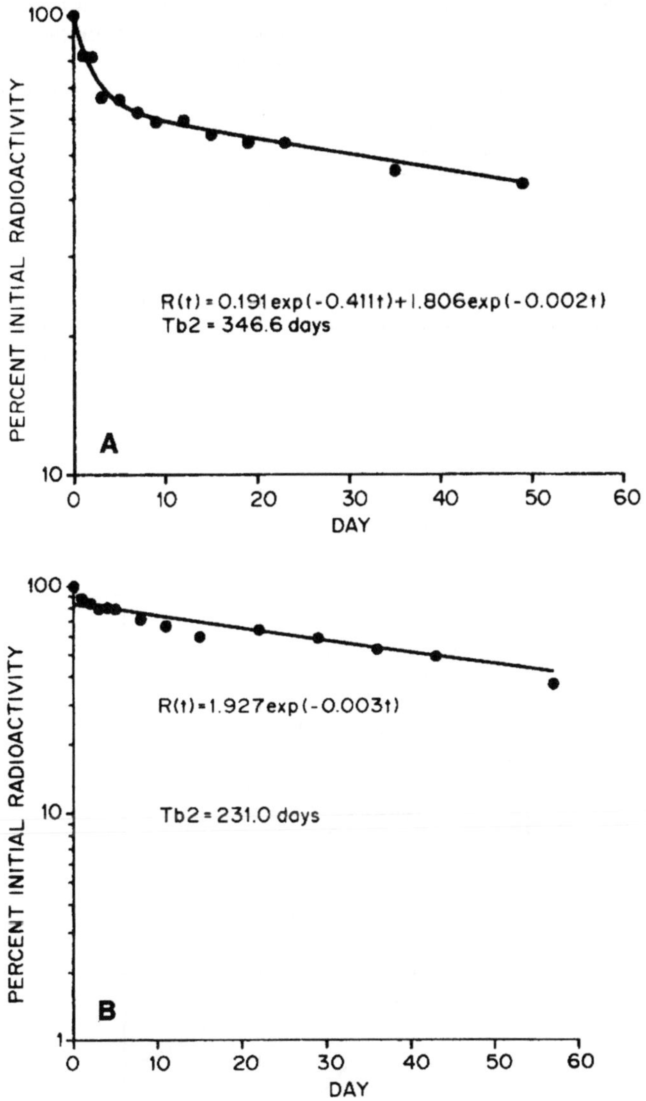

Figure 9-22. Whole-body retention curves for ^{46}Sc (A) and ^{88}Y (B), described by two- and one-compartment exponential elimination functions, respectively, and slow-component biological half-lives (Tb_2) of ^{46}Sc and ^{86}Y in selected meadow voles (*Microtus pennsylvanicus*). Data are based on individuals for which Tb_2 was the closest to the average calculated. Reprinted from Morimoto *et al.* (1985), with permission of the Health Physics Society.

Table 9-27. Biological Elimination Constants (k), Biological Half-lives (Tb), and Percent of Radioactivity in the Slow Component for Radioisotopes in Meadow Voles in the Laboratory[a]

	Biological elimination constant, (days^{-1})		Biological half-life (days)		Activity in slow component
Radioisotope	k_1	k_2	Tb_1	Tb_2	(%)
^{46}Sc	0.373	0.002	1.9	346.6	54
^{88}Y	—	0.003	—	231.0	72
^{85}Sr	1.194	0.021	0.7	33.0	12

[a] Values were derived from the retention functions for individual animals. Data reprinted from Morimoto *et al.* (1985), with permission of Williams and Wilkins.

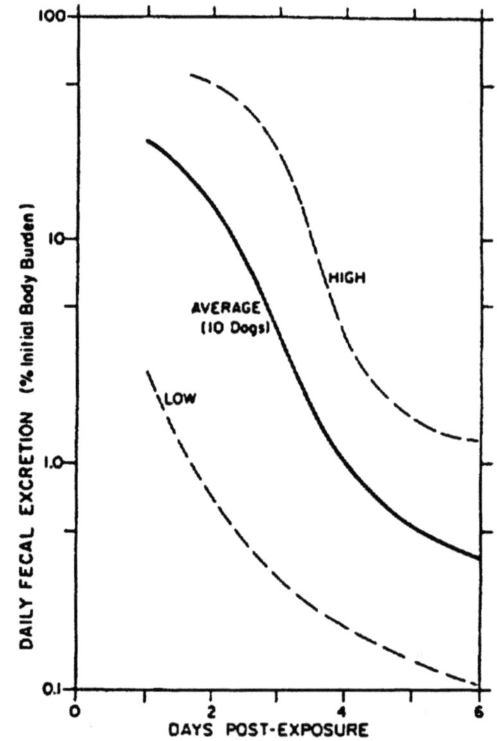

Figure 9-23. Daily fecal excretion ^{90}Y by dogs after inhalation of ^{90}Y in fused clay particles. Data were corrected for radioactive decay. Reprinted from Hahn *et al.* (1975), with permission of Academic Press.

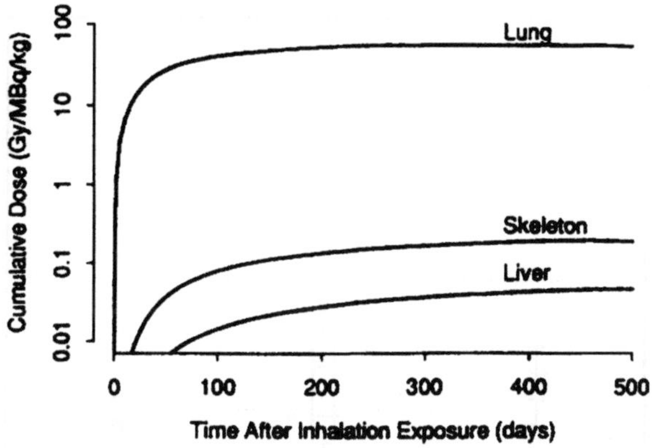

Figure 9-24. Doses to lung, skeleton, and liver in beagle dogs after inhalation of ^{91}Y in aluminosilicate particles, normalized to an initial lung burden of 1 MBq ^{90}Y/kg body weight and average retention. Reprinted from Hayn *et al.* (1994), with permission of the authors and the British Occupational Hygiene Society.

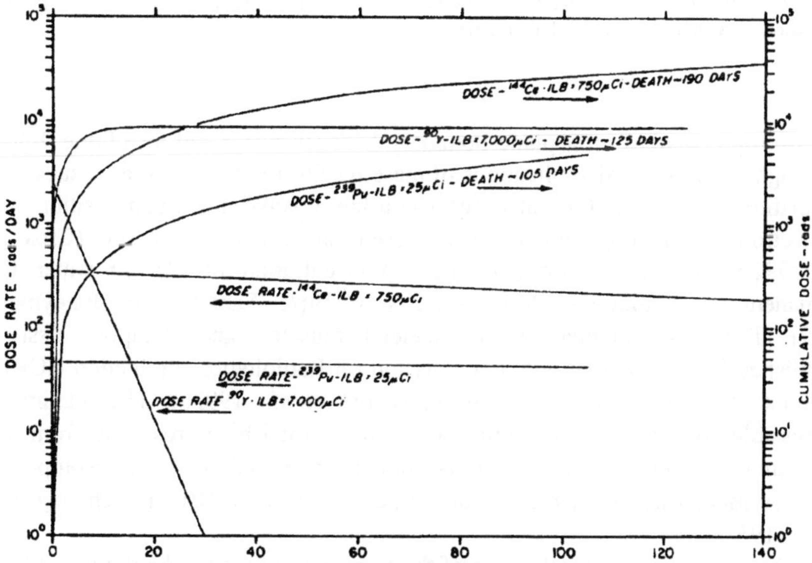

Figure 9-25. Dose and dose rate to a lung for 10-kg dogs after inhalation of an initial lung burden of 7000 μCi ^{90}Y, 25 μCi ^{239}Pu oxide, or 750 μCi ^{144}Ce in fused clay particles. Reprinted from Hobbs *et al.* (1972), with permission of the authors and Radiation Research Laboratories.

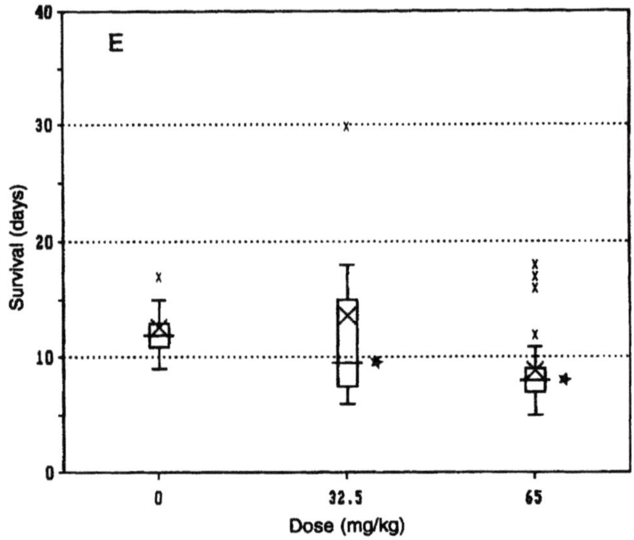

Figure 9-26. Survival of groups of mice irradiated with 10.5-Gy ^{60}Co γ-rays and treated with different doses of scandium sulfate. The horizontal bars correspond to the median survival time; the box shows the 25% and 75% quantities and contains 50% of the individual survival times, X, mean survival times; x, outliers. The stars show significant ($P < 0.01$) differences compared with the respective untreated control group according to the log-rank test. Reprinted from Floersheim (1995), with permission of Radiation Research Laboratories.

thin isotropic sources on a plane boundary between water and air (Table 9-30). Although most practical applications of these results are in relation to tissues, the calculations were made for water, because it has a well-defined composition and an electron dose attenuation that is very close to those of various tissues (Cross *et al.*, 1992b). The maximum permissible concentration of ^{90}Y in water for permanent consumption of food is 2.2×10^4 Bq/l (Santchi and Honeyman, 1989). The dose attenuation in "skeletal muscle" and "adipose tissue" established by the International Commission on Radiological Protection (ICRP) is about 2% lower than that in water (Cross *et al.*, 1992a). The maximum permissible concentration of radioisotopes in the total body, resulting from the exposure of the subject to maximum permitted concentrations of radioisotopes in water or food, when deposited in the critical organ is 9–10 μCi (Schwieg and Turba, 1961).

The dose absorbed by the skin of the fingers as a result of handling medical and laboratory containers (e.g., vials, and test tubes) was calculated by Beddoe and Kelly (1994). Beta doses ranging from 77 to 135 Gy/h per MBq/g emitted by ^{90}Y were estimated at a depth of 270 μm (a typical basal cell depth in the finger)

Table 9-28. Dose/Concentration Ratios Calculated from a Computer Model to Estimate Stochastic and Nonstochastic Effects of ^{46}Sc, ^{90}Y, and ^{91}Y in the Atmosphere and Water on Human Infants, and Adults[a]

			Dose/concentration ratio (Sv a^{-1}/Bg m^{-3})		
Medium	Adult or infant	Organ(s)	^{46}Sc	^{90}Y	^{91}Y
Atmosphere	Adult	Internal	0.45×10^2	0.21×10^2	0.80×10^2
Atmosphere	Adult	Internal	0.45×10^2	0.0	0.72×10^6
Atmosphere	Infant	Internal	0.16×10^2	0.13×10^2	0.40×10^2
Atmosphere	Infant	External	0.60×10^3	0.0	0.72×10^6
Water	Adult	Internal	0.29×10^5	0.42×10^5	0.53×10^5
Water	Adult	External	0.98×10^6	0.0	0.12×10^8
Water	Infant	Internal	0.73×10^6	0.12×10^5	0.18×10^5
Water	Infant	External	0.98×10^6	0.0	0.12×10^8
Atmosphere[b]	Adult	Large intestine	—	0.23×10^1	0.91×10^1
Water[b]	Adult	Large intestine	—	0.46×10^4	0.61×10^4
Atmosphere[b]	Adult	Skin	0.81×10^3	0.0	0.18×10^2
Water[b]	Adult	Skin	0.97×10^5	0.0	0.40×10^5

[a] Data reprinted from Laurens (1985), with permission of the Atomic Energy of Canada Ltd.
[b] Nonstochastic.

when finger protection is worn. Based on annual permissible dose rates, it was recommended that a minimum of 5-mm-thick perplex finger protection should be used by clinicians who inject specific activities of ^{90}Y and ^{32}P (Beddoe and Kelly, 1994). The ICRP groups all radioisotopes into four classes of danger; ^{46}Sc, ^{47}Sc, and ^{48}Sc are included in class C, with medium radiotoxicity. The grouping of radioisotopes according to their relative radiotoxicity puts ^{90}Sr/^{90}Y in group A (very highly toxic), whereas ^{48}Sc, ^{47}Sc, ^{90}Y, ^{92}Y, and ^{93}Y are put in group C (intermediate toxicity), and ^{91m}Y in group D (moderate toxicity) (Tölgyessy, 1993). The dose equivalent of ^{47}Sc used orally as a radiopharmaceutical was calculated by Johansson and Mattsson (1984). The total effective dose equivalent for this radioisotope is 0.42 mSv/MBq. These data may be useful in comparing the risk from various radiodiagnostic procedures and estimating the collective dose from diverse sources of radiation.

Lin (1994) reported that radiation interstitial pneumonitis effects were caused by the injection of ^{90}Y microspheres into a patient with liver carcinoma. Peters and Lee (1994) described radiation necrosis overlying the ankle joint of a female patient, 8 months after injection with ^{90}Y facilitate radiosynovectomy. Hematopoietic toxicity is the dose-limiting factor in radioimmunotherapeutic regimens, such as treatment with high doses of ^{90}Y-labeled antibodies to carcinoembryonic antigen. Murine γ-interferon administered after ^{90}Y treatment

Table 9-29. Dose–Rate Conversion Factors for Exposure of Various Human Organs to ^{46}Sc by Immersion in Contaminated Air or Water or Irradiation from a Contaminated Surface[a]

	Conversion factor (millirem yr^{-1} μCi^{-1})		
Organ	Air (per m^3)	Water (per m^3)	Ground surface (per cm^2)
Muscle	1.11×10^{10}	2.39×10^7	2.05×10^6
Skeleton	1.23×10^{10}	2.63×10^7	2.27×10^6
Skin	1.40×10^{10}	3.00×10^7	2.59×10^6
Pancreas	1.05×10^{10}	2.25×10^7	1.94×10^6
Spleen	1.14×10^{10}	2.44×10^7	2.09×10^6
Thymus	7.87×10^9	1.69×10^7	1.45×10^6
Thyroid	8.66×10^9	1.86×10^7	1.60×10^6
Uterus	7.71×10^9	1.65×10^7	1.42×10^6
Ovaries	6.99×10^9	1.50×10^7	1.29×10^6
Testes	9.95×10^9	2.13×10^7	1.84×10^6
Total body	1.11×10^{10}	2.38×10^7	2.04×10^6
Liver	—	—	1.74×10^6
Kidneys	—	—	1.84×10^6
Lung	—	—	1.93×10^6
Stomach	—	—	1.78×10^6
Small intestine	—	—	1.58×10^6
Large intestine	—	—	1.87×10^6
Bladder		1.84×10^6	—
Marrow	—	—	2.29×10^6

[a] Data reprinted from Kocher (1980), with permission of Williams and Wilkins.

Table 9-30. Beta-Ray Rates at Various Depths in Water on the Axis of a 1-Bq · cm Plane Isotropic Source Distributed Uniformly over a 1-cm^2 Circular Area on an Air–Water Boundary[a]

	β-Ray rate (nGy/h)				
Depth (mm)	^{46}Sc	^{47}Sc	^{49}Sc	^{90}Y	^{91}Y
0.07	1076	1294	1751	17,569	1669
0.4	57.7	214	1041	1049	897
3.0	—	—	148	200	64

[a] Data reprinted from Cross *et al.* (1992a), with permission of the authors and Atomic Energy of Canada Ltd. (By using a newer version of the ACCEPT computer code, slightly different values were generated. Cited from Cross, 1997, personal communication, with permission.)

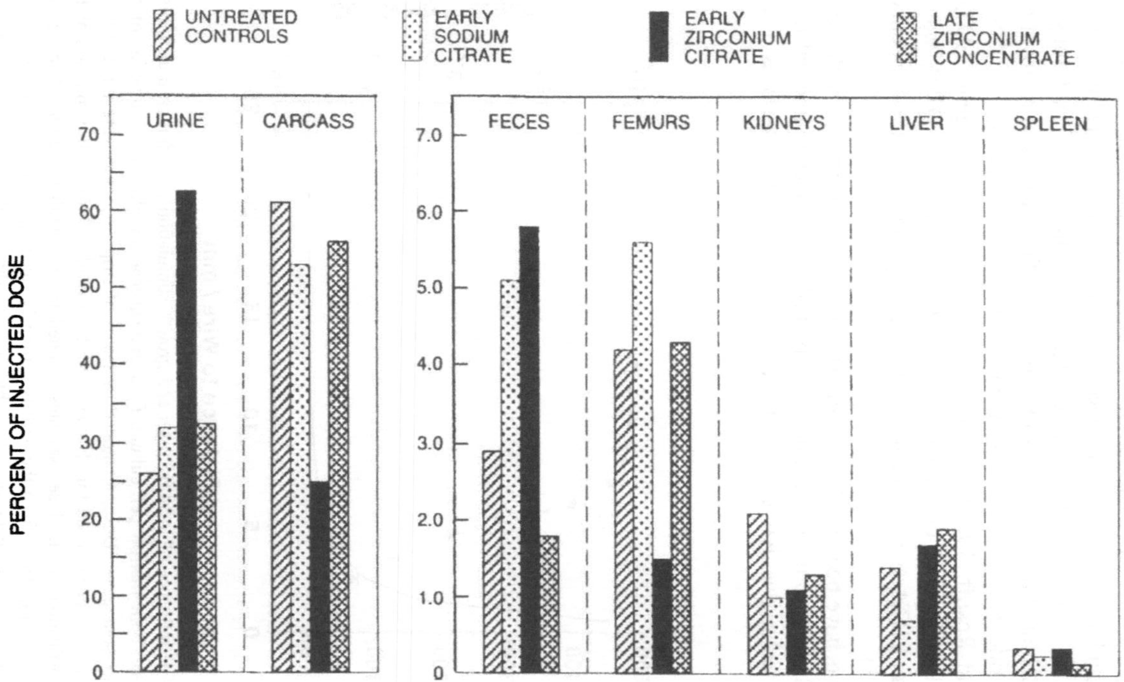

Figure 9-27. Effect of zirconium and sodium citrate on the distribution and excretion of ^{91}Y in rats, 6 days after i.p. administration of the radioisotope. Redrawn by Stanndard (1988) from Schubert and White (1950), with permisison of the American Society for Biochemistry and Molecular Biology.

reduced the hematopoietic toxicity and increased survival of affected mice (Thomas *et al.*, 1995).

In the postwar era, a large amount of experimental work was done to identify efficient agents for the elimination of radioactive elements from the body following accidental exposures. Experiments with rats established the efficacy of zirconium and sodium citrate in increasing the elimination of ^{90}Y through urine and feces (Schubert and White, 1950; Fig. 9-27). The successful removal of ^{46}Sc as the chloride was achieved in men following i.v. injection of ^{46}Sc-NTA (Spencer and Rosoff, 1965). The infusion of the chelating agents EDTA and DTPA resulted in an increased excretion of ^{46}Sc in urine, DTPA was more effective than EDTA as a decontaminating agent (Schmidtke, 1964; Spencer and Rosoff, 1965).

Attempts have been made to diminish the hazards associated with the use of ionizing radiation, to treat liver and other metastases. Selective internal radiation (SIR) is a technique that entails the treatment of liver metastases with ^{90}Y

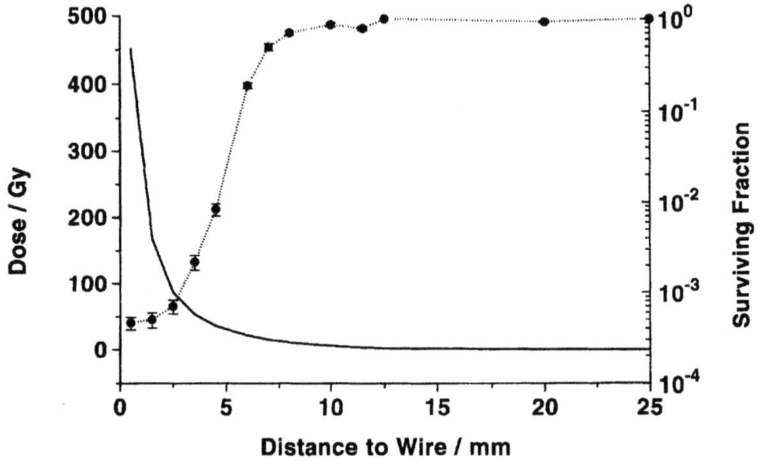

Figure 9-28. The influence of local necrosis and growth stimulation on radiation transformation *in vitro* of a hamster bone-marrow cell culture. Dose and survival are plotted as a function of distance from a ^{90}Y wire source. —, Dose, integrated over area of strips; ···, surviving fraction at specific distances (points have been joined for clarity). From Sigg *et al.* (1996), with permission of Elsevier Science. A reevaluation of the dose estimates resulted in an overall reduction of the doses. The decrease in dose with distance from the wire is extremely steep. Even the irradiation field over the whole distribution is inhomogeneous, but this has no direct influence on the basis for the transformation experiments. The ideal sublethal dose region for the transformation assay was based on measured survival data. This region, at a distance of between 9 and 11 mm from the wire, displays a constant toxicity. The dose is 6 Gy, administered at a dose rate of 4 m Gy/min. Cited from Crompton, personal communication, 1997, with permission.

micropheres. SIR therapy caused tumor regression in a high proportion of patients (Gray *et al.*, 1992).

Sigg *et al.* (1996) presented a model system for the assessment of the biological effects of inhomogeneous irradiated fields, which may result from exposure to particulate radioactive matter. The resulting harm per unit dose to tissue is qualitatively different from homogeneous irradiation sources because of specific hot particle effects, such as dose wasting of dose to necrotic tissues (overkill). Also, the formation of microlesions may occur, which leads to growth stimulation in adjacent tissue (Sigg *et al.*, 1996). Irradiation of hamster bone-marrow cells with ^{91}Y wires caused a strong dose-rate effect, which was observed with the colony-forming assay. Dose and survival as a function of distance from the ^{90}Y wire source were recorded (Fig. 9-28).

9.17 Hazard, Occupational, and Environmental Impacts

Data concerning the hazards of scandium and yttrium metals and compounds are presented in *Dangerous Properties of Industrial Materials* (Sax and Lewis, 1992), and in *Hazardous Chemicals Desk Reference* (Lewis, 1993). Adverse effects of scandium and yttrium metals and their compounds and radioisotopes may result from environmental, industrial, laboratory, or pharmacotherapeutic intoxication (Miller and Weidner, 1986; Hewitt, 1988; Goodman and Gilman, 1990; Calon, 1993).

Scandium metal should be handled carefully. In the form of dust, it is inflammable when exposed to heat or flame or by chemical reaction with oxidizing agents. It can react violently with halogens and air. Its hazard rating is 2 (medium toxicity). Scandium metal is reported as a chemical of interest in the 1990 edition of the U.S. Environmental Protection Agency (EPA) TSCA Inventory. *Scandium chloride* emits toxic fumes when heated to decomposition and its hazard rating is 3 (toxic, fire, explosive, or reactivity hazard). $ScCl_3$ is included in the EPA TSCA Inventory. Specified limits for assignment requirements of caution signs, labels, and signals and for disposal by burial in soil are: 10, 100, and 10 μCi, for ^{46}Sc, ^{47}Sc, and ^{48}Sc respectively, and 10, 10, 100, and 100 μCi, for ^{90}Y, ^{91}Y, ^{92}Y, and ^{93}Y, respectively (L'Annunziata, 1979). *Yttrium metal* is flammable in the form of dust when reacted with air and halogens. *Yttrium nitrate* and *yttrium citrate* are considered possible carcinogens, based on experimental tumorigenic data (Lewis, 1991).

Established criteria for metal toxicity to aquatic organisms are actute mortality, growth retardation, and impaired reproduction with mortality of offspring. Little is known about adaptive responses of aquatic organisms to metal toxicity (i.e., whole-animal responses, population dynamics, species abundances and diversity, cellular morphological changes, and biochemical reaction rates)

(Lelland and Kuwabara, 1985). Scandium and yttrium radioisotopes are generally retained strongly by soil and are taken up by terrestrial plants in very low amounts. Measurable levels may nevertheless accumulate on vegetation from aerosol-deposited fallout debris (Whicker and Schultz, 1982). For certain radioisotopes, such as those of strontium, cerium, and cesium, but also yttrium, the milk and meat pathways are of major significance. A transfer coefficient for yttrium and cerium of 0.002 kg^{-1} at an assumed transfer time of 20 days was estimated (Kathren, 1984). A review of the literature reassessed the burdens of radiation dose for 27 radioisotopes to consumers of fish (Poston and Klopfor, 1988). Factors of environmental partitioning, ecological trophic level discrimination, and specific tissue accumulation that may influence the dose commitment to humans from the consumption of fish were estimated. The recommended CF of ^{46}Sc freshwater fish is 100, a value similar to those of ^{95}Nb, ^{103,106}Ru, and ^{95}Zr, which exists at relatively high concentrations in nuclear effluents and have reasonably long half-lives. The recommended CF of ^{46}Sc for marine fish is 750, a value which is 25–700-fold higher than those recommended for niobium, ruthenium, technetium, and zirconium radioisotopes (Poston and Klopfer, 1988).

Scandium-46 is one of the major radioactive constituents in fly ash. Administration to rats of fly ash from fluidized-bed coal combustion caused a body burden of radioactivity in the lungs after the initial rapid clearance, with a concomitant increase in ^{46}Sc (Fig. 9-29). Estimates of the abundance of several chemical elements in car exhaust debris and gaseous effluent gave values of 18 and 2 mg/kg of dry deposit for scandium and yttrium, respectively (Hamilton, 1979). These values are relatively low when compared to those for lead, iron, bromine, silicon, and manganese, but higher than or equal to those for beryllium, cadmium, and other toxic TE.

High concentrations of yttrium and other TE were found in shells of *Dreissena polymorpha* from the reservoir of a Ukranian atomic power station (Vasiliev *et al.*, 1995). Twenty-seven fission products were measured in a spent reactor fuel, 1.13 ton of which contained 461 g of yttrium, with a total of 2522 TBq β activity of ^{90}Y (Adloff and Guillaumont, 1993). Scandium-46 and several other radioactive contaminants were found in the Yenisei River in Russia, resulting from sites of discharges from an industrial complex (Vakulovsky *et al.*, 1995).

Following the detonation of atomic bombs over Hiroshima and Nagasaki, Japan, in August 1945, measurements performed in November 1945 found γ radiation mainly from ^{46}Sc, ^{60}Co, and ^{134}Cs. An average of 10.5×10^{-5} μCi/g of ^{46}Sc was detected in soil and building materials. Also, calculations based on the doses inhaled by the American occupation forces found values of $0.9 \times 10^{-2}-1.8 \times 10^{-3}$ rem and $1.5 \times 10^{-2}-3.2 \times 10^{-3}$ rem for whole-body dose and bone dose, respectively. The 50-year dose-equivalent commitments to the bone, bone marrow, and whole body resulting from the inhalation of ^{90}Y, ^{91}Y,

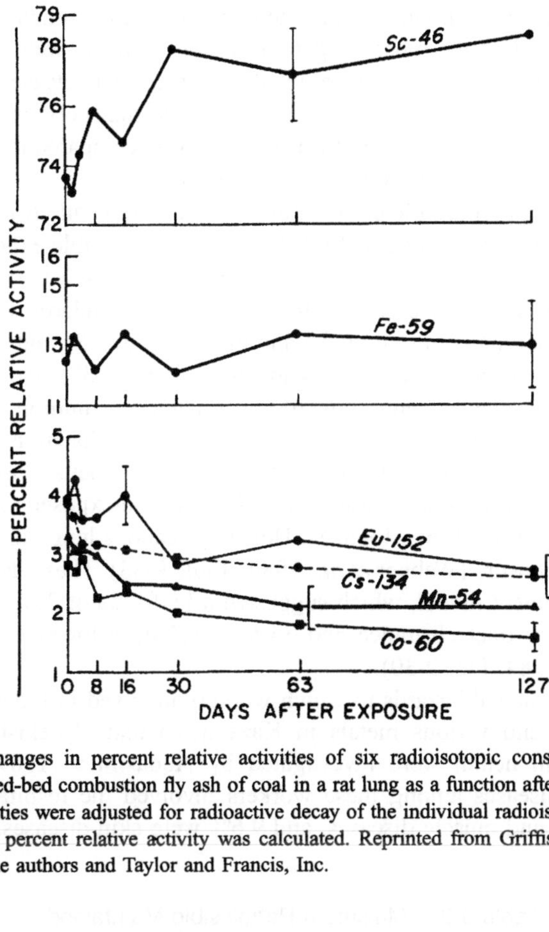

Figure 9-29. Changes in percent relative activities of six radioisotopic constituents of neutron-activated fluidized-bed combustion fly ash of coal in a rat lung as a function after a single inhalation exposure. Activities were adjusted for radioactive decay of the individual radioisotopes to the day of exposure before percent relative activity was calculated. Reprinted from Griffis *et al.* (1981), with permission of the authors and Taylor and Francis, Inc.

and other isotopes during the occupation period were estimated (McRaney and McGahan, 1980). Analysis of radioactive fallout of an atomic bomb explosion on Bikini Atoll in March 1954 detected a high specific activity ^{89}Sr, ^{90}Sr, ^{90}Y, and ^{91}Y (Kimura, 1954). Also, concentrations of ^{90}Y were detected in the liver and kidney of a deceased person who had been exposed to the radiation (Kimura *et al.*, 1956). As a result of the Marshall Island testing of atomic bombs, ^{90}Y and numerous other radioisotopes were detected in plants and animals (Luckey, 1980). Following a thermonuclear test that took place in Russia in October 1980, ^{88}Y and ^{155}Eu were measured in fallout samples in Romania (Dovlete *et al.*, 1983).

The explosion and resulting fire in the nuclear power station at Chernobyl, in the Ukraine, showered radiation debris over Eastern and even Western Europe. This provided a point source for radioactive contamination of atmospheric, terrestrial, aquatic, and urban compartments of the environment. Apart from $^{238,239,240}Pu$, several of the most dangerous fission products were ^{90}Sr, $^{128,131}I$, $^{134,137}Cs$, and ^{91}Y. The estimated value of the fractional release of REE and actinides, including yttrium, available for long-range transport from Chernobyl was 0.3, and a windscale value of 0.2 was estimated, as compared to a value of 80 estimated for cesium (Warner and Harrison, 1993).

The maximum permissible body burdens of scandium and yttrium with bone as reference organ are presented in Table 9-31. The permissible exposure limit in air for yttrium given in U.S.A. Federal standards is 1 mg Y^{3+}/m^3. However, several American states have set lower guidelines or standards for yttrium in ambient air. Annual limits of oral intake and inhalation of ^{47}Sc and ^{90}Y and other radioisotopes of concern have been established (Table 9-32). No criteria have been set for permissible concentrations of yttrium and scandium in water, as well as for permissible exposure limits in air (Sittig, 1991). Inhalation exposusre of laboratory animals to environmental or industrial air pollutants closely simulates the mode of human exposure to these agents. In rats nose-exposed to volcanic ash from the 1980 eruption of the Mount St. Helens volcano, a lung burden remained 128 days after exposure (Wehner *et al.*, 1984; Table 9-33). These data were fitted to a two-compartment model, which suggested a half-time of 2.6 ± 3.2 days for ciliated respiratory tract clearance and 34.5 ± 28.7 days for alveolar clearance (Wehner *et al.*, 1984; Fig. 9-30).

The environmental hazards to which workers involved in the production of scandium oxide and various metals in Kazakhstan and Uzbekistan and their effects on human health were investigated by Abdazimov (1989). Fifity-five percent of the diseases among these workers involved the respiratory system, 9.8% were bone and muscle diseases, and 9.6% were skin diseases. An analysis

Table 9-31. Maximum Permissible Maintained Body Burdens (MPBB) of Yttrium and Scandium Radioisotopes, with Bone as Organ of Reference

Radioisotope	Decay	MPBB Quantity per unit time
^{46}Sc	$\beta^-\gamma$	60
^{47}Sc	$\beta^-\gamma$	60
^{48}Sc	$\beta^-\gamma$	30
^{90}Y	β^-	3
^{91m}Y	$\beta^-\gamma$	5
^{91}Y	$\beta^-\gamma$	5
^{92}Y	$\beta^-\gamma$	2
^{93}Y	$\beta^-\gamma e^-$	2

[a] Data reprinted from McLean and Buddy (1964), with permission of Academic Press.

Table 9-32. Annual Limits of Intake and Derived Air Concentration for ^{47}Sc and ^{90}Y[a]

Radioisotope	Half-life	Chemical form	Oral intake (Bq)	Inhalation (Bq)	Derived air concentation (Bq/m^3)
^{47}Sc	3.4 d	All	8×10^7	10^8	5×10^3
^{90}Y	64.1 h	Oxides	2×10^7	2×10^7	9×10^3

[a] Data reprinted from Burkart (1988), by courtesy of Marcel Dekker, Inc.

of the working conditions in copper, zinc, and lead production plants in Uzbekistan found high levels of scandium and 17 other TE in polymetallic dust and sulfur dioxide emissions (Abdazimov, 1989). Proposals for improving the working conditions in these plants were worked out (Feighin, 1985), but it is doubtful that these measures were implemented. Occupational health standards, guidelines, and legislative regulations concerning the manufacturing and processing of metals and their compounds were recently prepared by the World Health Organization, the U.S. Environmental Protection Agency and European organizations. The guidelines refer mainly to high-risk hazardous metals such as arsenic, cadmium, mercury, manganese, lead, thallium, and a few others, but not to REE. The occupational exposure limits (maximum admissible time-weighted average concentrations for an 8-h work day and a 40-h work week) for yttrium metals and compounds range between 1 mg/m^3 in the United States and England and 5 mg/m^3 yttrium in Germany (Ewers, 1991; ACGIH, 1990–1991; Richardson and Gangolli, 1994). Under the European Community (EC) Directive

Table 9-33. Mount St. Helens Volcanic Ash Burden in Rat Lungs as a Function of Time after Exposure[a]

	Time of sacrifice	Mean ± SD
Exposed animals	15 min	119.2 ± 17.4
	1 h	114.4 ± 27.1
	2 days	96.4 ± 17.8
	8 days	81.3 ± 17.1
	33 days	61.3 ± 10.7
	128 days	12.5 ± 3.0
Control animals	0 days	< 0.9
	128 days	< 0.9

[a] Values for exposed animals are in micrograms of ash/lung for 5 lung samples, adjusted for differences in ash aerosol concentrations during the two exposures and for radioactive decay. Data reprinted from Wehner *et al.* (1984), with permission of Academic Press.

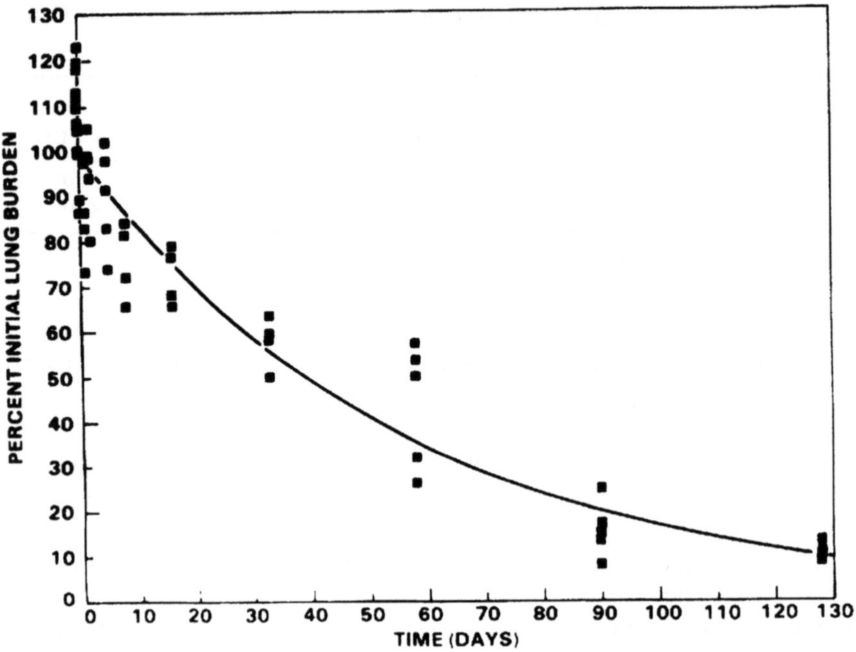

Figure 9-30. Alveolar deposition and clearance, in rats, of inhaled fly ash from the Mount St. Helens volcanic explosion, based on ^{46}Sc data. Reprinted from Wehner *et al.* (1984), with permission of Academic Press.

on Drinking Water Quality, a maximum admissible concentration of 50 mg yttrium chloride and yttrium nitrate per liter are permitted (Richardson and Gangolli, 1994).

The increasing use of high-T_c superconducting (HTS) materials requires hygienic regulations to limit the toxicological risks. Comparison of several HTS materials found that CuYBa ($YBa_2Cu_3O_{6-7}$) exhibited the most toxic effect at the site of i.p. administration to mice (Neizvestnova *et al.*, 1992). Further investigation of the fluorides of yttrium, ytterbium, and terbium showed a relatively low toxicity in rats, manifested by a minor irritation of the skin and eyes, but when these compounds were administered into the stomach, a specific intoxication resulted (Neizvestnova *et al.*, 1994).

The solid-state laser Nd : YAG, Ho : YSSG, Er : YSSG, and Er : YAG, which are used to perform laser sclerotomies, may cause thermal and focal damage to the iris and ciliary body. However, it is not clear if these effects are due to the laser irradiation alone, or possibly, are due to a combined effect of the irradiation and chemical composition of the device. Technical modifications have been investigated to minimize the tissue damage in human globes (Ozler *et al.*, 1991).

The use of uncertainty factors and modifying factors to calculate the reference data from experimentally no-observed adverse-effect levels was proposed; by Abernathy *et al.* (1993); changes in the activity of enzymes, convulsions, liver fat, and slight body weight changes were considered. The frank-effect level is a dose that induces very serious or life-threatening effects (Abernathy *et al.*, 1993).

9.18 Summary

In spite of their low abundance in the environment, at higher c scandium and yttrium may have some harmful effects on organisms, and, in particular, òn humans, because of their increasing industrial and other uses. Most of the existing data on their toxic or carcinogenic effects pertain to lower organisms or animals, and the extrapolation of this information to humans has to be done with reservations.

From a toxicological viewpoint, scandium and yttrium chlorides are less toxic, acutely or chronically, than several lanthanides. Their detrimental effects on ocular tissues and skin are only exceeded by niobium and hafnium. The overall pharmacological effects on the various body systems studied are of a depressant nature. The mode of death, respiratory paralysis coupled with cardiovascular collapse, appears similar to that observed for other chemical elements in the

Table 9-34. Future Directions for Studies of the Toxicology, Carcinogenicity, and Pharmacology of Scandium and Yttrium Salts and Compounds

Assay	Method	Reference
In vitro cytotoxicity	Comparison with *in vivo* LD_{50} of rat hepatocytes	Shrivastava *et al.* (1992)
Multispecies toxicity testing	Standardization in laboratory-scale and large-scale assay	Dickson *et al.* (1985)
Computational structure–activity relationship (SAR) methods	Structure–activity relationships in genotoxicity (DNA damage, chromosomal aberrations)	Richard *et al.* (1989)
Microtox test	Bacteria as assay agent	Qureshi *et al.* (1984)
Computer models	Membrane dialysis and lipid bilayers for metal speciation	Williams (1983)
		Morrison *et al.* (1989)
Toxicity dynamic models	Quantitative structure–activity relationships (QSAR)	Birdi (1993)
Molecular biology	Immunologic effects: alteration of humoral antibody titers; phagocytic activity of macrophages	Koller and Vos (1981)

lanthanide series. It seems that $ScCl_3$ exerts its greatest toxicity locally. It is considered that proper application of good industrial hygiene practices would prevent acute or chronic exposure.

Particular attention has to be devoted to scandium and especially yttrium radioisotopes, which are potentially harmful to organisms and their environment because of their long half-lives.

There is a need for further development of biological indices, appropriate to field investigations. These are a function of the stress acting on the individual or population and are sensitive over the range of low and ultralow concentrations that are characteristic of scandium, yttrium, and other dispersed elements (Table 9-34).

References

Abdazimov, A. D., 1989. Assessment of the working conditions and health susceptibility of workers in the production of copper, zinc, and lead, *Gig. Tr. Prof. Zabol.* 5:18–21. (Russian).

Abernathy, C. O., Cantilli, R., Du, J. T., and Levander, O. A., 1993. Essentially versus toxicity. Source considerations in risk assessment of essential trace elements, in: *Hazard Assessment of Chemicals* (J. Saxena, ed.), Taylor and Francis, Washington, D.C., pp. 81–113.

ACGIH (American Conference of Governmental and Industrial Hygienists), 1990–1991. Threshold limit value for chemical substances and physical agents and biological exposure. ACGIH, Cincinnati, Ohio. Cited by Ennever (1994).

Adloff, J.-P., and Guillaumont, R., 1993. *Fundamentals of Radiochemistry*, CRC Press, Boca Raton, Florida, pp. 126–135.

Ahrland, S., 1965. Strong and weak acids and bases, *Svensk. Kem. Tidskr.* 77:584–597.

Akabani, G., 1993. Absorbed dose calculations in Haversian canals for several beta-emitting radionuclides, *J. Nucl. Med.* 34:1361–1366.

Albert, A., 1985. *Selective Toxicity, The Physico-Chemical Basis of Therapy*, Chapman & Hall, London.

Apak, R., and Tor, I., 1988. Some correlations on acute metal toxicity, *Kim. Sanayi* 31:84–87.

Ariel, I. M., and Galey, D., 1964. The intralymphatic administration of radioactive isotopes and cancer chemotherapeutic drugs, *Surgery* 55:355–363.

Arvela, P., 1979. Toxicity of rare earths, *Progrs. Pharmacol.* 2:71–114.

Bachem, C., 1907. Pharmacology of several rare earths, *Arch. Int. Pharm. Ther.* 17:363–386 (German).

Baierl, T., Drosselmeyer, E., Polzer, G., Seidel, A., Thiele, H., and Wilczek, W., 1992. Cytotoxic effects of high T_c superconducting materials. Modifications by phospholipids and release of pro-inflammatory mediators, *J. Aerosol Sci.* 23(Suppl. 1):s527–s530.

Ballentyne, B., Marrs, T., and Turner, P., 1993. *General and Applied Toxicology*, Stockton Press, New York.

Baum, J. W., and Kaurin, D. G., 1991. Reassessment of data used in setting exposure limits for hot particles, *Radiat. Prot. Dosim.* 39:49–54.

Beddoe, A. H., and Kelly, M. T., 1994. Absorbed dose in the skin from beta emitters in medical and laboratory containers, *Br. J. Radiol.* 67:54–58.

Bendich, A., and Chandra. R. K., eds., 1990. Micronutrients and immune functions, *Ann N.Y. Acad. Sci.* 787:1–320.

Bienvenu, P., Nofre, C., and Cier, A., 1963. Comparative general toxicity of metal ions, *C. R. Acad. Sci. Paris* 256:1043–1044 (French).

Birdi, K. S., 1993. *Fractals in Chemistry, Geochemistry and Biophysics*, Plenum Press, New York, pp. 203–223.

Bland, M. R., Loutit, J. F., and Sansom, J. M., 1974. Histochemical phosphatases and metachromasia in murine tumors induced by bone seeking radionuclides, *Br. J. Cancer* 29:206–221.

Boecker, B. B., Hahn, F. F., Cuddihy, R. G., Snipes, M. B., and McClellan, R. O., 1986. Is the human nasal cavity at risk from inhaled radionuclies? In: *Life-Span Radiation Effects Studies in Animals. What Can They Tell Us?* (R. C. Thompson and J. A. Mahaffey, eds.), Office of Scientific and Technical Information, U.S. Department of Agriculture, Washington, D.C., pp. 564–577.

Bondy, S. C., and Prasad, K. N., eds. 1988. *Metal Neurotoxicity*, CRC Press, Boca Raton, Florida, pp. 61–86.

Brakhnova, I. T., 1975. *Environmental Hazards of Metals*, Consultants Bureau, New York, pp. 204–211.

Brooks, A. L., and McClellan, R. O., 1969. Chromosome aberrations and other effects produced by ^{90}Sr–^{90}Y in Chinese hamsters, *Int. J. Radiat. Biol.* 16:545–561.

Brown, D. B., O'Brien, W. J., and Schultz, R. O., 1990. Corneal ablations produced by the Nd doped Y–Li–F picosecond laser, *Cornea* 13:471–478.

Bruce, D. W., Hietbrink, B. E., and DuBois, K. P., 1963. The acute mammalian toxicity of rare earth nitrates and oxides, *Toxicol. Appl. Pharmacol.* 5:750–759.

Brusick, D., 1987. *Principles of Genetic Toxicity*, Plenum Press, New York.

Bryson, P. D., 1989. *Comprehensive Review in Toxicology*, 2nd ed., An Aspen Publication, Rockville, Maryland, pp. 457–461.

Bulman, R. A., 1988. Yttrium and the lanthanides, in: *Handbook on the Toxicity of Inorganic Compounds* (H. G. Seiler and H. Sigel, eds.), Marcel Dekker, New York, pp. 769–785.

Burkart, W., 1988. Radiotoxicity, in: *Handbook on the Toxicity of Inorganic Compounds* (H. G. Seiler and H. Sigel, eds.), Marcel Dekker, New York, pp. 805–826.

Calabrese, E. J., and Theil, L. S., 1989. How much soil do young children ingest. An epidemiologic study, *Regul. Toxicol. Pharmacol.* 10:123–130.

Calon, P., ed., 1993, *Handbook of Ecoxicology*. Blackwell Scientific Publ., Oxford.

Cannas, M., Beltrami, V., Lamberti, L., and Bigatti, P., 1994. Cytogenic study on the circulating lymphocytes of rheumatoid arthritis patients treated with ^{90}Y, *Med. Sci. Res.* 22:133–135.

Carson, B. L., Ellis, V., and McCann, J. L., 1986. *Toxicology and Biological Monitoring of Metals in Humans*, Lewis Publishers, Chelsea, Massachusetts, pp. 209–212.

Chargaff, E., and Green, C., 1948. On the inhibition of the thromboplastic effect, *J. Biol. Chem.* 173:263–270.

Chow, C. G., 1990. Mutagenesis and micronutrients relationships, *Food Addit. Contam.* 7:s44–s47.

Christie, N. T., and Costa, M., 1983. *In vitro* assessment of the toxicity of metal compounds. Effects of metal on DNA structure and function in intact cells, *Biol. Trace Elem. Res.* 5:55–71.

Clarkson, D. T., and Sanderson, J., 1969. The uptake of a polyvalent cation and its distribution in the root apices of *Allium cepa*, *Planta* 89:136–154.

Clarkson, D. T., and Sanderson, J., 1971. Inhibition of the uptake and long-distance transport of calcium by aluminum and other polyvalent cations, *J. Exp. Bot.* 22:837–851.

Cochran, K. W., Doull, R., Mazur, M., and DuBois, K. P., 1950. Acute toxicity of zirconium, columbium, strontium, lanthanum, cesium, tantalum and yttrium, *Arch. Ind. Hyg. Occup. Med.* 1:637–650.

Connell, D. W., and Miller, G. J., 1984. *Chemistry and Ecotoxicology of Pollution*, John Wiley and Sons, New York.

Cross, W. G., Freedman, N. O., and Wong, P. Y., 1992a. Tables of beta-ray dose distribution in water, AECL-10521, Atomic Energy of Canada Ltd., Chalk River, Ontario.

Cross, W. G., Freedman, N. O., and Wong, P. Y., 1992b. Beta ray dose distribution from skin contamination, *Radiat. Prot. Dosim.* 40:149–168.

Cross, WG, 1997, personal communication, with permission.

DeNardo, G. L., Kroger, L. A., DeNardo, S. J., Miers, L. A., Salako, Q., Kukis, D. L., Fand, I., Shen, S., Renn, O., and Mears, C. F., 1994. Comparative toxicity studies of ^{90}Y–90 MX-DTPA and 2-IT-BAD conjugated MAb, *Cancer* 73 Suppl. 3):1012–1022.

Dickson, K. L., Duke, T., and Loewengart, G 1985. A synopsis: Workshop on multispecies toxicity tests, in: *Multispecies Toxicity Testing* (J. Cairns, Jr., ed.), Pergamon Press, New York, pp. 248–253.

Ding, M., 1927, PhD thesis, Würzburg University. Cited by Steidle and Ding (1929).

Dolezal, V., Hajek, B., Kacl, K., Manova, I., and Petru, F., 1966. The toxicity of scandium, *Z. Chem.* 6:154–155 (German).

Dovlete, T. E., Sonoc, S., Cuculeanu, V., and Sima, O., 1983. Note on ^{155}Eu and ^{88}Y in fallout samples collected during 1981, *Atomkern. Kerntech.* 43:213–214.

Ducastaing, A., Mongeyron, C., Azanza, J. L., Creach, P., and Raymond, J., 1973. Action *in vitro* of the chlorides of lanthanum, yttrium and cerium on the first steps of the thrombin–fibrinogen reaction, *C. R. Acad. Sci. Paris* 266:262–269 (French).

Draize, J. H., Woodward, G., and Calvery, H. O., 1944. Methods for the study of irritation and toxicity of substances applied topically to the skin and mucous membrane, *J. Pharmacol.* 82:377–390.

Ellenhorn, M. J., and Barceloux, D. G., 1988. *Medical Toxicology*, Elsevier, New York, pp. 1009–1030.

England, M. W., Turner, J. E., and Hingerty, B. E., 1989. Ordering of metal-ion toxicities in different species. Extrapolation to man, *Health Phys.* 57(Suppl. 1):115–119.

Ennever, F. K., 1994. Metals, in: *Principles and Methods of Toxicology* (A. W. Hayes, ed.) Raven Press, New York, pp. 417–443.

Evans, C. H., 1990. Toxicology and pharmacology of the lanthanides, in: *Biochemistry of the Lanthanides*, Plenum Press, New York, pp. 339–389.

Ewers, U., 1991. Standards, guidelines, and legislative regulations concerning metals and their compounds, in: *Metals and Their Compounds in the Environment* (E. Merian, ed.), VCH, Weinheim, pp. 687–711.

Ewers, U., and Schlipköter, H. W., 1991. Chronic toxicity of metals and metal compounds, in: *Metals and Their Compounds in the Environment* (E. Merian, ed.), VCH, Weinheim, pp. 591–603.

Feighin, B. g., 1985. Working hygiene in the production of scandium oxide, *Gig. Tr. Prof. Zabol.* 4:39–40 (Russian).

Fekete, F. A., and Barton, L. L., 1991. Effects of Fe^{3+} analogs on growth and pseudobactin synthesis in a chromium-tolerant *Pseudomonas* isolate, *Biol. Metals* 4:211–216.

Filov, V. A., and Khudolei, V. V., 1993. Carcinogenic chemicals in the environment and their ecological significance. *Zh. Ekol. Khim*, 4:313–317 (Russian).

Fischer, N. S., 1986. On the reactivity of metals for marine phytoplankton, *Limnol. Oceanogr.* 31:443–449.

Floersheim, G. L., 1995. Modification of radiation sensitivity by salts of the metals of beryllium, indium and the rare earths cerium, lanthanum and scandium, *Radiat. Res.* 141:318–323.

Frankel, A. E., ed., 1988. *Immunotoxins*, Kluwer Academic Publishers, Boston.

Frouin, A., and Ledebt, S., 1913. Agglutinating and antihemolytic action of salts of the rare earth metals, *C. R. Soc. Biol.* 72:1038–1039 (French). CA 7:2625.

Furst, A., 1987. Toward mechanisms of metal carcinogenesis, in: *Genotoxic and Carcinogenic Metals. Environmental and Occupational Occurrence and Exposure* (L. Fishbein, A. Furst, and M. A. Mehlman, eds.), Princeton Scientific Publishing Co, Princeton, pp. 295–327.

Furst, A., and Radding, S. B., 1979. Unusual metals as carcinogens, *Biol. Trace Elem. Res.* 1:169–181.

Gabbiani, G., Jacqmin, M. L., and Selye, H., 1966. Thrombohemorrhagic lesions induced by combined treatment with REE and epinephrine, *J. Pharmacol. Exp. Ther.* 152:275–280.

Garg, A. N., Wankhade, H. K., Dogra, R. K. S., and Shanker, R., 1987. NAA of respirable mica samples and their pathological effects in lungs of rats, *Sci. Total Environ.* 67:165–176.

Garrett, I. R., and McClure, J., 1981. Lanthanide-induced calcergy, *J. Pathol.* 135:267–275.

Gebhart, E., and Rossman, T. G., 1991. Mutagenicity, carcinogenicity, teratogenicity, in: *Metals and Their Compounds in the Environment* (E. Merian, ed.), VCH, Weinheim, pp. 617–640.

Geldmacher-von Mallincrodt, M., 1991. Acute metal toxicology in humans, in: *Metals and Their Compounds in the Environment* (E. Merian, ed.), VCH, Weinheim, pp. 585–590.

Gensicke, F., Nitschke, H. W., and Hölzer, F., 1966. Comparative investigation of the metabolism of ^{91}Y, ^{144}Ce, ^{143}Pr and ^{147}Pm, *Stud. Biophys.* 1:347–352 (German).

Gilman, J. P. W., and Swierenga, S. H. H., 1984. Inorganic carcinogens, in: *Chemical Carcinogens* (C. E. Searle, ed.), American Chemical Society, Washington, D.C., pp. 577–630.

Goodman, L. S., and Gilman, A., 1990. *The Pharmacological Basis of Therapeutics*, Pergamon Press, New York.

Goyer, R. A., 1986. Toxic effects of metals, in: *Casarett and Doull's Toxicology*, Macmillan, New York, pp. 582–635.

Goyer, R. A., and Cherian, M. G., eds., 1995. *Toxicology of Metals, Biochemical Aspects*, Springer-Verlag, Berlin.

Graca, J. C., Davison, F. C., and Feavel, J. B., 1962. Comparative toxicity of stable rare earth compounds, *Arch. Environ. Health* 5:437–444.

Grant, W. M., and Kern, H. L., 1956a. Corneal bonding of rare earths and other cations, in: *Rare Earths in Biochemistry and Medical Research* (G. C. Kyker and E. R. Anderson, eds.), Institute of Nuclear Studies, Oak Ridge, Tennessee, Vol. 12, pp. 346–355.

Grant, W. M., and Kern, H. L., 1956b. Cations and the cornea. Toxicity of metals to the stroma, *Am. J. Ophthalmol.* 42:167–181.

Grant, W. M., and Schuman, J. S., 1993. *Toxcology of the Eye*, Thomas Publisher, Springfield, Illinois, pp. 1240–1256, 1518–1531.

Gray, B. N., Anderson, J. E., Burton, M. A., Van Hazel, G., Codde, J., Morgan, C., and Klemp, P., 1992. Regression of liver metastases following treatment with ^{90}Y microspheres, *Aust. N. Z. Surg.* 62:105–110.

Griffis, L. C., Snipes, M. B., and Brooks, A. L., 1981. Clearance by the rat of inhaled fly ash from fluidized-bed coal combustion, *J. Toxicol. Environ. Health* 7:117–124.

Gumpel, J. M., and Stevenson, A. C., 1975. Chromosome damage after intra-articular injection of different colloids of ^{90}Y, *Rheumatol. Rehabil.* 14:7–12.

Hahn, F. F., Barnes, J. E., Hobbs, C. H., and Mauderly, J. L., 1975. Effect of ^{90}Y inhaled in fused clay particles on the gastro-intestinal tract of beagles, *Radiat. Res.* 61:444–456.

Hahn, F. F., Muggenburg, B. A., Boecker, B. B., Cuddihy, R. G., Griffith, W. C., Guilmette, R. A., McClellan, R. O., and Mewhinney, J. A., 1986. Insights into radionuclide-induced lung cancer in people from life-span studies in beagle dogs, in: *Life-Span Radiation Effects Studies in Animals. What Can They Tell Us?* (R. C. Thompson, and J. A. Mahaffey, eds.), Office of Scientific and Technical Inforamtion, U.S. Department of Agriculture, Washington, D.C., pp. 521–534.

Hahn, F. F., Griffith, W. C., Hobbs, C. H., and McClellan, R. O., 1987. Toxicity of ^{90}Y inhaled in a relatively insoluble form by beagle dogs, in: Lovelance Inhalation Toxicology Research Institute Report 1986–1987. Springfield, Virginia, pp. 205–208.

Hahn, F. F., Griffith, W. C., Hobbs, C. H., Muggenburg, B. A., and Boecker, B. B., 1994. Biological effects of ^{90}Y in relatively insoluble particles inhaled by beagle dogs, *Ann. Occup. Hyg.* 38(Suppl):275–280.

Haley, P. J., 1991. Pulmonary toxicity of stable and radioactive lanthanides, *Health Phys.* 61:809–820.

Haley, T. J., 1965. Pharmacology and toxicology of the rare earth elements, *J. Pharmacol. Sci.* 54:663–670.

Haley, T. J., 1979. Toxicity (of the REE), in: *Handbook on the Physics and Chemistry of Rare Earths* (K. A. Gschneidner, Jr., and L. Eyring, eds.), North-Holland, Amsterdam, pp. 553–584.

Haley, T. J., Komesu, N., Mavis, L., Cawthorne, J., and Upham, H. C., 1962. Pharmacology and toxicology of scandium chloride, *J. Pharmacol. Sci.* 51:1043–1045.

Hamilton, E. I., 1979. Relations between metal elements in man's diet and environmental factors, in: *Trace Substances in Environmental Health* (D. D. Hemphill, ed.), University of Missouri, Columbia, Vol. 13, pp. 3–15.

Hamilton, J. G., 1949. The metabolism of the radioactive elements created by nuclear fission, *N. Engl. J. Med.* 240:863–870.

Hartung, R., 1986. Ranking the severity of toxic effects, in: *Trace Substances Environmental Health* (D. D. Hemphill, ed.), University of Missouri, Columbia, Vol. 20, pp. 204–214.

Hathway, D. E., 1984. *Molecular Aspects of Toxicology*, Royal Society of Chemistry, London.

Heeb, C. M., Gydesen, S. P., Simpson, J. C., and Bates, D. J., 1996. Reconstruction of radionuclide release from the Hanford site 1944–1972, *Health Phys.* 71:545–555.

Heflich, R. H., 1991. Chemical mutagens, in: *Genetic Toxicology* (A. P. Li and R. H. Heflich, eds.), CRC Press, Boca Raton, Florida, pp. 143–202.

Hewitt, P. J., 1988. Accumulation of metals in the tissues of occupationally exposed workers, *Environ. Geochem. Health* 10:113–116.

Hirano, S., and Suzuki, K. T., 1996. Exposure, metabolism, and toxicity of rare earths and related compounds, *Environ. Health Perspect.* 104(Suppl.1):85–95.

Hirano, S., Kodama, N., Shibata, K., and Suzuki, K. T., 1990. Distribution, localization and pulmonary effects of yttrium chloride following intra-tracheal instillation into the rat, *Toxicol. Appl. Pharmacol.* 104:301–311.

Hirano, S., Kodama, N., Shibata, K., and Suzuki, K. T., 1993. Metabolism and toxicity of i.v. injected YCl_3 in rats, *Toxicol. Appl. Pharmacol.* 121:224–232.

Hiroaki, I. O., Satoshi, T., Yoshiuki, N., Haruhiko, M., Kazunori, S., and Kato, Y., 1987. Mutagenicity of metal salts in combination with 9-aminoacridine in *Salmonella typhymurium*, *Jpn. J. Genet.* 62:159–162. CA 107:91632z.

Hobbs, C. H., Barnes, J. E., McClellan, R. O., Chiffelle, T. L., Jones, R. K., Lundgren, D. L., Mauderly, J. L., Pickrell, J. A., and Rypka, E. W., 1972. Toxicity in the dog of inhaled ^{90}Y in fused clay particles. Early biological effects, *Radiat. Res.* 49:430–460.

Höber, R., and Spaeth, R. A., 1914. The influence of rare earths on the contractability of the muscle, *Pflügers Arch. Physiol.* 159:433–456 (German).

Hrebinko, R. L., 1996. Severe injury from NdYAl garnet laser therapy for penile condylomata acuminata, *Urology* 48:155–156.

Hughes, M. N., and Poole, R. K., 1989. *Metals and Microorganisms*, Chapman and Hall, London, pp. 252–302.

Hutcheson, D. P., Gray, D. H., Venugopal, B., and Luckey, T. D., 1975. Safety of heavy metals as nutritional markers, *Environ. Qual. Safety*, 1975 (Suppl. 1):74–80.

Iskander, F. Y., Bauer, T. L., and Klein, D. E., 1986. Determination of 28 elements in American cigarette tobacco by NAA, *Analyst* 111:1207–109.

Iyehara, O. H., Hiroki, T., Satoshi, T., Niyitaki, Y., Mino, H., Sakata, K., and Kato, Y., 1987. Mutagenicity of metal salts in combination with 9-aminoacridine in *Salmonella thyphi, urinum, Japan. J. Genet.* 62:159–162.

Izraelson, Z. I., ed., 1967. *Toxicology of the Rare Metals*, Israel Program for Scientific Translations, Jerusalem.

Jenkins, C. E., 1969. Radionuclide distribution in Pacific salmon, *Health Phys.* 17:507–512.

Ji, Y. J., 1985. Toxicological study on safety evaluation of rare earths used in agriculture, *J. Chin. Rare Earth Soc.*, Special issue, pp. 4–10

Ji, Y. J., Cui, M. J., Dong, X. Y., and Zhu, S. F., 1985. Subchronic toxicity studies of rare earth nitrate in rats, in: *New Frontiers in Rare Earth Science and Applications* (G. X. Xu, and J. M. Xiao, eds.), Science Press, Beijing, pp. 697–699.

Johansson, L., and Mattsson, S., 1984. Effective dose equivalent in nuclear medicine investigations, in: *Radiation-Risk-Protection* (A. Kaul, ed.), TUV Verlag, Rheinland, Vol. 1, pp. 328–331.

Kaiser, K. L. E., 1980. Correlations and prediction of metal toxicity to aquatic biota. *Can. J. Fish. Aquat. Sci.* 37:211–218.

Kaiser, K. L. E., 1985. Correlation of metal ion toxicities to mice, *Sci. Total Environ.* 46:113–119.

Kaiser, K. L. E., 1997. Personal communication, with permission.

Kaplan, I. S., Tikhomirov, F. A., and Khvostova, V. V., 1973. Efficiency of the cytogenetic action of ^{90}Sr incorporated into *Crepis capillaris*, *Radiobiologiya* 13:686–689 (Russian). CA 80:24347w.

Kathren, R. L., 1984. *Radioactivity in the Environment*, Harwood Academic Publishers, Chur, Switzerland, pp. 248–270.

Kidman, B., Tutt, M. L., and Vaughan, J. M., 1951. The retention of radioactive strontium and yttrium (^{89}Sr, ^{90}Sr and ^{90}Y) in pregnant and lactating rabbits and their offspring, *J. Pathol. Bacteriol.* 63:253–268.

Kimura, K., 1954. Analysis of radioactive fallout of the atomic bomb explosion on Bikini, *Radioisotopes* 3:1–14.

Kimura, K., Ikeda, N., Kimura, K., Kawanishi, H., and Kimura, M., 1956. Radiochemical analysis of the body of the late Mr, Koboyama, *Radioisotopes* 4:22–27 (Japanese). CA 50:15247G.

Kinraide, T. B., 1991. Identity of the rhizotoxic aluminum species, *Plant Soil* 134:167–178.

Kinter W. B., and Pritchard, J. B., 1977. Altered permeability of cell membranes. In: *Handbook of Physiology*, Section 9 (D. H. K. Lee, ed.), American Physiological Society, Bethesda, Maryland, pp. 563–576.

Kocher, D. C., 1980. Dose-rate conversion factors for exposure to photon and electron radiation from radionuclies occurring in routine releases from nuclear fuel cycle facilities, *Health Phys.* 38:543–621.

Koller, L. D., and Vos, J. G., 1981. Immunologic effects of metals. In: *Immunologic Considerations in Toxicology* (R. P. Sharma, ed.), CRC Press, Boca Raton, Florida, Vol. 1, pp. 67–78.

Kulikov, N. V., and Molchanova, I. V., 1982. *Continental Radioecology*, Plenum Press and Nauka Publishers, Moscow, pp. 130–136.

Kuwabara, R., 1994. The effect of tritiated water and ^{90}Sr–^{90}Y on the egg formation of a brackish water amphypod, *Grandidierella japonica*, *Tokyo Nogyo Daigaku Aisotopu Senta Kenkya Hokoku* 8:25–29. CA 123:78504v.

Kyker, G. C., 1962. Rare earths, in: *Mineral Metabolism* (C. L. Comar and F. Bronner, eds.), Academic Press, New York, Vol. 2B, pp. 499–541.

Kyker, G. C., and Anderson, E. B., 1956. *Rare Earth Elements in Biochemical and Medical Research*, Proc. Conf. Oak Ridge, 1955, Oak Ridge Institute of Nuclear Studies, Oak Ridge, Tennessee.

Kyker, G. C., and Cress, E. A., 1957. Acute toxicity of yttrium, lanthanum and other REE, *Am. Med. Assoc. Ind. Health* 16:475–479.

Lachine, E. E., Noujaim, A. A., Ediss, C., and Wiebe, L. I., 1976. Toxicity, tissue distribution and excretion of $^{46}ScCl_3$ and ^{46}Sc-EDTA in mice, *Int. J. Appl. Radiat. Isot.* 27:373–377.

Lakin, K. M., Zimakov, Yu. A., Menkov, A. A., Bocharova, R. I., and Tzyu, N. P., 1970a. The influence of EDTA on the anticoagulant properties and dynamics of scandium excreted from the organism, *Farmakol. Toxikol.* 33:87–90 (Russian).

Lakin, K. M., Zimakov, Yu. A., Plyushchev, V. E., Nadezhdina, G. V., and Loseva, G. S., 1970b. Experimental investigations of the anticoagulative activity of REE salts, *Farmakol. Toxikol.* 33:205–210 (Russian).

Lambert, C. E., Barnum, E. C., and Shapiro, R., 1993. Acute toxicological evaluation of yttrium nitrate, yttrium oxide and bastnasite concentrate, *J. Am. Coll. Toxicol.* 12:612, 629, 630.

Lambotte-Vandeaer, M., and Duverger-van Bogaert, M., 1988. Scandium, in: *Handbook on Toxicity of Inorganic Compunds* (H. G. Seiler and H. Sigel, eds.), Marcel Dekker, New York, pp. 577–580.

L'Annunziata, M. F., 1979. *Radiotracers in Agriculture*, Academic Press, London.

Laufer, R. B., 1987. Paramagnetic metal complexes as water proton relaxation agents for NMR imaging, *Chem. Rev.* 87:910–927.

Laurens, J. M., 1985. A computer model of the biosphere to estimate stochastic and non-stochastic effects of radionuclides on humans, AECL-8645 Report, Atomic Energy of Canada Ltd., Chalk River, Ontario.

Lazo, V. V., and Makarenkova, E. P., 1995. Morphological change in the rabbit lens followinng 1.32 μm Nd:YAG laser irradiation, *Ark. Patol.* 57: 63–66.

Leitch, G. J., and Amer, M. S., 1975. Lanthanide inhibition of *Vibrio cholerae* and *Escherichia coli* enterotoxin-induced entero-sorption and its effects on intestinal mucosa cyclic adenosine 3′,5′-monophosphate and cyclic guanosine 3′,5′-monophosphate levels, *Infect. Immun.* 11:1038–1044.

Lelland, H. V., and Kuwabara, J. S., 1985. Trace metals, in: *Fundamental Aquatic Toxicology* (G. R. Rang and S. B. Petrocelli, eds.), Hemisphere Publishing Co., Washington, D.C., pp. 374–404.

Levan, A., 1945. Cytological reactions induced by inorganic salt solutions, *Nature* 156: 751–753.

Lewis, R. J., 1991. *Carcinogenically Active Chemicals*, Van Nostrand Reinhold, New York.

Lewis, R. J., 1993. *Hazardous Chemicals Desk Reference*, Van Nostrand Reinhold, New York.

Lin, M., 1994. Radiation pneumonitis caused by ^{90}Y microspheres. Radiologic findings, *Am. J. Roentgenol.* 162:1300–1302.

Lin, T. Y., Siemens, M. A., and Lam, K. W., 1988. The effect of YAG laser anterior capsulotomy on prostaglandin concentration in aqueous humor, *Ann. Ophthalmol.* 20:95–99.

Lloyd, E., 1961. The relative distribution of radioactive yttrium and stronium and the secondary deposition of ^{90}Y built up from ^{90}Sr, *Int. J. Radiat. Biol.* 3:475–492.

Luckey, T. D., 1980. *Hormesis with Ionizing Radiation*, CRC Press, Boca Raton, Florida.

Luckey, T. D., and Venugopal, B., 1977. *Metal Toxicity in Mammals*, Vol. 1, Plenum Press, New York.

Luckey, T. D., Venugopal, B., and Hutcheson, D., 1975. *Heavy Metal Toxicity, Safety and Hormology*, Thieme Publisher, Stuttgart.

Lundgren, D. L., Hahn, F. F., and McClellan, R. O., 1981. Toxicity of ^{90}Y in relatively insoluble fused alumino-silicate particles when inhaled by mice, *Radiat. Res.* 88:510–523.

Luo, G.-Y., 1991. Effects of yttrium nitrate on gastric mucosa and gastric acid secretion in the rat, *Zhongguo Yaolixue Yu Dulixue Zazhi* 5:50–52 (Chinese). CA 115:110855.

Magnusson, G., 1963. Acute toxicity effect of yttrium, cerium, terbium, holmiun and ytterbium on the liver, *Acta Pharmacol. Toxicol.* 20(Suppl. 3):1–95.

Maraveyas, A., Snock, D., Hird, V., Kosmas, C., Meares, C. F., Lambert, H. E., and Epenetos, A. A., 1994. Pharmacokinetics and toxicity of an ^{90}Y-CITC-DTPA-HMFG1 radioimmunoconjugate for i.p. radioimmunotherapy of ovarian cancer, *Cancer* 73(Suppl. 3):1067–1075.

Marinov, V. M., Lazarova, M. S., Mihajlov, M. I., and Ivanov, C. P., 1981. Dynamics of some micreoelement quantities in radiosensitive organs of rats subjected to sublethal dose of gamma rays, *Radiochem. Radioanal. Lett.* 46:87–94.

Masse, R., 1982. Metabolism and toxicity of radioactive rare earths, in: *Radionuclides, Metabolism and Toxicity* (P. Galle, and R. Mosse, eds.), Masson, Paris, pp. 143–155.

Maxwell, L. C., and Bischoff, F., 1931. Studies in cancer chemotherapy. The effects of thorium, cerium, erbium, yttrium, didymium, praseodymium, mangansese and lead upon transplantable rat tumors. *J. Pharmacol. Expr. Ther.* 43:61–70.

McConachie, N. S., and Jacobson, I., 1994. Bilateral aneurisms of the cavernous internal carotid arteries following ^{90}Y implanatation, *Neuroradiology* 36:611–612.

McFee, A. F., Robertson, S. D., and Washburn, L. C., 1994. Cytogenetic damage in marrow cells of mice after injection of ^{90}Y labeled MAb, *Nucl. Med. Biol.* 21:1109–1114.

McLean, F. C., and Budy, A. M., 1964. *Radiation, Isotopes and Bone*, Academic Press, New York, pp. 162–163.

McRaney, W., and McGahan, J., 1980. Radiation Dose Reconstruction, U.S. Occupation Forces in Hiroshima and Nagasaki, Japan, 1945–1946, Contract No. DNA 001-80-C-0052, Defense Nuclear Agency, Washington, D.C.

Migalovskaya, V. N., 1971. Chronic action of $^{90}Sr/^{90}Y$ on the frequency of chromosomal aberrations in salmon enbroyonic cells, *Tr. Polyar. Nautch, Issled. Preokt. Inst. Morsk. Ryb. Khoz. Okean.* 29:74–81 (Russian), CA 77:161583x.

Millar, W. T., Van den Aardweg, G. J. M. J., Hopewell, J. W., and Canney, P. A., 1996. Repair kinetics in pig epidermis. An analysis based on two separate rates of repair, *Int. J. Radiat. Biol.* 69:123–140.

Miller, K. L., and Weidner, W. A., 1986. *Handbook of Management of Radiation Protection Programs*, CRC Press, Boca Raton, Florida.

Mitchel, R. E. J., and Trivedi, A., 1992. Chronic exposure to ionizing radiation as tumor promoter in mouse skin, *Radiat. Res.* 129: 192–201.

Mogilevskaya, O. Ya., and Raikhlin, N. T., 1963. Biological effects of rare, dispersed, and other metals and their compounds used in industry of rare earth elements, *Toksikol. Redkhikh Metal.* 1963:195–208 (Russian). CA 69:2246b.

Mogilevskaya, O. Ya., and Raikhlin, N. T., 1967. Toxicology of rare earth metals. In: *Toxicology of the Rare Metals* (Z. I. Izrael's son, ed.), Israel Program of Scientific Translations, Jerusalem, pp. 132–141.

Morimoto, D. C., Tamarin, R. H., and Levy, C. K., 1985. Whole-body biological delimination rates of γ-emitting radionuclies in captive meadow voles *Microtus pennsylvanicus*, *Health Phys.* 49:1217–1228.

Morrison, G. M. P., Batley, G. E., and Florence, T. M., 1989. Metal speciation and toxicity, *Chem. Brit.* 1989:791–796.

Morrow, P. E., Gibb, F. R., Beiter, H., and Kilpper, R. W., 1979. Pulmonary retention of neutron-activated coal dust, *Arch. Environ. Health* 34:178–183.

Mu, C., and Shi, W., 1985. Investigation of mutagenicity and terratogenicity from rare earth compounds, in: *New Frontiers in Rare Earth Science and Application*, Science Press, Leipzig, pp. 1530–1534.

Muggenburg, B. A., Boecker, B. B., Hahn, F. F., Griffith, W. G., and McClellan, R. O., 1986. The risk of liver tumors in dogs and man from radioactive aerosols, in: *Life-Span Radiation Effects Studies in Animals. What Can They Tell Us?* (R. C. Thompson and J. A. Mahaffey, eds), Office of Scientific and Technical Information, U.S. Department of Agriculture, Washington, D.C. pp. 556–563.

Naruki, Y., Maloney, P. J., Reynolds, J. C., Larsons, S. M., and Carrasquillo, J. A., 1995. Selective cytotoxicity of ^{90}Y labeled T101 MAb on human malignant T-cell line, *Antib. Immunoconjug. Radiopharm.* 8:81–92.

Nasisse, M. P., McGahan, M. C., Shields, M. B., Echelman, D., and Fleisher, L. N., 1992. Inflammatory effects of continuous-wave Nd:YAG laser cyclophotocoagulation, *Invest. Ophthalmol. Visual Sci.* 33:2216–2223.

Neizvestnova, E. M., Katsnel'son, B. A., Davydova, V. I., Malysheva, L. G., and Konovalova, N. E., 1992. Toxicological characteristics and regulation of certain polymetallic oxides, *Gig. Tr. Prof. Zabol.* 6:23–26 (Russian). BA 96:34969.

Neizvestnova, E. M., Grehova, T. D., Privalova, L. I., Babakova, O. M., Konovalovan, N. E., and Petelina, E. V., 1994. Problems of hygienic regulation of yttrium fluorides, terbium, ytterbium and lutetium in the air of workplace, *Med. Tr. Prom. Ekol.* 7:32–35 (Russian).

Olefir, A. I., 1967. Effect of oxides, hexaborides and rare earth sulfides on some factors of natural immunity, *Vrach. Delo* 11:95–97 (Russian). CA 68:28153e.

Ootsuyama, A., Makino, H., Nagao, M., Ochiai, A., Yamauchi, Y., and Tanooka, H., 1994. Frequent p53 mutation in mouse tumors induced by repeated β-irradiation, *Mol. Carcinog.* 11:236–242.

Ozler, S. A., Hill, R. A., Andrews, J. J., Baerveldt, G., and Berns, M. W., 1991. Infrared laser sclerostomies, *Invest. Ophthalmol. Visual Sci.* 32:2498–2503.

Pavlyushchik, A. V., Golubovich, I. A., Makkaeva, M. Yu, and Stankevich, M. P., 1970. Level of sulfhydryl groups and morphological changes in rat organs after interperitoneal injection of colloidal ^{90}Y and ^{196}Au, *Med. Radiol.* 15:54–60 (Russian). Cited by Haley, (1979).

Peters, W., and Lee, P., 1994. Radiation necrosis overlying the ankle joint after injection with ^{90}Y, *Ann. Plast. Surg.* 32:542–543.

Polikarpov, G. G., 1966. *Radioecology of Aquatic Organisms*, Reinhold, New York, pp. 153–160.

Pöllänen, R., and Toivonen, H., 1995. Skin calculations from uranium fuel particles below 500 μm in diameter, *Health Phys.* 68:401–405.

Polzer, G., Matejka, M., Lind, I., Wiener-Schmuck, M., and Seidel, A., 1993. *In vitro* effects of $CeCl_3$ and $LaCl_3$ on macrophages, in: *Endomyocardial Fibrosis* (M. S. Valiathan *et al.*, eds.), Oxford University Press, Bombay, pp. 254–265.

Polzer, G., Baierl, T., Drosselmeyer, E., and Seidel, A., 1994. *In vitro* effects of high T_c superconducting materials and their components on some immuno-functions of macrophages, in: *Toxic and Carcinogenic Effects of Solid Particles on the Respiratory Tract* (D. L. Dungworth *et al.*, eds.), DILSI Press, Washington, D.C.

Poston, T. M., and Klopfer, D. D., 1988. Concentration factors used in the assessment of radiation dose to consumers of fish: A review of 27 radionuclides, *Health Phys.* 55:751–766.

Pounds, J. G., 1985. The toxic effects of metals, in: *Industrial Toxicology* (P. L. Williams and J. L. Burson, eds.), Van Nostrand Reinhold, New York, pp. 197–210.

Quereshi, A. A., Coleman, R. N., and Paran, J. H., 1984. Evaluation and refinement of the Microtox test for use in toxicity screening, in: *Toxicity Screening Procedures Using Bacterial Systems* (D. Liu, and B. J. Dutka, eds.), Marcel Dekker, New York, pp. 1–22.

Ragaini, R. C., and Ondow, J. M., 1976. Trace elements emissions from Western US coal-fired power plants, in: *Modern Trends in Activation Analysis*, Proc. Conf., Münich, pp. 654–663.

Ramsden, E. N., 1961. A review of experimental work on radio-yttrium comprising: the tissue distribution, the mechanism of deposition in bone, and the state in the blood, *Ind. J. Rad. Biol.* 3:399–410.

Rayner, B., Tutt, M., and Vaughan, J., 1953. The deposition of ^{91}Y in rabbit bones, *Brit. J. Exp. Pathol.* 34:138–145.

Reece, W. O. 1965. Effects of exercise and inhaled yttrium oxide on blood lactic acid, erythrocyte volume and lung histology in dogs, PhD thesis, Iowa State University.

Reid, R. J., Rengel, Z., and Smith, F. A., 1996. Membrane fluxes and comparative toxicities of aluminum, scandium and gallium, *J. Exp. Bot.* 47:1881–1888.

Reller, H. H., and Luedders, W. L., 1977. Pharmacologic and toxicologic effects of topically applied agents on the eccrine sweat glands. In: *Dermato-toxicology and Pharmacology* (F. N. Marzulli and H. I. Maibach, eds.), John Wiley and Sons, New York, pp. 18–35.

Rezvani, M., Hamlet, R., Hopewell, J. W., and Sieber, V. K., 1994. Time and dose-related changes in the thickness of pig skin after irradiation with single doses of $^{90}Sr/^{90}Y$ β-rays, *Int. J. Radiat. Biol.* 65:497–502.

Richard, A. M., Rabinowitz, J. R., and Waters, M. D., 1989. Strategies for the use of computational SAR methods in assessing geno-toxicity, *Mutat. Res.* 221:181–196.

Richardson, M. L., and Gangolli, S., eds., 1994. *The Dictionary of Substances and Their Effects*, Royal Society of Chemistry, London, Vol. 7, pp. 869–872.

Romney, E. M., Steen, A. J., Wood, R. A., and Rhoads, W. A., 1967. Concentration of radionuclides by plants grown on ejecta from the Sedan thermonuclear cratering detonation, in: *Radioecological*

Concentration Processes (B. Aberg and F. P. Hungate, eds), Pergamon Press, Oxford, pp. 391–398.

Santschi, P. H., and Honeyman, B. D., 1989. Radionuclides in aquatic environment, *Radiat. Phys. Chem.* 34:213–240.

Sasser, L. B., Bell, M. C., and West, J. L., 1972. Gastrointestinal irradiation of sheep with an insoluble ^{90}Y-tagged fallout simulant, *Health Phys.* 22:467–470.

Sax, S., and Lewis, R. J., 1992. *Dangerous Properties of Industrial Materials*, Van Nostrand Reinhold, New York.

Schalnus, R. W., Ohrloff, C., and Magone, T., 1995. The aqueous vitreous barrier and blood aqueous barrier after YAG laser posterior capsulotomy in sulcus versus capsular fixation of posterior chamber lens, *Ophthalmologe* 92:289–292 (German).

Schardein, J. L., 1993. *Chemically Induced Birth Defects*, Marcel Dekker, New York, pp. 722–741.

Schlatter, C., 1985. Speculations on mechanisms of metal carcinogenesis, in: *Carcinogenic and Mutagenic Metal Compounds* (E. Merian *et al.*, eds.), Gordon & Breach Science Publishers, New York, pp. 529–539.

Schmidtke, I., 1964. Removal of inhaled radioactive yttrium by the use of diethylenetriaminepenta-acetic acid, *Health Phys.* 10:1235–1241.

Schroeder, H. A., and Mitchener, M., 1971. Scandium, chromium^{6+}, gallium, yttrium, rhenium, palladium and indium in mice. Effects on growth and life span, *J. Nutr.* 101:1431–1438.

Schubert, J., and White, M. R., 1950. The effect of different dose levels of zirconium on the excretion and distribution of plutonium and yttrium, *J. Biol. Chem.* 184:191–196.

Schwieg, H., and Turba, F., eds., 1961. *Artificial Radioactive Isotopes. Physiology, Diagnosis and Therapy*, Springer-Verlag, Berlin, pp. 50–63, 270–273 (German).

Sedman, R. M., and Mahmood, R. J., 1994. Soil ingestion by children and adults reconsidered using the results of recent tracer studies, *Air Waste Manage. Assoc.* 44:141–144.

Selye, H., Tuchweber, B., and Rohan, P., 1965. Thrombosis in large veins and pulmonary embolism induced by catecholamine or serotonin, *Nature* 208:900–901.

Selye, H., Garga, S., and Winandy, G., 1966a. Production and prevention of intra-cranial venous thromboses in the rat, *Exp. Neurol.* 15:329–337.

Selye, H., Tuchweber, B., and Rohan, P., 1966b. Metal-induced sensitization for the production of different types of tissue reactions by adrenaline, *Br. J. Exp. Pathol.* 47:281–285.

Sharma, S. N., Kamboj, V. P., and Kar, A. B., 1972. Calcification of the rat testis by some rare earth and radioactive metallic salts, *Exp. Pathol.* 7:176–181.

Shevchenko, V. V., 1970. Effect of incorporated ^{90}Sr–^{90}Y radionuclides on *Crepis capillaris* chromosomes, *Radiobiologiya* 10:575–579 (Russian). CA 74:28485p.

Shrivastava, R., Delomenie, C., Chevalier, A., John, G., Ekwall, B., Walum, E., and Massingham, R., 1992. Comparison of *in vivo* acute lethal potency and *in vitro* cytotoxicity of 48 chemicals, *Cell Biol. Toxicol.* 8:157–170.

Shvydko, N. S., Ilin, L. A., Norets, R. A., and Antonova, V. A., 1972. Fixation mechanism for isotopes of the rare earth elements by individual skin structure. *Gig. Sanit.* 37:43–46 (Russian). CA 73:73557s.

Sigg, M., Crompton, N. E. A., and Burkart, W., 1996. A pure beta-like source to assess hot particle effects *in vitro*, *Health Phys.* 71:135–140.

Sittig, M., 1991. *Handbook of Toxic and Hazardous Chemicals and Carcinogens*, Noyes Publications, Park Ridge, Illinois, pp. 936–937.

Snipes, M. B., Olson, T. R., and Yeh, H. C., 1988. Deposition and retention patterns for 3-, 9-, and 15-μm latex microspheres inhaled by rats and guinea pigs, *Exp. Lung Res.* 14:37–50.

Snyder, F., 1960. Rare earth fatty liver, *Nature* 185:480–481.

Solymoss, B., Selye, H., and Gabbiani, G., 1966. Predisposition to thrombosis not reflected by the blood coagulogram, *J. Clin. Pathol.* 19:331–333.

Somers, E., 1961. The fungi-toxicity of metal ions, *Ann. Appl. Biol.* 49:246–253.

Speer, R. J., Hill, J. M., and Denton, A. D., 1961. Colloidal ^{90}Y hydroxide and ^{90}Y phosphate. Preparation and tissue distribution followiing i.v. administration, *Proc. Soc. Exp. Biol.* 106:444–448.

Spencer, H., and Rosoff, B., 1965. Removal of ^{46}Sc in man, *Health Phys.* 11:1181–1185.

Stannard, J. N., 1988. *Radioactivity and Health. A History*, Pacific Northwest Laboratory, San Diego.

Steffee, C. H., 1959. Histopathologic effects of REE administered i.p. to rats, *Arch. Ind. Health* 20:414–419.

Steidle, H., and Ding, M., 1929. Contribution to the pharmacology of rare earth elements, *Naunyn-Schmiedeberg's Arch. Exp. Pathol. Pharmacol.* 141:273–279 (German).

Stettler, L. E., Proctor, J. E., Platek, S. F., Carolan, R. J., Smith, R. J., and Donaldson, H. M., 1988. Fibrogenicity and carcinogenic potential of smelter slags used as abrasive blasting substitutes, *J. Toxicol. Environ. Health* 25:35–56.

Takamura, K., Hayashi, K., Ishinishi, N., Yamada, T., and Sugioka, Y., 1994. Evaluation of carcinogenicity and chronic toxicity associated with orthopedic implants in mice, *J. Biomed. Mater. Res.* 28:583–589.

Talbot, R. B., Davidson, F. C., Green, J. W., Jr., Reece, W. O., and Vangelder, G., 1968. Effects of subcutaneous implantation of REE in mice, *Bull. Ga. Acad. Sci.* 26:1–9.

Talburt, D. E., and Johnson, G. T., 1967. Some effects of rare earth elements and yttrium on microbial growth, *Mycologia* 59:492–503.

Taylor, D. M., 1983. The comparative retention of bone-seeking radionuclides in the skeletons of rats, *Health Phys.* 45:768–772.

Thies, H. A., Oppelt, W., Noddack, W., and Noddack, I., 1962. Experiments with animals for studying the anti-coagulative effects of twelve REE, *Med. Welt* 16:891–896 (German).

Thomas, G. E., Esteban, J. M., Raubitschek, A., and Wong, J. Y. C., 1995. Gamma-interferon administration after ^{90}Y radiolabeled antibody therapy: Survival and hematopoietic toxicity studies, *Int. J. Radiat. Oncol. Biol. Phys.* 31:529–534.

Thomassen, P. R., and Leicester, H. M., 1964. Uptake of radioactive berryllium, vanadium, selenium, cerium and yttrium in the tissue and teeth of rats, *J. Dent. Res.* 43:346–352.

Timofeeva-Resovskaya, E. A., 1963. Distribution of radioisotopes in basic components of fresh waterbodies, *Tr. Inst. Biol. Akad. Nauk SSR Ural.* 30:1–79. English transl. JPR Software, Winter Park, Florida 21816.

Tölgyessy, J., ed., 1993. *Chemistry and Biology of Water, Air and Soil, Environmental Aspects*, Elsevier, Amsterdam.

Tuchweber, B., and Savoie, L., 1968. REE and soft tissue calcification, *Proc. Soc. Exp. Biol. Med.* 128:473–476.

Turner, J. E., Williams, M. W., Jacobson, K. B., and Hingerty, B. E., 1985. Correlations of acute toxicity of metal ions and the covalent/ionic character of their bonds, *Pharmacochem. Libr.* 8:171–178.

Vakulovsky, S. M., Kryshev, I., I., Nikitin, A. I., Savitsky, Y. V., Malyshev, S. V., and Tertyshnik, E. G., 1995. Radioactive contamination of the Yenisei River, *J. Environ. Radioact.* 29:225–226.

Van Middlesworth, L., 1947. Study of plutonium metabolism in bone, PhD thesis. Cited by J. N. Stannard (1988).

Varga, S., Solymoss, B., and Rohan, P., 1967. The alterations of venous endothelium in experimental thrombosis produced by catecholamine and metals, *Pathol. Microbiol.* 30:139–145.

Vasiliev, A. N., Zhuravl', N. E., Onis'ko, T. G., 1995. Biogeochemistry of *Dreissena* shells from a reservoir with anomalous thermal conditions, *Dop. Nats. Akad. Nauk Ukr.* 8:145–147.

Venugopal, B., and Luckey, T. D., 1978. *Metal Toxicity in Mammals*, Plenum Press, New York, Vol. 2.

Vriesendorp, H. M., Quadri, S. M., Anderson, B. S., and Dicke, K. A., 1996. Hematologic side effects of radiolabeled immunoglobulin therapy, *Exp. Hematol.* 24:1183–1190.

Waldmann, T. A., White, J. D., Carrasquillo, J. A., Reynolds, J. C., Paic, H. C., Gansow, O. A., Brechbiel, M. W., and Nelson, D. L., 1995. Radioimmunotherapy of interleukin-2Rα-expressing adult T-cell leukemia with ^{90}Y-labeled anti-Tac, *Blood* 86:4063–4075.

Warner, F., and Harrison, R. M., eds., 1993. *Radioecology after Chernobyl*, John Wiley and Sons, Chichester, pp. 16–27.

Wehner, A. P., Wilkerson, C. L., Mahaffey, J. A., and Milliman, E. M., 1980. Fate of inhaled fly ash in hamsters, *Environ. Res.* 22:485–498.

Wehner, A. P., Wilkerson, C. L., and Stevens, D. L., 1984. Lung clearance of neutron-activated Mount St. Helens volcanic ash in the rat, *Environ. Res.* 35:211–217.

Wen, H., Guo, G., and Wang, J., 1995. Effect of rare earth compounds on alveolar macrophage *in vitro*, *Wuji Huaxue Xuebao* 11:389–393 (Chinese). CA. 124:138136j.

Whicker, F. W., and Schultz, V., 1982. *Radioecology. Nuclear Energy and the Environment*, CRC Press, Boca Raton, Florida, Vol. 1, pp. 157–158.

Wiener-Schmuck, M., Lind, I., Polzer, G., and Seidel, A., 1990. *In vivo*, and *in vitro* effects of rare earth compounds, *J. Aerosol Sci.* 21:s505–s508.

Wilczek, W., Drosselmeyer, E., and Seidel, A., 1989. The *in vivo* effects of high-T_c-superconduting particles (YBa_2Cu_{6-7}) and quartz (SiO_2) on bovine alveolar macrophages, *Exp. Pathol.* 37:269–272.

Wilczek, W., Drosselmeyer, E., Krug, H. F., and Seidel, A., 1990. Effects of high T_c-superconducting materials on alveolar macrophages, *J. Aerosol Sci.* 21:s513–s516.

Williams, D. R., 1983. Computer model of metal biochemistry and metabolism, in: *Chemical Toxicology and Clinical Chemistry of Metals* (S. S. Brown and J. Savory, eds.), Academic Press, London, pp. 167–182.

Xiao, B., Li, Q., and Xiao, B., 1992. Lung granuloma induced by the powder of yttria and its mineral in rats. Observations on histopathology and enzymes, *Gongye Weisheng Yu Zhiyebing* 18:79–81 (Chinese). CA 118:53787y.

Yang, C. S., Schaedle, M., and Tepper, H. H., 1989. Phytotoxicity of scandium in solution culture of loblolly pine (*Pinus taeda* L.) and honey locust (*Gleditsia triacanthos* L.), *Environ. Exp. Bot.* 29:155–164.

Zhang, K., Tian, Z., Tan, C., Huang, S., and Huang, R., 1985. Study on teratogenic effect of rare earth nitrate on rats, *J. Chin. Rare Earth Soc.*, Special Issue, pp. 33–39.

Zhao, G., Li, S., and Ji, Y., 1992. Experimental pathologic observation of rare earth element dusts, *Gongye Weisheng Yu Zhiyebing* 18:86–89 (Chinese). CA 118:53789a.

Zhong, N. C., Deng, R. W., and Wu, J. G., 1992. Synthesis, characterization and anti-inflammatory activity of naproxen complexes with REE, *J. Inorg. Biochem.* 47:81–87.

Zhou, T., Shen, H., Zou, S., Duan, M., Song, L., and Dong, Z., 1992. Cytotoxicity of five rare earth dusts, *Gongye Weisheng Yu Zhiyebing* 18:82–85 (Chinese). CA 118:53788z.

Zimakov, Yu. A., and Zimakova, I. E., 1971. Effect of rare earth element salts on antibody formation during primary immunological response, *Tr. 2nd Mosk, Med. Inst.* 3:93–99 (Russian). CA 79:51706y.

Zimmerman, A., Schubiger, P. A., Mettler, D., Geiger, L., Triller, J., and Rösler, H., 1995. Renal pathology after arterial ^{90}Y microspheres administered in pigs, *Invest. Radiol.* 30:716–723.

Biochemical and Other Techniques That Employ Yttrium and Scandium. Outlook

10

Les belles actions cachées sont les plus estimables.

Blaise Pascal (1623–1662)

10.1 Introduction

Much information has been accumulated in the last two decades about the abilities of scandium and yttrium to serve as inert markers in nutrition research and the investigation of metabolic processes; also, they serve as informative probes of the interactions of calcium, magnesium, aluminum, and other metals with molecules of biochemical interest.

In this chapter, we give emphasis to the applications of stable and radioactive scandium and yttrium in biochemistry, physiology, and toxicology, as well as in medicine, agriculture, and environmental processes. Their numerous applications in technology and industry are only briefly considered insofar as these may have health and environmental implications (Himmelstein and Levy, 1986; Sax and Lewis, 1992).

10.2 Stable and Radioactive Isotopes as Markers in Nutrition Research

Pioneering work with radioisotopes of scandium, yttrium, Ln, and heavy metals as nutrient markers was done by Hutcheson and Luckey at the University of Missouri (Gray and Vogt, 1974; Hutcheson *et al.*, 1979, 1983) and Miller at the University of Tennessee (Miller and Byrne, 1970; Miller *et al.*, 1972). Investigations in nutrition and biochemistry of animals and humans used ^{46}Sc, ^{90}Y, and ^{47}Sc with efficiency (Table 10-1). Scandium-46 and several Ln radioisotopes found applications in nutrition studies with an irradiation of approximately 20 min and a decay of approximately 1 wk (Gray and Vogt, 1974). Scandium-47 is a suitable fecal marker, when the excretion of ^{51}Cr, used to tag hemoglobin, is being studied (Pearson, 1966).

Table 10-1. Use of Scandium Radioisotopes as Nutrition Markers

Radioisotope(s)	Application	Reference
^{46}Sc	Monitoring of nutrient intake in laboratory animals	Gray and Vogt (1974)
^{46}Sc chloride	Determination of fecal fat excretion in patients	Barnard (1973)
^{46}Sc	Measurement of rumen dry matter passage in buffalo	Kishan *et al.* (1982)
^{46}Sc, ^{90}Y, and REE isotopes	Measurement of multiple markers by neutron activation analysis	Pond *et al.* (1985)
^{47}Sc	Measurement of calcium absorption in humans	McCredie *et al.* (1984)
^{47}Sc	Inert marker of ^{47}Ca absorption	Ogg *et al.* (1968)
^{47}Sc	Inert fecal marker	Pearson (1966)

Multiple markers (^{46}Sc and seven Ln radioisotopes) were employed to measure several processes simultaneously in the same animal (cattle or sheep). It was found that 4 ng of Sc^{3+} introduced in the diet was required for the quantitative measurement of this marker within a digest or fecal matrix of 1 g by neutron activation analysis (Pond *et al.*, 1985). The measurement of the rumen dry matter passage and turnover rates in buffalo fed on various ration combinations was achieved with ^{46}Sc (Kishan *et al.*, 1982). Multiple nutrient markers in a single sample, consisting of oxides of scandium and other Ln, were employed for the determination of intake and apparent utilization of nutrients in animals and in humans (Miller *et al.*, 1972). Scandium oxide was given as an intake marker at the rate of 8.5 μg Sc^{3+}/g of fat, whereas ytterbium, europium, and samarium were employed as markers of calcium and protein metabolism and energy values, respectively. Among the five subjects involved in the study, the marker concentrations of scandium that were detected in feces varied between 37.5 and 47.8 mg Sc^{3+}/kg. These concentrations were similar to those of lanthanum, but fivefold higher than those of samarium. The method permitted the monitoring of any of several nutrients (protein, fat, calcium, and energy), without direct monitoring of food intakes or complete fecal collection (Hutcheson *et al.*, 1979).

Recognition of biochemical similarities between pigs and humans has prompted studies into the nutrition and physiological states of the pig. A kinetic study of scandium and other Ln oxides as markers in the alimentary tract of the pig showed that the rate of passage of scandium through the entire g.i. tract was 3.8 % per hour; values of 1.9–2.7 % were obtained for europium, samarium, terbium, and ytterbium, and 4.8 % as found for lanthanum (Hutcheson *et al.*, 1983). Scandium and the other markers were found to have similar alimentary kinetics in each individual. Fecal fat excretion was measured in humans with ^{125}I-labeled triolein and ^{46}Sc as an unabsorbable marker. A disadvantage of this routine total collection is that most of the activity is excreted almost as bolus in some samples (Barnard, 1973). Other data on the uses of radioactive isotopes as markers in nutrition research are presented in section 8.9.

10.3 Mimicking Action of Calcium Cellular Fluxes and Metabolism

The use of ^{48}Ca as a stable isotope tracer in biological systems was accomplished by proton activation via the (p,n) reaction with subsequent measurement of γ-ray spectra from ^{48}Sc ($t_{1/2} = 44$ h) and ^{44}Sc ($t_{1/2} = 4$ h). With protons of about 5.5 MeV and a Ge(Li) γ-ray detector, and no chemical processing of the samples, the ratios of these two radioisotopes were determined with good precision. This method allowed ^{48}Ca tracer studies with a sensitivity at least equivalent to that obtained with the stable ^{45}Ca tracer (Edwards *et al.*, 1977).

A fecal recovery method with ^{47}Sc was developed for the measurement of the intestinal absorption of calcium in patients with renal disorders (Ogg *et al.*, 1968; Fig. 10-1). A good correlation was found between the values for calcium absorption obtained by the single-tracer fecal ^{47}Ca recovery method and the ^{47}Sc/^{47}Ca ratio technique for studying intestinal absorption of calcium in animals and humans. This latter method offers the advantage that it requires neither blood sampling nor complete faecal collection. Calcium-47 provides its own marker by its decay to ^{47}Sc. Furthermore, the effect of agents that inhibit or stimulate intestinal calcium absorption was more apparent in treated rats with the ^{47}Sc/^{47}Ca ratio method than with the fecal ^{47}Ca recovery technique (Fig. 10-2). A mean recovery of 96 % was obtained after intragastric administration of scandium (McCredie *et al.*, 1984). Because ^{47}Sc is not absorbed by the human intestine (Pearson, 1966), an accurate measurement of the intestinal absorption of ^{47}Ca was achieved with ^{47}Sc. The high correlation between the ^{47}Sc method and the unreliable fecal recovery method showed that the latter method is unnecessary. Scandium-47 was also used for bone scintigraphy and was an adequate tracer for calcium in the studying of osteopathic disorders. Owing to the 3.35-day half-life of ^{47}Sc, when injected intravenously, it was deposited in liver, kidney, spleen and liver, whereas ^{47}Ca was mostly fixed in bone (Basse-Cathalinat *et al.*, 1968). The image obtained from rats and rabbits after 8 days showed a better definition and a higher contrast than usually obtained with ^{47}Ca or ^{85}Sr.

Intrasellar implantation of ^{90}Y in patients with Cushing's syndrome caused a drop in urinary excretion of adrenal steroids and changes in carbohydrate and calcium metabolism (Molinatti *et al.*, 1967). In all patients, a notable decrease of calciuria and a clear decrease in urine calcium during an infused calcium load

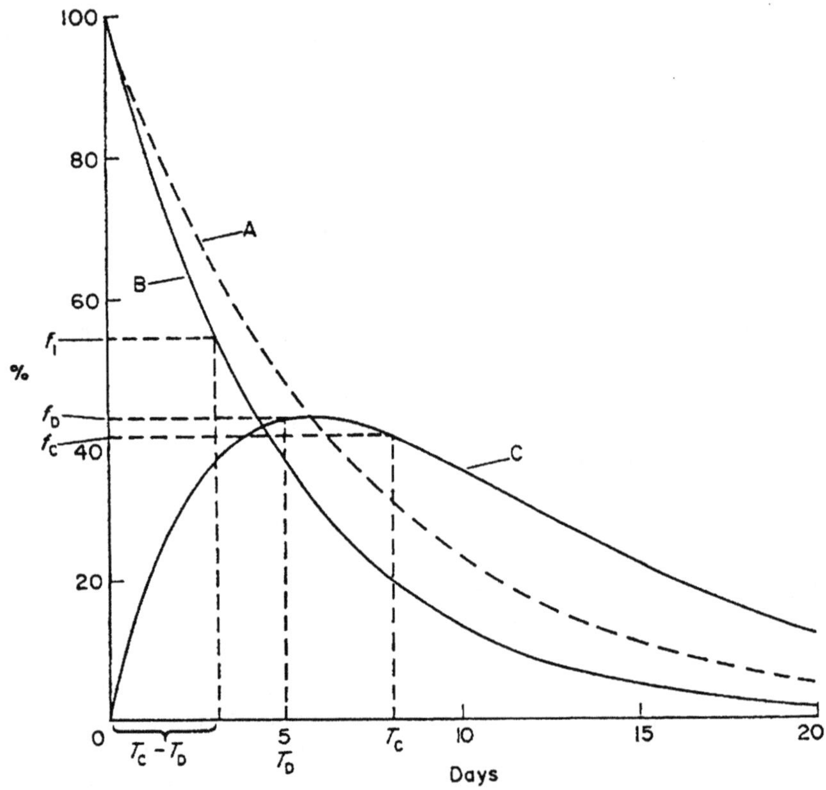

Figure 10-1. Correlation between results obtained by the ^{47}Sc inert marker technique (B) and the fecal recovery method (A) of measuring the intestinal absorption of ^{47}Ca in humans. From these two curves, the growth and decay curves for ^{47}Sc and ^{47}Ca are plotted (C). The vertical and horizontal dashed lines represent the 95 % confidence limits of the observations. Reprinted from Ogg *et al.* (1968), with permission of the Biochemical Society and Medical Research Society, London.

was observed, in contrast to the preexisting pattern (Fig. 10-3). The correct implantation of ^{90}Y in the pituitary caused complete regression of Cushing's syndrome (Molinatti *et al.*, 1967).

10.4 Blood Flow and Microsphere Therapy

Microspheres (MS) represent a convenient way of delivering radiation to tumors. The administration of 58 mCi of ^{90}Y attached to MS allowed roughly 10,000 rads of irradiation to be delivered to the liver (Ariel and Padula, 1984). The 2.18-meV pure beta emission, with a maximum penetration in tissue of

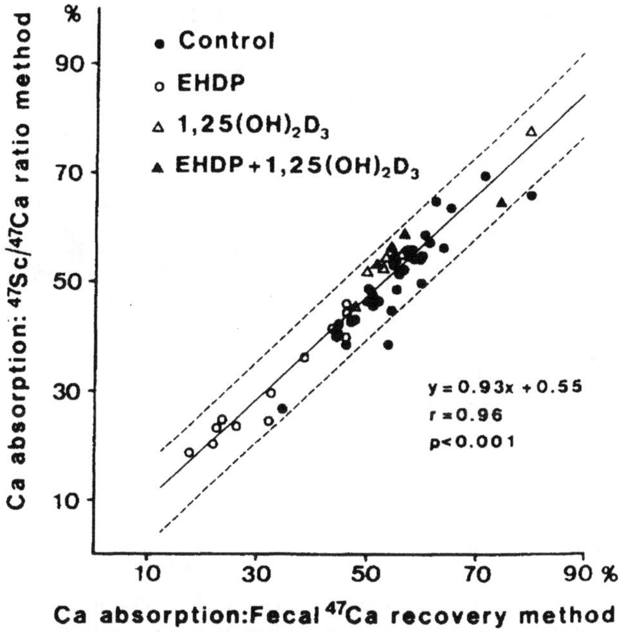

Figure 10-2. Correlation between the *in vivo* Ca^{2+} absorption in rats determined by the $^{47}Sc/^{47}Ca$ ratio method and that determined by the fecal ^{47}Ca recovery method for controls and animals treated with 1-hydroxyethane-1,1-diphosphonate (EHDP) and $1,25(OH)_2D_3$. EHDP and $1,25(OH)_2D_3$ 1.25-dihydroxychole calciferol are agents that respectively decrease or increase Ca^{2+} absorption. The dashed lines represent the 95 % confidence limits of the regression line. Reprinted from McCredie *et al.* (1984), with permission of the authors and Mosby-Year Book, Inc.

8 mm, limits the irradiation to the immediate vicinity of the MS. Combined chemotherapy and ^{90}Y MS radioisotope therapy yielded good results in the treatment of liver and brain metastases, asymptomatic cancers, and radiation splenectomy (Ariel and Padula, 1984; Fig. 10-4).

The measurement of cardiac output or regional blood flow with the use of ^{46}Sc-labeled MS in many animal species and in humans approached 100 % extraction efficiency of labeled MS. Hamlin *et al.* (1962) measured the cardiac output in the lung of a dog with ^{46}Sc-labeled MS. Table 10-2 summarizes investigations of the blood flow in various organs of animals that were conducted with ^{46}Sc or tracer imaging with double or multiple tracers (Medvedev *et al.*, 1987). By using batches of MS labeled with ^{46}Sc and other radioisotopes, several flow measurements can be made in the same animal (Fig. 10-5).

A computer program based on a stripping method for analyzing data from multiple infusion MS blood flow and cardiac output experiments was developed by Irion *et al.* (1988). This program measured the actual number of counts per

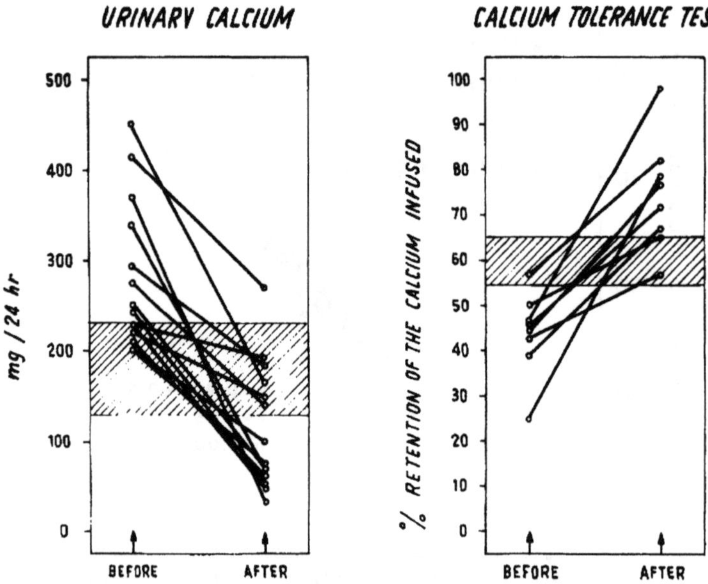

Figure 10-3. Changes in calciuria and in calcium tolerance test values following pituitary implantation of ^{90}Y in patients with Cushing's syndrome. Reprinted from Molinatti *et al.* (1967), with permission of the Endocrine Society.

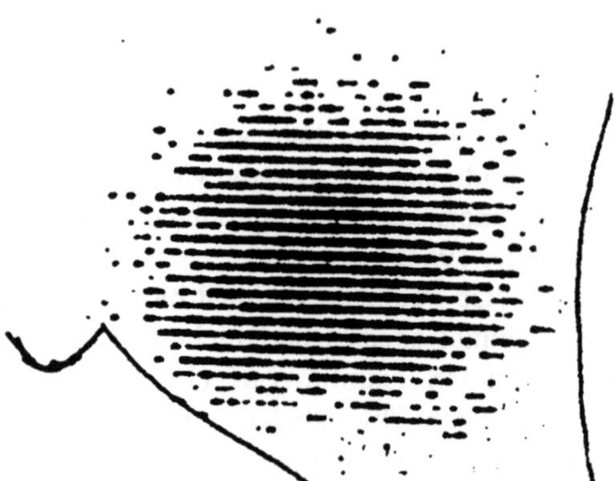

Figure 10-4. Scan of the spleen of a patient with liver cancer, which had spread to other vital organs. The bremsstrahlung of ^{90}Y, shortly after the administration of 50 mCi of ^{90}Y, shows a uniform distribution. Reprinted from Ariel and Padula (1984), with permission of Grune & Stratton.

Table 10-2. Studies of Blood Flow in Animals with ^{46}Sc

Blood flow	Animal	Method	Reference
Brain	Dogs	Multiple imaging	Skarstein (1979)
Brain cerebellum	Baboons	Cerebral blood flow	McCalden (1979)
Cardiac output	Rats	^{46}Sc microspheres	Flaim *et al.* (1984)
Carotid artery	Dogs	^{46}Sc–^{57}Co imaging	Kim *et al.* (1988)
Heart	Dogs	Autoradiography	L'Abbate *et al.* (1961)
Kidneys	Rabbits	^{46}Sc–^{169}Yb imaging	Axen and Janson (1979)
Placenta	Guinea pigs	Isoproterenol infusion	Myers *et al.* (1986)
Muscle tumor	Rabbits	$C^{15}O_2$ steady state	Kairento *et al.* (1983)
Myocardial ischemia	Dogs	^{46}Sc microspheres	Caldwell *et al.* (1990)
Myocardial ischemia	Dogs	Uptake of [^{14}C]deoxyglucose	Bianco *et al.* (1988)
Follicular ovary	Rabbits	^{46}Sc–^{113}Sn imaging	Janson (1975)
Pancreas	Piglets	^{46}Sc–^{169}Yb imaging	Lehtola *et al.* (1986)
Stomach	Cats	Multiple imaging	Gross *et al.* (1981)
Urinary bladder	Dogs	^{46}Sc–^{99m}Tc imaging	Kroyer *et al.* (1990)
Blood–brain barrier	Rats	Diffusion technique	Sawada *et al.* (1989)
Various tissues and organs	Humans	Nonradioactive Y_2O_3 microspheres	Mori and Takanashi (1992)

minute (cpm) in a preset window for a given isotope; only radioisotopes that spill over from a higher to a lower energy range were employed. Specific activities of ^{46}Sc-labeled MS were 19.7 cpm. The protocol was designed to reduce the amount of arterial blood that must be withdrawn during MS distribution; this avoids contamination of reference samples by radioisotopes used in previous flow measurements (Irion *et al.*, 1988).

Intralymphatic administration of ^{90}Y- and ^{46}Sc-labeled MS produced promising results for the treatment of patients suffering from cancer (Mantravadi *et al.*, 1982; Ariel and Padula, 1984; Wollner *et al.*, 1988; Schubiger *et al.*, 1991). An improved method was worked out, which avoids the nonuniformity of hepatic deposition for ^{90}Y-labelled glass-MS used in cancer therapy (Roberson *et al.*, 1992). A three-dimensional tumor dose calculation and treatment planning system was investigated in experiments with rabbits (Fig. 10-6).

A matrix-type polymeric drug delivery system employing poly(lactic acid) (PLA) MS as carriers for the ionic form of ^{90}Y was developed (Häfeli *et al.*, 1997), with application in magnetically enhanced radionuclide therapy (MERIT) (Rayman *et al.*, 1997). The PLA MS could be selectively delivered to the target site in cancer treatment. Incorporation of 10 % Fe_3O_4 (magnetite) made the magnetic MS responsive to an external magnetic field (Fig. 10-7). Stability studies showed that approximately 95 % of added ^{90}Y was retained within the PLA matrix after 28 days (Häfeli *et al.*, 1994).

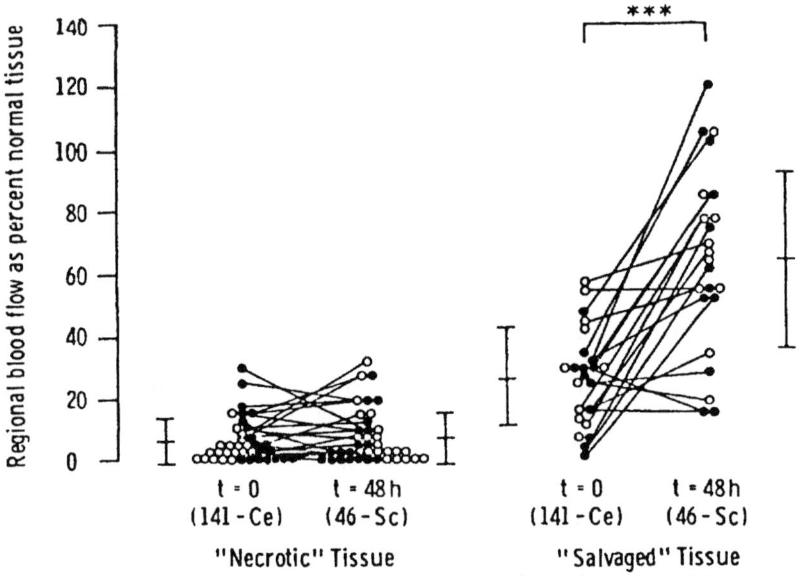

Figure 10-5. Changes in blood flow in dogs in a 48-h period following the onset of ischemia ($t = 0$ on the left-hand side of each group of results) in various reactions of the necrotic zone. Blood flow values are expressed as percentages of that in normal well-perfused tissue. Preferential reflow in "salvaged" tissue (right), as compared to "necrotic" tissue (left), is observed. ○ Control dogs; ● nifedipine-treated dogs. Reprinted from Kudoh *et al.* (1986), with permission of the authors and Academic Press.

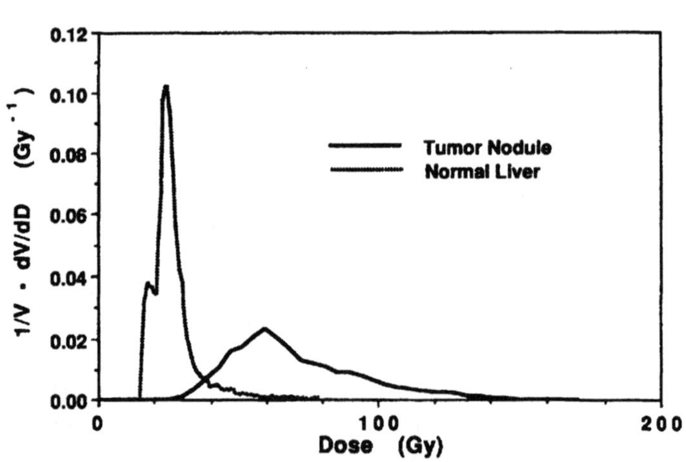

Figure 10-6. A differential dose–volume histogram for a normal rabbit liver and a tumor nodule treated with ^{90}Y-labeled microspheres. The ordinate values are the differential fractional volumes (dV/V) containing the differential dose increment (dD). The curves are normalised so that their integrated areas are unity, representing the full volume. Reprinted from Roberson *et al.* (1992), with permission of the authors and the Society of Nuclear Medicine.

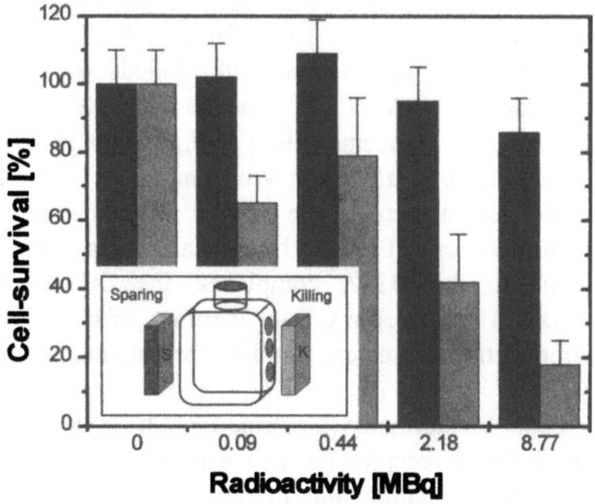

Figure 10-7. Survival of SK-N-SH neuroblastoma cells after incubation with ^{90}Y-labeled magnetic microspheres (MMS) for 24 h, as measured by the MTT [3-(4,5-dimethylthiazol-2-yl)-2,5-diphenyltetrazolium bromide] assay. The inset shows the position of the cells growing on the flask wall. After ^{90}Y-MMS were added to the flask, a magnet was placed either adjacent to the cells ("killing magnet," box "K"; cell survival data represented by shaded bars) or on the side opposite to the cell layer ("sparing magnet," box "S"; cell survival data represented by solid bars). The data are expressed as percentages of the control cell survival (100 %) with no added radioactivity. Reprinted from Häfeli *et al.* (1994), with permission of John Wiley and Sons, Inc.

10.5 Radiolabeling of Antibodies and Other Compounds

Much current research is focused on imaging and cancer therapy with suitable radioisotopes. The versatility of ^{46}Sc, and especially ^{90}Y and ^{212}Bi, for radiolabeling of MAbs was demonstrated by Gansow (1991). The development of many classes of MAbs (Fig. 7-7) and polyclonal antibodies (PAbs) and newer coupling chelators has stimulated interest in the use of radionuclide metals in cancer therapy (Order *et al.*, 1986; White *et al.*, 1996) and in radiolabeled immunoglobulin therapy (Vriesendorp *et al.*, 1996). Yttrium-90 is among the frequently selected radioisotopes. Among its desirable properties are (a) availability from a long-lived parent ($t_{1/2} = 28$ years), (b) decay to a stable daughter, (c) pure β emitter, (d) suitable half-life of 64 h, and (e) good chelation properties (Coursey *et al.*, 1993). The production of ^{90}Y for clinical radiolabeling requires a low level of TE contamination, because some TE, such as iron, compete with ^{90}Y for antibody binding sites, and a low $^{90}Sr/^{90}Y$ ratio (Wike *et al.*, 1990).

Chelating agents such as cyclic anhydride diethylenetriaminepentaacetic acid (DTPA) (Hnatowich *et al.*, 1985, 1988; Brechbiel and Gansow, 1991; Mardirossian *et al.*, 1993) and 1,4,7,10-tetraazacyclododecane-*N*-[Gly$_3$ (L-(*p*-isothiocyanate)-Phe-amide)acetyl]-*N'*,*N''*,*N'''*-triacetic acid (Meares *et al.*, 1995) have been synthesized and used (Anderson-Berg *et al.*, 1987; Hyams *et al.*, 1987, Snook *et al.*, 1991; Deshpande *et al.*, 1990; Wu *et al.*, 1992). The use of these chelating agents is intended to increase the percent injected dose per gram of tissue, while decreasing the percent uptake by critical organs, such as the bone marrow. As compared to the DTPA complexes, the DOTA-^{90}Y complex conjugated to MAB B72.3 displayed a greater inertness of blood circulating in the marrow and, by implication, a greater suitability for immunotherapy (Harrison *et al.*, 1991).

The ^{90}Y-labeled anti-Tac antibody (a murine IgG 2a class MAb) specific to the high-affinity interleukin-2 receptor (IL-2R) was reported to have potential applications in organ transplantation and in the treatment of Tac-expressing neoplastic diseases (Cooper *et al.*, 1990). Anti-Ly1 MAb (the murine homolog of human CD5) labeled with ^{90}Y exhibited a prolonged and selective protection against the T-cell mouse lymphoma EL4 (Schmidberger *et al.*, 1991) and graft-versus-host disease in mice (Vallera *et al.*, 1991).

When antibodies produced for tumor scanning or therapy and labeled with ^{47}Sc were administered, high radioactivity was found in the liver, associated with the blood clearance (Vaughan *et al.*, 1987). High-resolution camera images of mouse erythroid tumors labeled with bifunctional radioactive scandium and other metal chelates were achieved with leukemia-cell-specific MAbs. Chelate-conjugated antibodies (2 mg/ml) labeled to high specific activity with 1 mμCi of ^{46}Sc were also found to be versatile; ^{47}Sc and other radioisotopes having different half-lives and emission spectra yield superior imaging. Both IgGs and IgMs were labeled with scandium, indium, or gallium radioisotopes. The advantage of labeling with radioactive metal chelates is that if the chelate becomes unconjugated from the antibody during metabolism, the protected radioisotope is excreted immediately. The chelate-conjugated MAbs show therapeutic potential when chelated with ^{47}Sc, (Scheinberg *et al.*, 1982).

Static and dynamic imaging of tumor slices with negatron and positron-emitting radioisotopes was achieved with the use of a new detection system, employing a beta camera based on a plastic scintillator and a light-sensitive detector (Ljunggren and Strand, 1991). The ability of this system to image the energetic β-particles ($E_{max} = 2.27$ MeV) emitted by ^{90}Y-labeled MAbs in human ovarian tumors in mice was demonstrated (Fig. 10-8).

More examples of applications of yttrium and scandium radioisotopes for labeling MAbs are presented in Table 10-3. The biochemical and physiological aspects of complexes of these radioisotopes with MAbs are presented in more detail in Section 7.4.

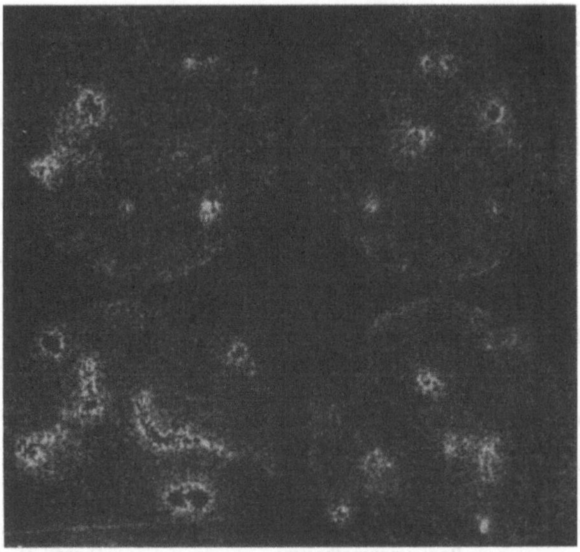

Figure 10-8. A beta-camera image of human ovarian tumors in mice injected with ^{90}Y-labeled monoclonal antibodies. Reprinted from Ljunggren and Strand (1991), with permission of Elsevier Science.

10.6 Other Biochemical and Medical Applications

In addition to ^{46}Sc and ^{90}Y, which have found a large number of applications in various fields of medicine (Evans 1990; Wike *et al.*, 1990; Rakovic *et al.*, 1991; Tilyou, 1991; Horwitz and Dietz, 1994; Salako and DeNardo, 1997; Franssen *et al.*, 1997), other radioisotopes and the stable isotope ^{45}Sc are also employed (Herzog *et al.*, 1993; Rowlinson-Busza *et al.*, 1994; Macklis *et al.*, 1994; Andrews *et al.*, 1994; Hopkins *et al.*, 1995; Dillehay *et al.*, 1995; Tian *et al.*, 1996; Table 10-4). Yttrium-87 was complexed with ethylenediaminetetramethylenephosphonate (EDTMP) for *in vivo* imaging and dosimetry of soft tissues (Keeling and Beaney, 1991). Implanted ^{90}Y rods were used for interstitial irradiation of hypertrophic soft tissues of a patient with active acromegaly. Decrease of the levels of growth hormone (GH) to 50 % of baseline values within 12 months of this kind of therapy suggested a potential cure of acromegaly (MacSweeney *et al.*, 1990).

Koral *et al.* (1993) utilized a fast Fourier transform method to calculate the three-dimensional dose-rate distribution for murine, human tumor xenografts. The ^{90}Y dose-rate distribution was found generally more uniform than either the activity distribution of the ^{131}I dose-rate distribution. Given the long range of the

Table 10-3. Use of Yttrium and Scandium Radioisotopes to Label Chelators

Radiolabeled chelator	Application	Reference
^{90}Y-CITC-DTPA	Ovarian cancer therapy	Kosmas *et al.* (1995)
^{88}Y-SCN-Bz-DTPA	Minimization of marrow toxicity	Roselli *et al.* (1989)
^{90}Y-chimeric L6	Human breast cancer therapy	DeNardo *et al.* (1995)
^{90}Y-EDTMP	Treatment of leukemias	Keeling *et al.* (1989)
^{90}Y-anti-Ly1	Therapy of T-cell lymphoma EL4	Schmidberger *et al.* (1991)
^{90}Y-antifibrin Fab	Intratumoral radioimmunotherapy	Kairemo *et al.* (1993)
^{90}Y-antiferritin	Treatment of resistant Hodgkin's disease	Herpst *et al.* (1995)
^{88}Y-Janus 2-DOTA	Model of radioimmunotherapy	Goodwin *et al.* (1994)
^{90}Y-RS7	Radioimmunotherapy of lung carcinoma	Stein *et al.* (1994)
^{90}Y-DOTA-avidin, biotin	Three-step tumor targeting	Chinol *et al.* (1997)
^{90}Y-A33 IgG	Recognition of tumor-associated antigens on human colorectal carcinoma	Antoniw *et al.* (1996)
^{90}Y-HMFG1-CITC-DTPA	Treatment of head and neck cancer	Maraveyas *et al.* (1995)
Long-lived ^{91}Y	Radioimmunotherapy	Howell *et al.* (1994)
^{47}Sc-EDTA derivatives	Tumor imaging	Scheinberrg *et al.* (1982)
Sc-L-Bz-EDTA	Marking MAb with dual antigen specificity	Reardan *et al.* (1985)
^{46}Sc-DTPA	Optimization of targeting virus-infected spleen	Anderson and Strand (1985)
^{47}Sc-DTPA	Cytotoxic radiolabel	Vaughan *et al.* (1987)
^{47}Sc-MOPC-21	Diagnosis of breast and colon cancer	St. Raymond (1990)

[a]Chelators: A33 IgG, MAb which recognizes a tumor-associated antigen in human colorectal carcinoma; CITC, (S)-4-[2,3-bis(carboxymethyl)aminoprolyl]isothiocyanate; DTPA, diethylenetriaminepentaacetic acid; HMFG1-CITC-DTPA, a radioimmunoconjugate; Janus 2-DOTA, bivalent hapten of 1,4,7,10-tetraazacyclododecanetetraacetic acid; EDTA, ethylenediaminetetraacetic acid; EDTMP, ethylenediaminetetramethylene phosphonate; MOPC-21, a nonspecific mouse antibody; RS7, murine MAb IgG1 with pancarcinoma reactivity; SCN-Bz-DTPA, chelate conjugate with isothiocyanatobenzyl group covalently linked to the carbon backbone of DTPA.

^{90}Y β-particles, the results showed the ^{90}Y dose-rate distribution to be the least nonuniform (Fig. 10-9).

Since aluminum has no radioisotopes suitable for use in imaging, ^{46}Sc was applied for the investigation of localization of aluminum in parathyroid adenomas in rats. Unlike the 440-keV β emission of ^{47}Sc, which may limit the allowable trace dose, ^{46}Sc may be useful in multiplane tomographic scanning for visualizing small and larger adenomas (Cann and Prussin, 1980).

When scandium citrate was employed as an agent for the detection of malignancies, blood clearance following i.v. administration in humans was greater than that seen with ^{67}Ga (Hayes and Hübner, 1983). Tracheobronchial deposition, retention, and clearance of inhaled particles in rabbits and rats was

Table 10-4. Biochemical and Medical Uses of Yttrium and Scandium

Material or method	Application	Reference
^{90}Y-binding prostate-specific antigen	Therapy of nonprostatic endocrine cancer	Diamandis and Redshaw (1995)
^{90}Y bremsstrahlung	Radioimmunotherapy	Shen *et al.* (1994)
^{90}Y-EDTA solution	Protein surface mapping	Greiner *et al.* (1996)
^{90}Y-lipidol	Carrier in a model of liver cancer	Lin *et al.* (1995)
^{90}Y seeds	Implantation into pituitary gland of Cushing's disease patients	Burke *et al.* (1973)
^{90}Y implantation	Treatment of diabetic retinopathy	Kohner *et al.* (1969)
^{90}Y irradiation	Treatment of cystic craniopharyngiomas	Lange *et al.* (1995)
^{90}Y-glass microspheres	Toxicity tests with rabbits	Yan *et al.* (1993)
^{90}Y microspheres	Treatment of liver metastases	Ho *et al.* (1996)
^{90}Y irradiation	Treatment of epilepsy	Talairach and Szikla (1965)
^{90}Y implantation	Treatment of endocrine malfunction	Nother (1959)
^{90}Y pituitary	Treatment of Nelson's syndrome	Cassar *et al.* (1976)
^{90}Y-Ca oxalate	Treatment of hemophilic arthropathy	Dawson *et al.* (1994)
^{165}Dy-^{90}Y silicate	Radiation synovectomy in knee	Edmonds *et al.* (1994)
Yttrium-modified hapten	Imaging and therapeutic agent	Anderson *et al.* (1989)
$^{90}Y_2O_3$ powders with poly(vinyl alcohol)	Selective internal radiation therapy	Gray *et al.* (1995)
^{90}Y embolization	Treatment of congestive hypersplenism	Becker *et al.* (1995)
^{86}Y citrate	Radiation in prostate cancer	Rösch *et al.* (1996)
Y oxysulfide	Visualization of multicolor luminescent immunophosphors	Beverloo *et al.* (1992)
Y-TCMT[a]	Catalyst for hydrolytic cleavage of RNA	Morrow (1996)
^{89}Y NMR	Probe for Ca-binding protein	Holz and Horrocks (1990)
Mixture of Y and REE	Antimicrobial thermoplastic articles	Shimizu (1993)
^{46}Sc-irradiated talc and fly ash	Pulmonary clearance of toxic inhaled materials	Wehner *et al.* (1980)
Short-lived ^{44}Sc	Bone scanning	Syed and Hosain (1975)
^{47}Sc injection	Bone scintigraphy in rats	Basse-Cathalinet *et al.* (1968)
^{47}Sc	Radiolabeling of antibodies for tumor therapy	Humm (1986)
^{46}Sc microspheres	Perfusion measurement in lungs during respiratory distress	Wiener *et al.* (1990)
Sc–phospholipid complexes	Biomediators with multiple medical activities	Maurel and Masse (1991)
Sc and La metals	Nucleic acid amplification	Filler and Lever (1996)

[a]TCMT, 1,4,7-Tris(carbamoylmethyl)-1,4,7-triazacyclononane.

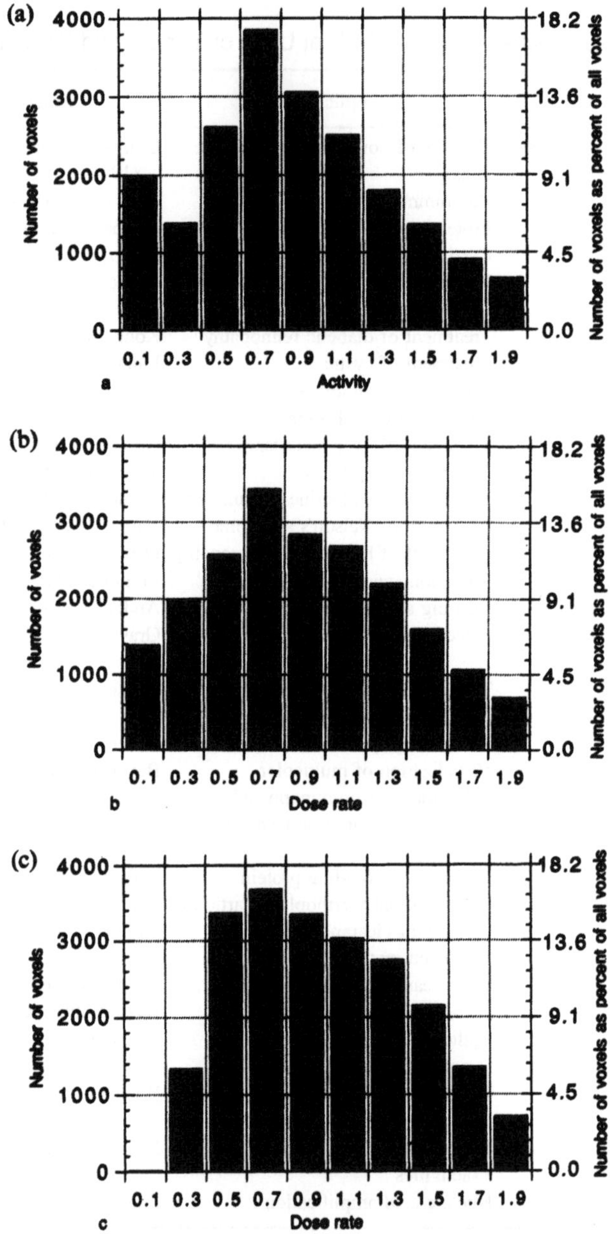

Figure 10-9. Histogram comparison of calculated three-dimensional dose-rate distributions with ^{131}I- (a), ^{125}I- (b), and ^{90}Y-labeled compounds (c) for a human ovarian carcinoma xenograft. The number of voxels within a range is plotted against the midpoint of the range. Each midpoint is a fraction of the mean value for the entire tumor. Reprinted from Koral *et al.* (1993), with permission of the authors and Elsevier Science.

followed by using fly ash and other aerosol particles labeled with ^{46}Sc (Tomenius, 1973).

Based on measurement of the characteristic K series of x-rays, scandium was employed as a target material in microradiography. A special demountable x-ray apparatus with a titanium window was equipped with a target of high-purity scandium metal and used for microradiography of biopsy specimens (Fig. 10-10). A microradiograph of the enamel of a human molar tooth obtained with a scandium target revealed small differences in mineralization not readily detectable with a copper source (Gwinnett, 1967).

Scandium-47 has found applications in bone scintigraphy (Basse-Cathalinat *et al.*, 1968), in radiodiagnostics and radiotherapy, and in pharmaceutical design (Smith *et al.*, 1982; Mirzadeh *et al.*, 1992). The utility of carrier-free ^{47}Sc for *in vitro* radioassays in tumor-bearing mice and rabbits and as a tumor or bone-marrow scanning agent was demonstrated (Hara and Freed, 1973; Fig. 10-11). Improved methods were proposed for the preparation of a source of high-purity ^{47}Sc (Gladney and Goode, 1979) and for rapid preparation of curie amounts of ^{49}Sc (Soloway *et al.*, 1963). Radioisotope generators are efficient in the rapid separation of short-lived daughter radioisotopes; these are applied in angiography and metabolic studies and in other medical and life sciences. Scandium-44 can be obtained in this manner, and, owing to its short half-life of 3.93 h and the long

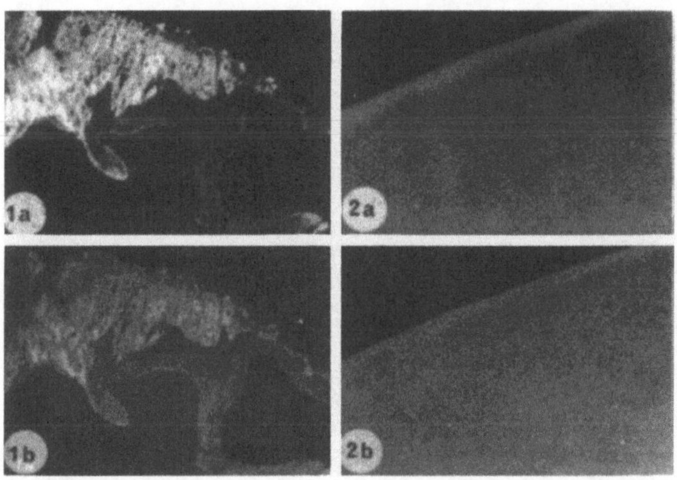

Figure 10-10. Microradiographs obtained with a scandium or copper target. 1, Biopsy specimen: (1a) Scandium target, exposed 150 min; (1b) copper target, exposed 10 min (original magnification, ×50). (2a) Outer enamel of human permanent molar tooth; (2a) Scandium target, exposed 240 min; (2b) copper target, exposed 15 min (original magnification, ×250). Reprinted from Gwinnett (1967). We made efforts without results, to trace the copyright holders.

half-life of its parent, ^{44}Ti, this radioisotope has promising possibilities; it is considered a well-behaved source of positrons in combination with positron annihilation spectroscopy (Mirza and Aziz, 1969). Syed and Hosain (1975) proposed the use of ^{44}Sc as a positron-emitting source for the study of bone disease by positron tomography.

The holmium:yttrium:scandium garnet (HoYSG) laser dye lasers have found use in surgery of salivary stones (Sterenborg *et al.*, 1990), in tracheobronchial endoscopy (Shapshay *et al.*, 1990), and in other medical domains (Table 10-5). *Yttrium* oxide-partially-stabilized zirconia (YPSZ) belongs to a new class of ceramics used in orthopedic surgery. Cylindrical YPSZ samples were implanted on the paraspinal muscles of rats for up to 12 weeks. The tissue reaction was evaluated by measurements of both the membrane thickness and the cell distribution surrounding the implants. Whereas at 1 wk inflammation type cells were present within the loose connective tissue, at 4 wk a decrease of inflammatory cells was observed. To overcome the brittleness of alumina ceramics, the use of YPSZ, which exhibits an improved fracture toughness was

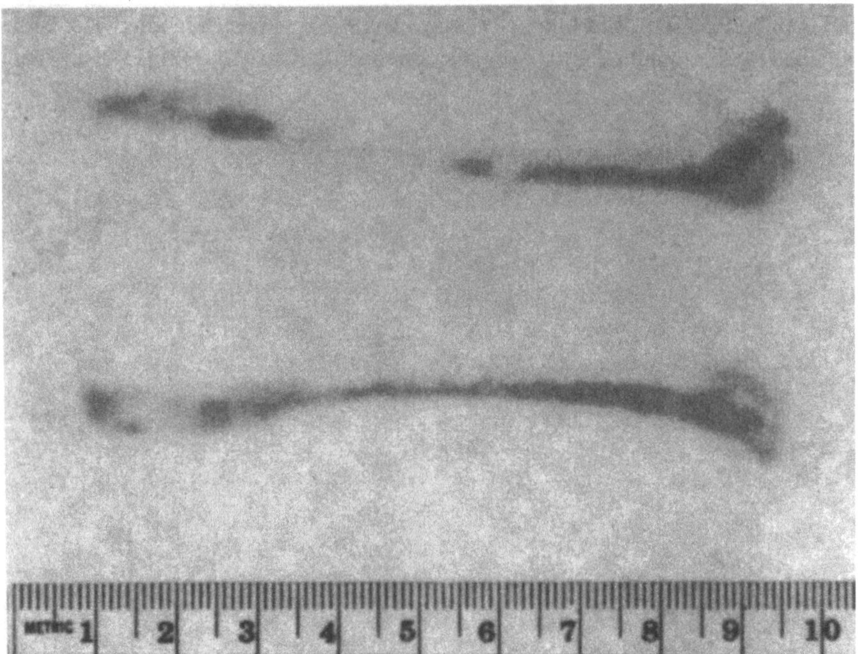

Figure 10-11. Autoradiography of ^{47}Sc uptake in rabbit femur. The rabbit was sacrificed 18 h after injection of 100 μCi of ^{47}Sc in the citrate form. The autoradiograph was obtained by exposure for 4 days on x-ray film. Reprinted from Hara and Freed (1973) with permission of the authors and Elsevier Science.

Table 10-5. Examples of Laser Applications in Medicine and Biology

Type of laser	Application	Reference
Nd:YLiF picosecond	Corneal ablation	Brown *et al.* (1994)
Nd:YAG[a]	Pulmonary metastasectomy	Harvey *et al.* (1993)
Nd:YAG	Strength of tissue bonds in sheep hearts	Fenner *et al.* (1992)
Nd:YAG	Tympany treatment	Tetens *et al.* (1994)
Nd;YAG and diode	Complementary treatment of ciliary bodies	Brancato *et al.* (1991)
Low-level Nd:YAG	Stimulation of cartilage repair	Spivak *et al.* (1992)
$YAl_3(BO_3)_4$		
Ho:YAG	Coronary angioplasty	Geschwind *et al.* (1993)
Ho:YScGd	Lithotripsy of salivary stones	Sterenborg *et al.* (1990)
Er:YAG	Endotracheal tube surgery	Gonzales *et al.* (1990)

[a]YAG, yttrium aluminum garnet.

proposed. Further evaluation of the long-term biological and mechanical behaviour of YPSZ as for alumina did not show any significant difference between the induced tissue reactions (Christel *et al.*, 1989).

The potential interaction of the radiation-enhancing agent 5-fluorouracil (5-FU) with the radioimmunotherapeutic effect of a ^{90}Y-labeled MAb in a human adenocarcinoma xenograft was evaluated (Remmenga *et al.*, 1994). A thymic mouse inoculated with the human colon carcinoma showed a significant delay of tumor growth under the influence of ^{90}Y-marked 5-FU; the radiation-enhancing effect by 5-FU may increase the ability to treat solid tumors (Fig. 10-12).

More examples of the application of scandium and yttrium radioisotopes in biological sciences are depicted in Figs. 10-13 through 10-17.

10.7 Biological, Agricultural, and Environmental Applications

A differential external radioassay technique permitted separate estimation of tail and body radioactivity of mice and snails previously exposed to radioisotope-labeled *Schistosoma mansoni* cercariae (Georgi, 1982). However, as compared to ^{75}Se, ^{49}Sc yielded fewer satisfactory results in labeling cercariae. Scandium-46 showed 13 % retention by microfilaria of *Dirofilaria immitis*, and 100 % motility of microfilaria; these levels were higher than those obtained with other radioisotopes investigated (Lengemann *et al.*, 1986). Although ^{203}Hg exhibited the highest retention levels, its toxicity to the microfilaria presents an impediment in studying the migration and distribution of parasites in the host. The movement of the insect *Pantorhytes szentivanyi* in a cacao plantation was investigated with a solution of acetoacetone labeled with ^{46}Sc (Lamb *et al.*, 1971). The diffusion

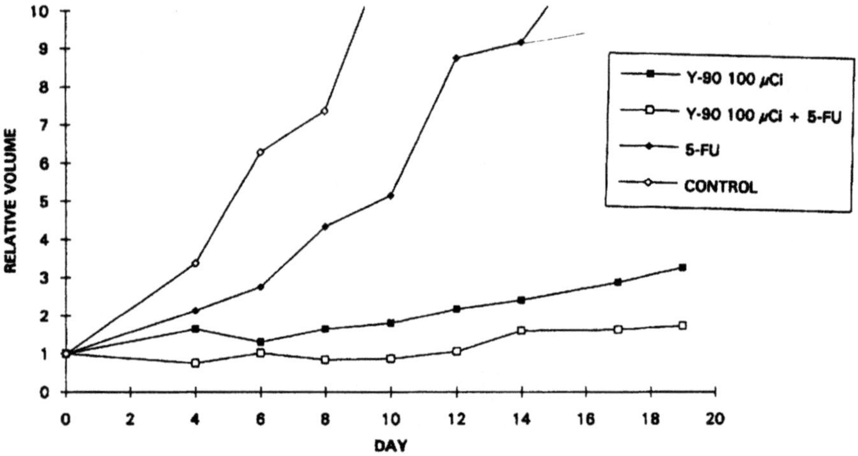

Figure 10-12. Relative volumes of human tumor xenografts inoculated into mice treated with 100 μCi of ^{90}Y with and without 5-fluorouracil (5-FU) compared with controls. Reprinted from Remmenga *et al.* (1994), with permission of the authors and the American Association for Cancer Research.

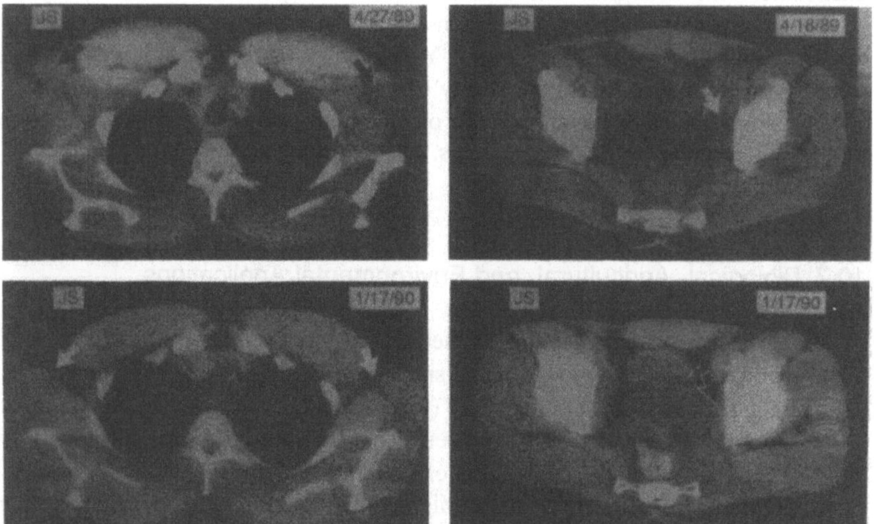

Figure 10-13. CT scans of chest (*left*) and pelvis (*right*) of a lymphoma patient before (April, 1989) and after (January 1990) three courses of ^{90}Y anti-idiotype monoclonal antibody therapy. Reprinted from White *et al.* (1996) with permission of W. B. Saunders, Co.

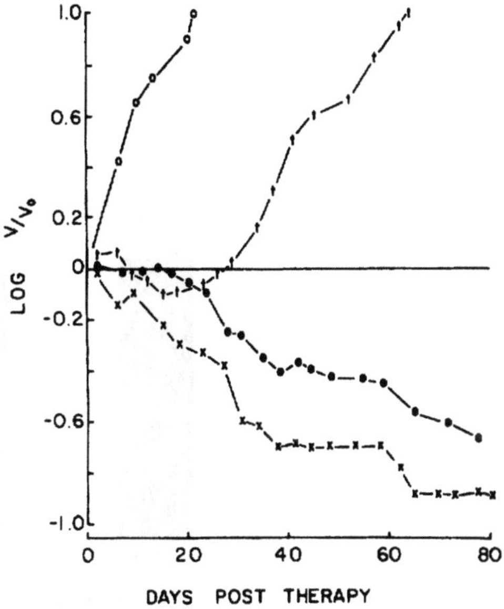

Figure 10-14. Effects of treatment with ^{90}Y-labeled monoclonal antiferritin (QCI054) on tumor growth *in vivo*. Results are expressed as the log of the ratio of tumor size to that at the start of therapy. Therapy was initiated when tumors reached approximately 0.3–0.4 g. O, Control saline solution injected; +, 100 μCi of ^{90}Y -QCI054; ×, 200 μCi of ^{90}Y-QCI054; ●, 300 μCi of ^{90}Y-QCI054. At the conclusion of the experiment at 140 days, 75% of the animals treated with 300 uCi and 50 % of the animals treated with 200 % had no evidence of viable tumor upon pathological examination. Reprinted from Klein *et al.* (1989) with permission of the authors and the American Association for Cancer Research.

model seemed useful for comparative treatment of insect dispersal occurring in closed and open canopies. Caches of eastern white pine (*Pinus strobus* L.) seeds made by small mammals were located by exposing seeds treated with ^{46}Sc and using a portable survey meter and scintillation probe to find them after the animals buried then in the forest floor (Quink *et al.*, 1970).

Despite the extensive uses of Ln^{3+} (mainly europium and terbium, but also dysprosium, gadolinium, and samarium) in luminescent probes in the life and chemical sciences and in technological applications (Bünzli and Choppin, 1989), scandium and yttrium have found few such applications.

Several dissertations (St. Raymond, 1990; Hoffman, 1991; Yang, 1991) and patents (Rubin and Strauss, 1987; Thakur, 1990; Xu, 1992; Zamora and Griffith, 1995) concern various applications of scandium and yttrium in biochemistry and medicine. Examples of biological and environmental applications are presented in Table 10-6.

10.8 Applications in Industry and Technology

Scandium and yttrium metals and compounds have a broad variety of applications in industry and technology (Xu and Xiao, 1985; Kudo, 1988; Gupta and Krishnan, 1988; Gschneidner, Jr., 1991; Hedrick, 1993, 1996; Kobayashi *et*

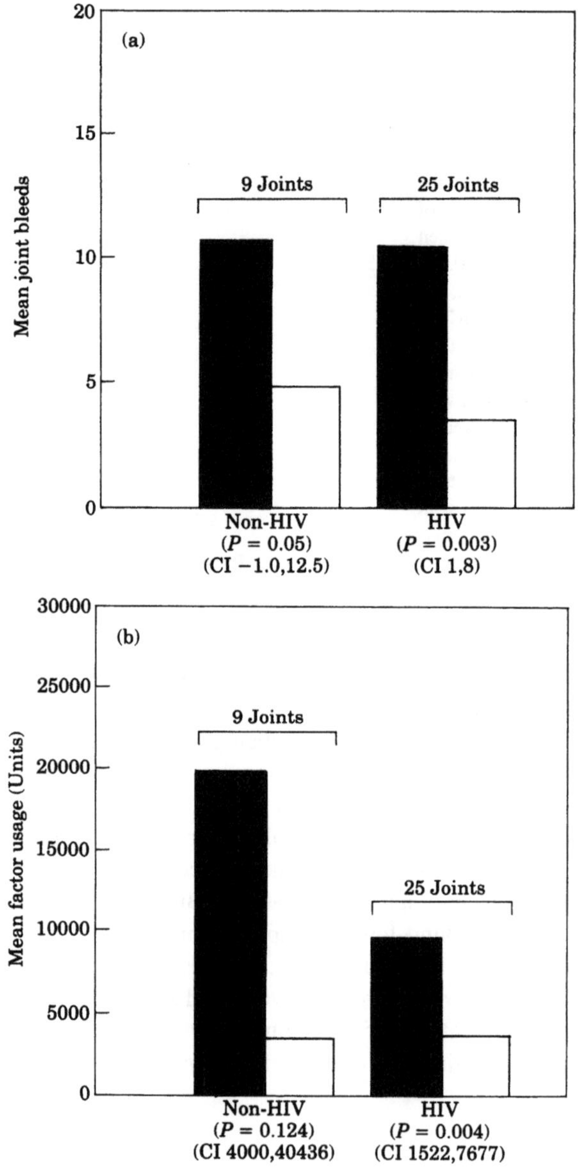

Figure 10-15. Effect of ^{90}Y silicate synovectomy in patients with hemophilic arthropathy. (a) Mean number of joint bleeds in HIV-negative and HIV-positive patients, 6 months before synovectomy (■), and 12 months after synovectomy (□). (b) Mean amount of factor usage in HIV-negative and HIV-positive patients, 6 months before synovectomy (■) and 12 months after synovectomy. Reprinted from Dawson *et al.* (1994), with permission of the authors and Bailliere Tindall.

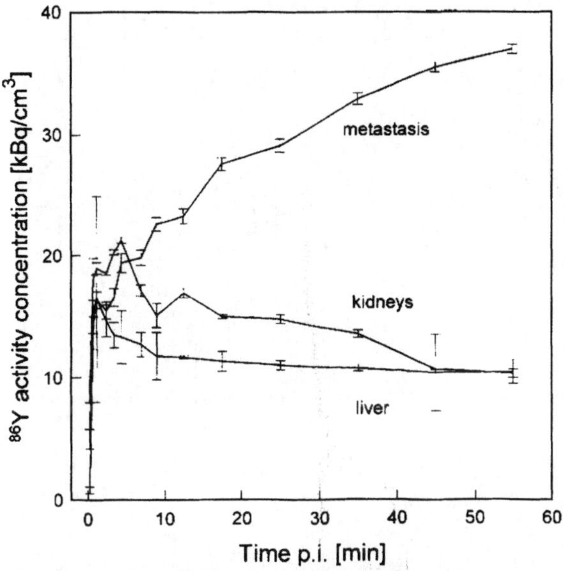

Figure 10-16. Decay-corrected initial time–activity curves of the early uptake kinetics of ^{86}Y citrate in bone metastases, liver, and kidney of a patient with a prostatic carcinoma. Reprinted from Rösch *et al.* (1996), with permission of Springer-Verlag.

al., 1997). Some examples are given in Tables 10-7 and 10-8, respectively. A considerable number of patents dealing with these types of applications have been filed in recent years in the United States, Japan, Russia, France, and other countries (Mirzoyan *et al.*, 1995; Oshikawa *et al.*, 1995; Huguenin, 1996; Yamamoto and Ishihara, 1996, 1998).

Scandium oxide, the main compound used in industry and in other fields, is now produced in high-purity grades, up to 99.995 %. The purity of scandium oxide is very important for growing high-quality laser crystals and for other purposes. Gd:Sc:Ga garnets (GSGG) doped with chromium and neodymium are twice as efficient as the Y:Al garnets (YAG) and much easier to grow as large-size crystals. GSGG is one of the most efficient solid-state laser material known at this time (ASM International Panel, 1988).

Besides the 29 known superconducting elements, some 14 other elements, including yttrium and scandium, only become superconducting under pressure or in high-pressure phases. In a state of high purity, certain crystal forms of these elements would become superconducting at a sufficiently low temperature. In the 1990s, new alloys and superconducting oxides, such as the "1,2,3 compound" in the Y:Ba:Cu:O system, with transition temperatures above 100 K were discovered (Buckel, 1991).

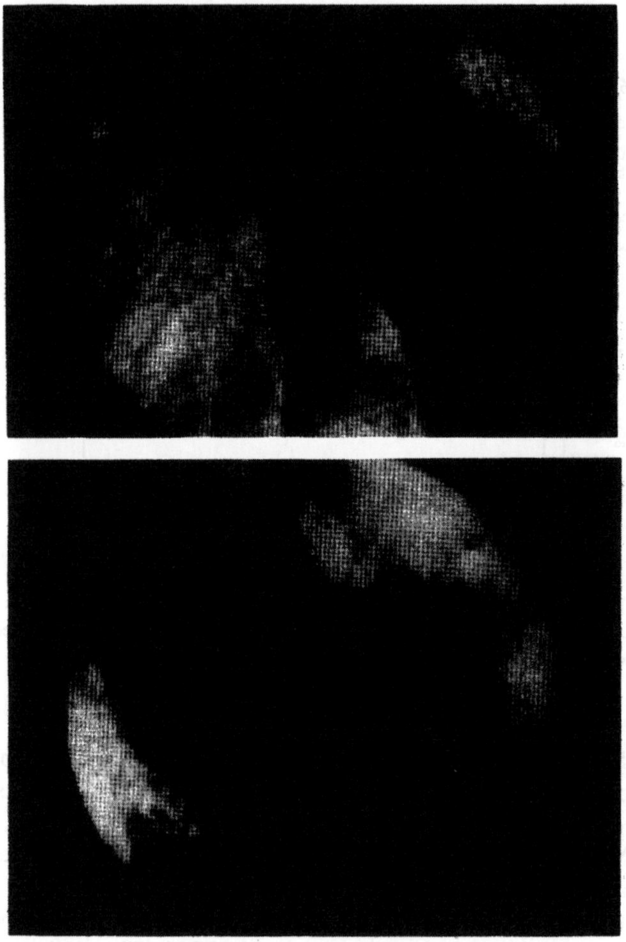

Figure 10-17. Cystoscopic view of right lobe of the human prostate prior to (*top*) and after extensive Nd:YAG photoirradiation treatment (*bottom*). Reprinted from Shanberg *et al.* (1994). Efforts were made without success to trace the copyright holders.

The production of catalysts for the oligomerization and polymerization of alkenes has led to the growth of a multibillion dollar industry. These are particularly useful in the production of plastics, elastomers, and inexpensive feedstock chemicals for copolymers, surfactants, and other useful additives. Such catalysts include the dimeric hydrides of scandium and yttrium catalysts, which exhibit a very rapid ethylene polymerization behavior (Hlatky, 1994).

Table 10-6. Biological and Environmental Applications of Scandium and Yttrium

Material or method	Application	Reference
^{49}Sc labeling	Quantification of skin penetration by cercariae	Georgi (1982)
$^{46}ScCL_3$ labeling	Measurement of water drinking rates in crustacea	Dall and Smith (1977)
^{46}Sc labeling	Investigation of dynamics of seed caches	Vander and Stephen (1995)
^{46}Sc labeling	Investigation of dispersion of cacao tree weevils	Lamb *et al.* (1971)
^{46}Sc labeling	Detection of endoparasites in humans and animals	Lengemann *et al.* (1986)
^{46}Sc labeling	Study of *in vivo* solubility of asbestos in thorax	Morgan *et al.* (1971)
^{46}Sc	Tracer in sugar industry	Fernandes (1993)
^{46}Sc	Tracer of soil adhesion to vegetation	Hinton *et al.* (1995)
^{46}Sc, ^{88}Y, and other radioisotopes	Study of kinships in a meadow vole population	Sheridan and Tamarin (1986)
REE-labeled fungi	Extraction of scandium from minerals	Komissarova *et al.* (1995)
Sc, and other TE	Nutrient supplement for animals and plants	Baisden (1982)
Sc, Y, and other TE	Production of healthy, tasty, and stable mineral water	Aburai (1991)
Y sintering of apatite ceramics	Corrosion-resistant bone implants	Xi *et al.* (1996)
Y-DTPA complex	Tracer in water systems	Dzata and Loveland (1982)
Y metaphosphate	Packing material for biochromatography	Inoue and Ohtaki (1994)
Y, Sc, and other metals	Plant virucide, insecticide	Wu (1989)
Yttria ultrafiltration membrane	Food additive polymer	U.S.A. FDA (1993)
Y-REE labeling	Seed identification	Bazin *et al.* (1994)
^{90}Sr–^{90}Y β-gauging	Measurement of water content in plants	Pandey *et al.* (1996)
High-yield Sc esters (RCO_2R')	Drugs, agrochemicals, food additives, cosmetic bases	Yamamoto and Ishihara (1996)

10.9 Outlook

Scandium, and its close relative yttrium are relatively rare in nature. However, they occur everywhere; they are part of us, and part of the land on which we live, the food we eat, the water we drink, and the air we breath. These elements are widely distributed in living matter, and highly positive or negative correlations with many other elements are found.

Scandium and yttrium are similar in their chemical and physical properties. However, differences exist in their relative body distribution and their actions in

Table 10-7. Technological Applications of Scandium

Material	Applications	Reference
^{46}Sc	Tracer in petroleum production	Roskill Information Services (1985)
^{46}Sc	Radioactive electron emitter	Singelton and Yannopoulos (1975)
High-purity scandia	High-energy laser crystals, etc	ASM (1988)
Sc iodide	High-intensity lamps	Roskill Information Services (1985)
Scandia	Dispersed cathode-ray tubes for large screens	Ishiwata *et al.* (1990)
Scandia	Ceramics for ultrahigh-power aerospace engines	Jones (1990)
Scandia film	Antireflection coatings in lasers	Ladany (1986)
Scandia	Cathode used in oxygen-microwave devices	Yan *et al.* (1985)
Sc metal	Luminescent SiC films	Ballandovich (1991)
Top-layer Sc cathode	High-resolution display and projection TV	Gärtner *et al.* (1989)
GdScAl and YScGa garnets	Solid-state lasers	Zharikov (1991)
Sc trifluoromethanesulfonate	Extremely active acylation catalyst	Ishihara *et al.* (1995)
CrSc multilayer	Multimirror for x-ray microscopy and astronomy	Salashchenko and Shamov (1997)
CrGdScAlO	Phonon-terminated laser	Murray (1987)
GdScGa	High-energy, free-electron laser	Hecht (1987)

animals and humans. Scandium tends to accumulate in the liver and kidney, whereas yttrium prefers bone, spleen, lung, and endocrine glands. Increased or decreased concentrations of scandium have been found in some disease and pathological conditions; whether these are a cause or an effect is unclear.

Several investigations have found specific stimulatory effects on some fungi and on higher organisms. Inhibitory effects observed in physiological processes have been found to be reversible with changes in pH and other conditions. Scandium and yttrium are "hard" acceptors, with a predominant preference for oxygen donor groups; they interact with a variety of biological compounds, mainly through the carboxylic or phosphate groups. Mixed-ligand chelates with many primary and secondary ligands are formed; the stability of these complexes increases in the order La > Y > Sc. It is believed that scandium and yttrium bind to the surfaces of cells and do not penetrate bilayers; however, in several microorganisms they penetrate through the membrane and become apparently involved in the functions of the cell. Multiple and long-range interactions with carbohydrates, carboxylic acids, amino acids, proteins, nucleic acids, and polymeric systems have so far been scarcely investigated.

Table 10-8. Technological Applications of Yttrium

Material	Applications	Reference
Yttria	Ceramics for ultrahigh-power aerospace engines	Jones (1990)
Yttria	Stabilization of zirconia refractors	Kudo (1988)
Yttria antireflection film	Photolithography	Oshikawa *et al.* (1995)
YEu and YTaNb phosphors	High-voltage illumination tubes and radiology	Franz *et al.* (1989)
YVO_4, yttria	Phosphor for color TV	Kudo (1988)
$YBa_2Cu_3O_7$ thin film	High superconductor in microelectronics	Habermeier (1992)
High-yield Y esters (RCO_2R')	Drugs, agrochemicals, food additives, cosmetic bases	Yamamoto and Ishihara (1996)
Y metal	Alloyed steel	Mirzoyan *et al.* (1995)
Y_2O_3–ZrO_2	Synthetic gems	Su (1991)
$Y_2Ru_2O_7$	Magnetic insulator	Cox (1992)
Y boride	Refractory materials	Perry and Phillips (1995)
Y fluoride	Sputtering targets for multilayers	Perry and Phillips (1995)
Y tantalate	Low-voltage luminescence	Huguenin (1996)
^{90}Y	Radioactive electron emitter	Singleton and Yannopoulos (1975)
Nd:doped YAl_3 $(BO_3)_4$	Self-doubling laser	Cheetham (1994)

When compared with several other chemical elements, scandium and yttrium are relatively minor in terms of both research and applications. The lack of a viable source, high prices, and limitations of the methods of analysis for minute quantities have hindered a more thorough knowledge of the properties of these elements. Scandium and yttrium metals are of low toxicity, even at higher concentrations. Their metabolic behavior can be manipulated by the presence of specific chelators, by varying the site or method of injection, or by varying chemical, physical, and environmental factors.

Scandium and yttrium show abilities to serve as substitutes for several elements that have well-established physiological functions. The most prominent of these is calcium, which is spectroscopically silent yet chemically similar to scandium. Scandium and yttrium have antimicrobial and anticoagulant properties, which suppress many calcium-dependent cellular activation processes.

The long-lived (^{46}Sc, ^{88}Y, ^{91}Y) and short-lived (^{44}Sc, ^{47}Sc, ^{86}Y, ^{90}Y) radioisotopes have found extensive applications as inert markers in nutrition research, blood flow investigations, and studies of metabolic processes, medical imaging, radiodiagnostics, and radioimmunotherapy. Yttrium-90 is increasingly being employed for radioassays required in nuclear medicine at gigabecquerel activity levels for measuring injection activity, and at becquerel levels for measuring tissue samples. Scandium and yttrium radioisotopes may also be useful for enhanced magnetic resonance imaging and in the separation, preparation, and purification of molecules and cells, in immunoassay, and in histochemistry. The various industrial and technological applications of scandium and yttrium may stimulate further perspectives in biological, medical, and environmental domains (Fig. 10-18).

The production of high-purity scandium and yttrium metals and compounds is important in technology and biochemistry, because even minute impurities may alter their properties and the expected effects. The synthesis of early metal porphyrins, such as scandium porphyrin derivatives, has uncovered completely unknown compounds (Hoffman, 1991).

Today, the periodic table of elements is like an iceberg floating in the ocean of an incomprehensible human perception of nature (Fig. 10-19). Only the tip of the iceberg is seen, and almost only the tip is investigated (the essential elements). The largest part of the iceberg (the "not essential" elements) is, practically and theoretically, hardly known. In the figure, the man in the boat is occasionally fishing something out of the ocean, but the whale at the bottom is, for the time being, not retrieved.

The concept that scandium, and yttrium, like lanthanum, could behave as super-calcium in certain circumstances, but were not selected due to their low availability (Frausto da Silva and Williams, 1991), awaits further experimental confirmation. We may assume that further developments in inorganic biochemistry, molecular and cellular biology, developmental and comparative holistic biology, biotechnology, medicine, and other fields of the natural sciences will lead to the discovery of new, unexpected biochemical properties and roles of the

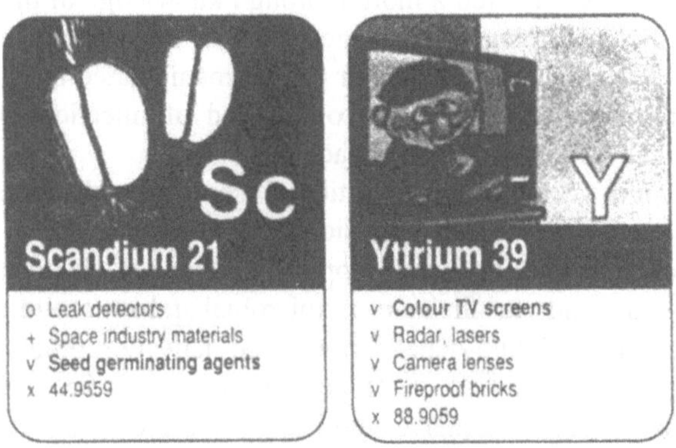

Figure 10-18. Pictorial representation of some properties and uses of scandium and yttrium. Reprinted from *The Periodic Table of Elements*, with permission of the Association of the Dutch Chemical Industry.

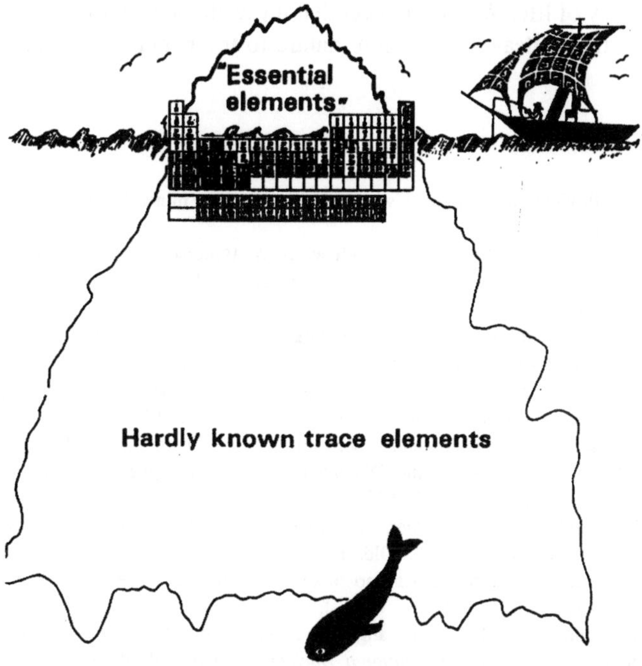

Figure 10-19. The "essential elements" at the tip of the iceberg. The current belief is that some 18–22 chemical elements are essential for life, the remainder of the considerable number of chemical elements being considered either ballast, not essential, neutral, or occasionally, beneficial elements. It may be expected that better knowledge of the biochemical properties of the periodic table of elements will change these views in the future.

chemical elements. The perspective of supramolecular chemistry, which is based on intermolecular interactions, may be exemplified with the siderophores, with which scandium and yttrium form stable complexes (Vögtle, 1991). Advancements in the life sciences may yield cures for combating disease-causing organisms, remedy enzyme deficiencies, a deeper understanding of the origin of life, a reasonable insight into how the brain works, and the emergence of the development of organisms (O'Neill *et al.*, 1994). However, the thought that we can do better than nature is provocative and should evoke ecological concerns (Hoffmann, 1993). Wilson (1998) emphasizes the advent of a true theory of biology, as opposed to the thick descriptions of particular living processes. Wilson argues for the fundamental unity of all knowledge and the need to search for consilience. This term describes the concept that everything in our world is organized in terms of a small number of fundamental natural laws.

Even though they are not as well known as the established essential elements, such as zinc or selenium, scandium and yttrium are good supporting actors in the play of life. We may expect that these elements will find an increased importance in our endeavor to explore nature to the benefit of mankind.

References

Aburai, Y., 1991. Production of mineral water containing trace minerals for purposes of health, Japanese patent 03224690.

Anderson, L. D., Frincke, J. M., Meyer, D. L., Sullivan, B. W., Battersby, T. R., and Kusnierz, E., 1989. Improvements relating to modified haptens as imaging and therapeutic agents, European patent 327365. CA 113:187358q.

Anderson, W. T., and Strand, M., 1985. Stability, targeting and bio-distribution of ^{46}Sc- and ^{67}Ga-labeled MAb in erythroleukemic mice, *Cancer Res.* 45:2154–2158.

Anderson-Berg, W. T., Squire, R. E., and Strand, M., 1987. Specific radioimmunotherapy using ^{90}Y-labeled MAb in erythroleukemic mice, *Cancer Res.* 47:1905–1912.

Andrews, J. C., Walker, S. C., Ackermann, R. J., Cotton, L. A., Ensminger, W. D., and Shapiro, B., 1994. Hepatic radioembolization with ^{90}Y containing glass microspheres. Preliminary results and clinical follow-up, *J. Nucl. Med.* 35:1637–1644.

Antoniw, P., Farnsworth, A. P. H., Turner, A., Haines, A. M. R., Mountain, A., Mackintosh, J., Shochat, D., Humm, J., Welt, S., Old, L. J., Yarranton, G. T., and King, D. J., 1996. Radioimmunotherapy of colorectal carcinoma xenografts in nude mice with ^{90}Y A33Ig and Tri-Fab, *Br. J. Cancer* 714:513–524.

Ariel, I. M., and Padula, G., 1984. Selective angiography as a means of delivering ^{90}Y microspheres in the treatment of cancer, in: *Interventional Nuclear Medicine* (R. P. Spencer, ed.), Grune & Stratton, Orlando, pp. 15–34.

ASM International Panel, 1988. Assessment of quality and material form of minor metals for the national defense stockpile, Final report, ASM International Panel, Washington, D.C.

Association of the Dutch Chemical Industry and the Royal Society of Chemistry, Great Britain, 1991. *The Periodic Table of Elements.*

Axen, C., and Janson, P. O., 1979. Diaphragmatic blood flow at various levels of ventilation in the rabbit, *Acta Physiol. Scand.* 106:1–3.

Baisden, C. R., 1982. Nutrient compound, U.S. patent 4337245. CA 97:71338e.

Ballandovich, V. S., 1991. Scandium acceptor in GH-SiC, *Sov. Phys. Semicond.* 25:174–178.

Barnard, D. J., 1973. A simpler radio-isotope method for the determination of faecal fat, *Med. Lab. Technol.* 30:239–244.

Basse-Cathalinat, B., Blanquet, P., Collignon, G., and Mirande-Iriberry, J., 1968. Bone scintigraphy with ^{47}Sc and the scintillation camera, *J. Nucl. Med.* 9:436–438.

Bazin, M., Levaire, W., and Riglet, G., 1994. Seeds labeled with rare earth elements, Canadian patent 2120634. CA 122:128139g.

Becker, C. D., Rüsler, H., Biasiutti, F. D., and Baer, H. U., 1995. Congestive hypersplenism. Treatment by means of radioembolization of the spleen with ^{90}Y, *Radiology* 195:183–186.

Beverlo, H. B., Vanschadewijk, A., Bonnet, J., Vander Geest, R., Runia, R., Verwoerd, N. P., Vrolijk, J., Ploem, J. S., and Tanke, H. J., 1992. Preparation and microscopic visualization of multicolor luminescent immunophosphors, *Cytometry* 13:561–570.

Bianco, J. A., Bakanauskas, J., Carlson, M., Jones, S., Moring, A., Alpert, J. S., and Klassen, V., 1988. Augmented uptake of 2-^{14}C-D-deoxyglucose in reversibly-injured myocardium, *Eur. J. Nucl. Med.* 13:557–562.

Brancato, R., Leoni, G., Trabucchi, G., and Cappelini, A., 1991. Histopathology of continuous wave Nd:YAG and diode laser cataract transscleral lesions in rabbit ciliary body, *Invest. Ophthalmol. Visual Sci.* 32:1586–1592.

Brechbiel, M. W., and Gansow, O. A., 1991. Backbone-substituted DTPA ligands for ^{90}Y, *Bioconjug. Chem.* 2:187–194.

Brown, D. B., Obrien, W. J., and Schultz, R. O., 1994. The mechanism of ablation of corneal tissue by the Nd doped YLiF picosecond laser. *Cornea* 13:479–486.

Buckel, W., 1991. *Superconductivity*, VCH, Weinheim.

Bünzli, J. C., and Choppin, G. R., eds., 1989. *Lanthanide Probes in Life, Chemical and Earth Sciences*, Elsevier, Amsterdam.

Burke, C. W., Doyle, F. H., Joplin, G. F., Arnot, R. N., Macerlean, D. P., and Fraser, T. R., 1973. Treatment of Cushing's disease by pituitary implantation of radioactive gold and yttrium seeds, *Quart. J. Med.* 42:693–714.

Caldwell, J. H., Martin, G. V., Link, J. M., Krohn, K. A., and Bassingthwaighte, J. B., 1990. Iodophenylpentadecanoic acid myocardial blood flow relationship during maximal exercise with coronary occlusion, *J. Nucl. Med.* 30:99–105.

Cann, C. E., and Prussin, S. G., 1980. Possible parathyroid imaging using ^{67}Ga and other aluminum analogues, *J. Nucl. Med.* 21:471–474.

Cassar, J., Doyle, F. H., Lewis, P. D., Mashiter, K., Van Noorden, S., and Joplin, G. F., 1976. Treatment of Nelson's syndrome by pituitary implantation of ^{90}Y or ^{108}Au, *Br. Med. J.* 2:269–272.

Cheetham, A. K., 1994. Advanced inorganic materials. An open horizon. *Science* 264:794–795.

Chinol, M., Paganelli, G., Sudati, F., Meares, C., and Fazio, F., 1997. Biodistribution in tumor-bearing mice of two ^{90}Y-labeled biotins using three-step tumor targeting, *Nucl. Med. Commun.* 18:176–182.

Christel, P., Meunier, A., Heller, M., Torre, J. P., and Peille, C. N., 1989. Mechanical properties and short-term *in vivo* evaluation of Y_2O_3-partially stabilized zirconia. *J. Biomed. Mater. Res.* 23:45–61.

Cooper, M. M., Robbins, R. C., Goldman, C. K., Mirzadeh, S., Brechbiel, M. W., Stone, C. D., Gansow, O. A., Clark, R. E., and Waldmann, T. A., 1990. Use of ^{90}Y-labeled anti-Tac antibody in primate xenograft transplantation, *Transplantation* 50:760–765.

Coursey, B. M., Calhoun, J. M., and Cessna, J. T., 1993. Radioassays of ^{90}Y used in nuclear medicine, *Nucl. Med. Biol.* 20:693–699.

Cox, P. A., 1982. *Transition Metal Oxides*, Clarendon Press, Oxford.

Dall, W., and Smith, D. M., 1977. Measurement of water drinking rates in marine Crustacea, *J. Exp. Mar. Biol. Ecol.* 30:199–208.

Dawson, T. M., Ryan, P. F. J., Street, A. M., Robertson, R. L., Kalff, V., Kellys, M. J., and Cicuttini, F. M., 1994. Yttrium synovectomy in hemophilic arthropathy, *Br. J. Rheumatol.* 33:351–356.

DeNardo, S. J., Gumerlock, P. H., Winthrop, M. D., Mack, P. C., Chi, S.-G., Lamborn, K. R., Shen, S., Miers, L. A., White, R. W., and DeNardo, G. L., 1995. ^{90}Y chimeric L6 therapy of human breast cancer in nude mice, and apoptosis-related messenger RNA expression, *Cancer Res.* 55:5837s–5841s.

Deshpande, S. V., DeNardo, S. J., Kukis, D. L., Moi, M. K., McCall, M. J., DeNardo, G. L., and Meares, C. F., 1990. ^{90}Y-labeled MAb for therapy. Labeling by a new macrocyclic bifunctional chelating agent, *J. Nucl. Med.* 31:473–479.

Diamandis, E. P., and Redshaw, R., 1995. Localization and therapy of nonprostatic endocrine cancer with agents directed against prostate-specific antigen, PCT patent 9502,424. CA 122:208793f.

Dillehay, L. E., Mayer, R., Zhang, Y., Shay, Y., Song, S., Mackensen, D., and Williams, J., 1995. Prediction of tumor response to experimental radioimmuno-therapy with ^{90}Y in nude mice, *Biol. Phys.* 33:417–427.

Dzata, F. K. A., and Loveland, W. D., 1982. Development of stable fluorescence tracers for use in small water systems, *Nucl. Instrum. Methods* 193:319–322.

Edmonds, J., Smarts, R., Laurent, R., Butler, P., Brooks, P., Hoschl, R., Wiseman, J., George, S., Lovegrove, F., and Lambrecht, R., 1994. A comparative study of the safety and efficacy of ^{165}Dy hydroxide macro-aggregate and ^{90}Y silicate colloid in radiation synovectomy. *Br. J. Rheumatol.* 33:947–953.

Edwards, F. M., Kraushaar, J. J., Ristinen, R. A., and Rudolph, H., 1977. Use of ^{48}Ca as stable isotope tracer analyzed by proton activation, *Med. Phys.* 4:58–65.

Evans, C. H., 1990. Past, present and possible future clinical applications of the lanthanides, in: *Biochemistry of the Lanthanides*, Plenum Press, New York, pp. 391–425.

Fenner, J., Moseley, H., Martin, W., and Wheatley, D. J., 1992. Strength of tissue bonds as a function of surface apposition, *Lasers Med. Sci.* 7:375–379.

Fernandes, E. A. N., 1993. Scandium as tracer in the sugar and alcohol agro-industry, *J. Radioanal. Nucl. Chem.* 168:41–46.

Filler, A. G., and Lever, A. M. L., 1996. Nucleic acid amplification using scandium and lanthanum ions, U.S.A. patent 5554498. BA 48:166119.

Flaim, S. F., Nellis, S. H., Toggart, E. J., Drexler, H., Kanda, K., and Newman, E. D., 1984. Multiple simultaneous determination of hemodynamics and flow distribution in conscious rat, *J. Pharmacol. Methods* 11:1–39.

Franssen, M. J. A. M., Koenders, E. B., Boerbooms, A. M. T., Buijs, W. C. A. M., Lemmens, J. A. M., and Van De Putte, L. B. A., 1997. Does application of radiographic contrast medium in radiation synovectomy influence the stability of ^{90}Y colloid? *Br. J. Rheumatol.* 36:506–508.

Franz, K. A., Kehr, W. G., Siggel, A., Wieczoreck, J., and Adam, W., 1989. Luminescent materials, in *Ullmann's Encyclopedia of Industrial Chemistry* (B. Elvers *et al.*, eds.), VCH, New York, Vol. A13, pp. 519–555.

Frausto da Silva, J. J. R., and Williams, R. J. P., 1991. The Biological Chemistry of the Elements, The Inorganic Chemistry of Life, Clarendon Press, Oxford, pp. 545–551.

Gansow, O. A., 1991. Newer approaches to the radiolabeling of monoclonal antibodies by use of metal chelates, *Nucl. Med. Biol.* 18:369–381.

Gärtner, G., Janiel, P., Crombeen, J. E., and Hasner, J., 1989. Top-layer scandate cathode by plasma activated CVD, *Inst. Phys. Conf. Ser.*, No. 99, pp. 25–28.

Georgi, J. R., 1982. *Schistosoma mansoni*: Quantification of skin penetration and early migration by differential external radioassay and autoradiography, *Parasitology* 84:263–281.

Geschwind, H. J., Nakamura, F., Kvasnicka, J., and Dubois-Rande, J. L., 1993. Excimer and Ho:YAG laser coronary angioplasty. *Am. Heart J.* 125:510–522.

Gladney, E. S., and Goode, W. E., 1979. Preparation of carrier free ^{47}Sc by ^{47}Ti(n,p) reaction with epithermal neutrons, *Int. J. Appl. Radiat. Isot.* 30:65.

Gonzales, M. J., Smith, M., and Reinisch, L., 1990. Endotracheal tube safety with the Er:YAG laser, *Ann. Otol. Rhinol. Laryngol.* 991:553–556.

Goodwin, D. A., Meares, C. F., Watanabe, N., McTigue, M., Chaova Pong, W., Ransone, C. M., Renn, O., Greiner, D. P., Kukis, D. L., and Kronenberger, S. I., 1994. Pharmacokinetics of pretargeted MAb 2D12.5 and ^{88}Y-Janus-2-(*p*-nitrobenzyl)-1,4,7-10-tetraazacyclododecanetetraacetic acid (DOTA) in BALB/c mice with KHJJ mouse adenocarcinoma. A model for ^{90}Y radioimmunotherapy, *Cancer Res.* 54:5937–5946.

Gray, B. N., Anderson, J. E., Burton, M. A., Van Hazel, G., Codde, J., Morgan, C., and Klemp, P., 1995. Regression of liver metastases following treatment with ^{90}Y microspheres, *Aus. N. Z. Surg.* 62:105–110.

Gray, D. H., and Vogt, J. R., 1974. NAA of stable heavy metals as multiple markers in nutritional monitoring, *J. Agric. Food Chem.* 22:144–146.

Greiner, D. P., Hughes, K. A., and Meares, C. F., 1996. Radiolytic protein surface mapping, *Biochem. Biophys. Res. Commun.* 225:1006–1018.

Gross, P. M., Marcus, M. L., and Heistad, D. D., 1981. Measurement of blood flow to bone and marrow in experimental animals by means of the microsphere technique, *J. Bone Joint Surg.* 63A:1028–1031.

Gschneidner, K. A., Jr., 1991. Rare earths in the electronic industry, in: *Fine Chemicals for the Electronic Industry. Chemical Applications for the 1990's.* (D. J. Ando and M. G. Pellat, eds.), Royal Society of Chemistry, London, pp. 63–94.

Gupta, C. K., and Krishnan, T. S., eds., 1988. *Rare Earths—Applications and Technology*, Trans Tech Publ., Bombay.

Gwinnett, A. J., 1967. Scandium as a target material in microradiography, *J. Dent. Res.* 46:1479.

Habermeier, H.-U., 1992. YBeCuO high temperature superconductor thin film preparation by pulsed-laser deposition. Recent developments, *Mater. Sci. Eng.* B13:1–7.

Häfeli, U. O., Pauer, G. J., Roberts, W. K., Humm, J. L., and Macklis, R. M., 1997. Magnetically targeted microspheres for intracavitary and intraspinal ^{90}Y radiotherapy, *Sci. Clin. Appl. Magnetic Carriers*, Proc. Int. Conf., 1st, 1996 (U. Häfeli, ed.), Plenum Press, New York, pp. 501–516.

Hamlin, R.L., Marsland, W.P., Smith, C.R., and Sapirstein, L.A., 1962. Fractional distribution of right ventricular output in the lungs of dogs, *Circ. Res.* 10:763–766

Hara, T., and Freed, B. R., 1973. Preparation of carrier-free ^{47}Sc by chemical separation from ^{47}Ca and its distribution in tumor bearing mice, *Int. J. Appl. Radiat. Isot.* 24:373–376.

Harrison, A., Walker, C., Parker, D., Jankowski, K., Cox, J., Craig, A., Sansom, J., Beeley, N., Boyce, R., Chaplin, L., Eaton, M., Farnsworth, A., Millar, K., Millican, A., Randall, A., Rhind, S., Secher, D., and Turner, A., 1991. The *in vivo* release of ^{90}Y from cyclic and acyclic ligand–antibody conjugates, *Nucl. Med. Biol.* 18:469–476.

Harvey, J. C., Lee, K., and Beattie, E. J., 1993. Utility of the Nd:YAG laser for extensive pulmonary metastasectomy, *J. Surg. Oncol.* 54:175–179.

Hayes, R. L., and Hübner, K. F., 1983. Basis for the clinical use of gallium and indium radionuclides, in: *Metal Ions in Biological Systems* (H. Sigel, ed.), Marcel Dekker, New York, Vol. 16, pp. 279–315.

Hecht, J., 1987. Why scandium rose in value, *New Scientist* 114:26.

Hedrick, J. B., 1993. Rare earths. The lanthanides, yttrium and scandium, US-DI Bureau of Mines, Washington, D.C.

Hedrick, J. B., 1996. Rare earths, *Am. Ceram. Soc. Bull.* 75:147–150.

Herpst, J. M., Klein, J. L., Leichner, P. K., Quadri, S. M., and Vriesendorp, H. M., 1995. Survival of patients with resistant Hodgkin's disease after polyclonal ^{90}Y-labeled antiferritin treatment, *J. Clin. Oncol.* 13:2394–2400.

Herzog, H., Rösch, F., Stöcklin, G., Lueders, C., Qaim, S. M., and Feinendegen, L. E., 1993. Measurement of pharmacokinetics of ^{86}Y radiopharmaceuticals with PET and radiation dose calculation of analogous ^{90}Y radiotherapeutics, *J. Nucl. Med.* 34:2222–2226.

Himmelstein, J. S., and Levy, B. S., 1986. Environmental health aspects of the microelectronics industry. *Occup. Med.* 1:175–194.

Hinton, T. G., Stoll, J. M., and Tobler, L., 1995. Soil contamination of plant surfaces from grazing and rainfall interactions, *J. Environ. Radioact.* 29:11–26.

Hlatky, G. G., 1994. Oligomerization and polymerization by homogeneous catalysts, in: *Encyclopedia of Inorganic Chemistry* (R. B. King, ed.), John Wiley, Chichester, Vol., 5, pp. 2728–2740.

Hnatowich, D. J., Virzi, F., and Doherty, P. W., 1985. DTPA-coupled antibodies labeled with ^{90}Y, *J. Nucl. Med.* 26:503–509.

Ho, S., Lau, W. Y., Leung, T. W., Chan, M., Ngar, Y. K., Johnson, P. J., and Li, A. K. C., 1996. Partition model for estimating radiation doses from ^{90}Y microspheres in treating hepatic tumors, *Eur. J. Nucl. Med.* 23:948–952.

Hoffman, C. G., 1991. Some alkali metals and scandium porphyrin derivatives, Thesis, University of California, Berkeley.

Hoffmann, R., 1993. How should chemists think, *Sci. Am.* 2:66–73.

Holz, R. C., and Horrocks, De W., Jr., 1990. ^{89}Y NMR spectrometry, a new probe for calcium-binding proteins, *J. Magn. Reson.* 89:627–631.

Hopkins, K., Chandler, C., Bullimore, J., Sandeman, D., Coakham, H., and Kemshead, J. T., 1995. A pilot study of the treatment of patients with recurrent malignant gliomas with intratumoral ^{90}Y radioimmunoconjugates, *Radiother. Oncol.* 34:21–31.

Horwitz, P. E., and Dietz, M. L., 1994. Process for the separation and purification of ^{90}Y for medical applications, U.S. patent 5368736. CA 122:64489m.

Howell, R. W., Goddu, S. M., and Rao, D. V., 1994. Application of the linear-quadratic model to radioimmunotherapy. Further support for the advantage of longer-lived radionuclides. *J. Nucl. Med.* 35:1861–1869.

Huguenin, D., 1996. Use of doped yttrium, lutetium, or gadolinium tantalate in low voltage luminescence, European patent 690115. CA 124:130957g.

Humm, J. L., 1986. Dosimetric aspects of radiolabeled antibodies for tumor therapy. *J. Nucl. Med.* 27:1490–1497.

Hutcheson, D. P., Venugopal, B., Gray, D. H., and Luckey, T. D., 1979. Lanthanide markers in a single sample for nutrient studies in humans, *J. Nutr.* 109:702–707.

Hutcheson, D. P., Kay, M., Enderle, M., Francis, R., and Luckey, T. D., 1983. The kinetics of lanthanide oxide markers in the alimentary tract of the pig, *Nutr. Res.* 3:913–921.

Hyams, D. M., Esteban, J. M., Lollo, C. P., Beatty, B. G., and Beatty, J. D., 1987. Therapy of peritoneal carcinomatosis of human colon cancer xenografts with ^{90}Y-labeled anti-CEA antibody ZCE 025, *Arch. Surg.* 122:1333–1337.

Inoue, S., and Ohtaki, N., 1994. Metaphosphates as packing materials for biochromatography, *J. Chromatogr.* 679A:247–253.

Irion, G. L., Wolfson, M. R., and Alsberge, M. S., 1988. Computerized stripping method for multiple nuclide microsphere studies, *Comput. Biomed. Res.* 21:27–37.

Ishihara, K., Kubota, M., Kurihara, H., and Yamamoto, H., 1995. Scandium trifluoromethanesulfonate as an extremely active acylation catalyst, *J. Am. Chem. Soc.* 117:4413–4414.

Ishiwata, S., Kimura, E., and Nakanishi, H., 1990. Preparation and applications of scandium compounds, in: *Rare Metals '90*, Proc. Symp. Osaka, pp. 53–56.

Janson, P. O., 1975. Effects of the luteinizing hormone on blood flow in the follicular rabbit ovary, as measured by radioactive microspheres, *Acta Endocrin.* 79:122–133.

Jones, R. L., 1990. Oxide acid–base reaction in ceramic corrosion, *High Temp. Sci.* 27:369–380.

Kairemo, K., Ljunggren, K., Strand, S. E., Hiltunen, J., Penttila, P., Nikula, T., Laine, A., and Wahlström, T., 1993. Radioimmunotherapy with ^{90}Y-labeled MAb in a nude mouse ovarian cancer model, *Acta Oncol.* 32:801–805.

Kairento, A. L., Brownell, G. L., Schluederberg, J., and Elmaleh, D. R., 1983. Regional blood-flow measurement in rabbit soft-tissue tumor with positron imaging using the $C^{15}O_2$ steady-state and labeled microspheres, *J. Nucl. Med.* 24:1135–1142.

Keeling, A. A., and Beaney, R. P., 1991. ^{87}Y for *in vivo* imaging, *Nucl. Med. Biol.* 18:261–262.

Keeling, A. A., Vaughan, A. T. M., and Beaney, R. P., 1989. ^{90}Y-*EDTMP*: A radiotherapeutic agent in the treatment of leukaemias, *Br. J. Cancer* 60:74–78.

Kim, S., Dorscher-Kim, J., Liu, M. T., and Trowbridge, H. O., 1988. Biphasic pulp blood-flow response to substance P in the dog as measured with a radiolabelled microsphere injection method, *Arch. Oral Biol.* 33:305–309.

Kishan, J., Gupta, B. N., Arora, S. P., and Chopra, R. C., 1982. Measurement of rumen dry matter passage and turnover rate in buffaloes fed on various ration combination using ^{46}Sc, *J. Nucl. Agric. Biol.* 11:5–7.

Klein, J. L., *et al*, 1989. ^{90}Y and ^{131}I radioimmunoglobulin therapy of an experimental human hepatoma, *Cancer Res.* 49:6383–6389.

Kobayashi, S., Hachiya, I., Ishitani, H., and Araki, M., 1997. Scandium trifluoromethanesulfonate as a novel reusable Lewis acid catalyst in aldol and Michael reactions, *Synlett* 7:472–474.

Kohner, E. M., Dollery, C. T., and Bulpitt, C. J., 1969. Cotton-wool spots in diabetic retinopathy, *Diabetes* 18:691–704.

Komissarova, L. N., Mel'nikov, P. P., Pushkina, G. Ya., Karavajko, G. I., Avakyan, Z. A., and Voronchikhin, D. A., 1995. Use of fungi for extracting rare earth elements and especially scandium from phosphorus-containing mineral raw materials, Russian patent 2036.148. CA 123:346249e.

Koral, K. F., Kwok, C. S., Yang, F. E., Brown, R. S., Sisson, J. C., and Wahl, R. L., 1993. Autoradiography-based, three-dimensional calculation of dose rate for murine, human-tumor xenographts, *Nucl. Med. Biol.* 20:901–909.

Kosmas, C., Maraveyas, A., Gooden, C. S., Snook, D., and Epenetos, A. A., 1995. Anti-chelate antibodies after intraperitoneal ^{90}Y-labeled MAb immunoconjugates for ovarian cancer therapy, *J. Nucl. Med.* 36:746–753.

Kroyer, K., Bülow, J., Nielsen, S. L., and Kromann-Andersen, B., 1990. Urinary bladder blood flow. 1. Comparison of clearance of locally injected ^{99m}Tc pertechnate and radioactive microsphere technique in dogs, *Urol. Res.* 18:223–226.

Kudo, Y., 1988. Industrial applications of rare earths in Japan, in: *Rare Earths. Applications and Technology* (C. K. Gupta and T. S. Krishnan, eds.), Translations of Technical Publications, Switzerland, pp. 243–258.

Kudoh, Y., Hearse, D. J., Maxwell, M. P., Yoshida, S., Downey, J. M., and Yellon, D. M., 1986. Calcium antagonists and myocardial infarction, *J. Mol. Cell. Cardiol.* 18 (Suppl. 4):77–92.

L'Abbate, A. Camici, P., Pelosi, G., Trivella, G., Valli, G., Marzilli, M., Levantesi, D., and Taddei, L., 1961. A new technique for multitracer macroscopic autoradiography, *J. Nucl. Med. Allied Sci.* 25:27–33.

Ladany, I., Zanzucchi, P. J., Andrews, J. T., Kane, J., and De Piano, E., 1986. Scandium oxide antireflection coatings for superluminescent LEDS, *Appl. Opt.* 25:472–473.

Lamb, K. P., Hassan, E., and Scotter, D. R., 1971. Dispersal of ^{46}Sc-labelled *Pantorhytes* weevils in Papuan cacao plantations, *Ecology* 52:178–182.

Lange, M., Kirsch, C. M., Steude, U., and Oeckler, R., 1995. Intracavitary treatment of intrasellar cystic craniopharyngiomas with ^{90}Y by trans-spheroidal approach, *Acta Neurochir.* 135:206–209.

Lehtola, A., Kivilaakso, E., Puolakkainen, P., Karonen, S.-L., Lempinen, M., and Schröder, T., 1986. Effects of dextran 70 versus crystalloids in the microcirculation of porcine hemorrhagic pancreatitis, *Surg. Gynecol. Obstet.* 162:556–562.

Lengemann, F. W., Grieve, R. B., Chmielewicz, M., and Georgi, J. R., 1986. ^{203}Hg and other gamma-emitting radio-isotopes as labels for *Dirofilaria immitis* microfilaria, *Parasitology* 92:451–461.

Lin, W. Y., Wang, S. J., Chen, M. N., Shen, L. H., Tsai, Z. T., and Ting, G., 1995. Preparation and biodistribution of ^{90}Y lipidol in rats following hepatic arterial injection, *J. Nucl. Med.* 36:163P.

Ljunggren, K., and Strand, S. E., 1991. Imaging detector for static and dynamic studies in biological samples, *Nucl. Instrum. Methods Phys. Res.* 31A:475–478.

Luckey, T. D., Venugopal, B., Gray, D., and Hutcheson, D., 1977. Lanthanide marker evidence for one and two physiologic compartments in the human alimentary tract, *Nutr. Rep. Int.* 16:339–347.

Macklis, R. M., Beresford, B. A., and Humm, J. L., 1994. Radiobiologic studies of low-dose-rate ^{90}Y-lymphoma therapy, *Cancer* 73 (Suppl. 3):966–973.

MacSweeney, J. E., Baxter, M. A., and Joplin, G. F., 1990. Heel pad thickness is an insensitive index of biochemical remission in acromegaly, *Clin. Radiol.* 42:348–350.

Mantravadi, R. V., Spigos, D. G., Tan, W. S., and Felix, E. L., 1982. Intraarterial ^{90}Y in the treatment of hepatic malignancy, *Radiology* 142:783–776.

Maraveyas, A., Myers, M., Stafford, N., Rowlinson-Busza, G., and Stewart, J. S. W., 1995. Radiolabeled antibody combined with external radiotherapy for the treatment of head and neck cancer. Reconstruction of a theoretical phantom of the larynx for radiation dose calculation to local tissues, *Cancer Res.* 55:1020–1027.

Mardirossian, G., Wu, C., and Hnatowich, D. J., 1993. The stability in liver homogenates of ^{111}In and ^{90}Y attached to antibody via two popular chelators, *Nucl. Med. Biol.* 20:65–74.

Maurel, J. C., and Masse, J. P., 1991. New organic derivatives of metals with phospholipids, use of these derivatives, process for isolating the derivatives from plants, and pharmaceutical compositions containing them, PCT patent. CA 116:28114c.

McCalden, T. A., 1979. Serial measurement of cerebral blood flow using external counting of microspheres, *Stroke* 10:29–33.

McCredie, D. A. Troehler, U., and Bonjour, J. P., 1984. *In vivo* determination of intestinal calcium absorption with ^{47}Sc used as marker, *J. Lab. Clin. Med.* 103:354–362.

Meares, C. F., Li, M., and DeNardo, S. J., 1995. Method for preparing radionuclide-labeled chelating agents for ESR and NMR, PCT patent 9526206. CA 124:49701x.

Medvedev, O. S., Martynova, E. R., Akchurin, R. S., and Khalatov, V. Yu., 1987. Experimental estimation of chronic microsphere loss from the rat myocardium, *Bull. Exp. Biol. Med.* 103:7–9.

Miller, J. K., and Byrne, W. F., 1970. Comparison of ^{46}Sc and ^{144}Ce as nonabsorbed reference materials in studies with cattle, *J. Nutr.* 100:1287–1291.

Miller, J. K., Byrne, W. F., and Lyke, W. A., 1972. Comparison of fecal excretions of ^{46}Sc tagged sand and soluble ^{144}Ce by calves, *Health Phys.* 22:461–465.

Mirza, M. Y., and Aziz, A., 1969. A scandium generator, *Radiochim. Acta* 11:43–44.

Mirzadeh, S., Schenter, R. E., Callahan, A. P., and (Russ) Knapp, F. F., Jr., 1992. Production capabilities in US nuclear reactors for medical radioisotopes, Oak Ridge National Laboratory, Report ORNL/TM-12010, Oak Ridge, Tennessee.

Mirzoyan, G. S., *et al.*, 1995. Steel alloyed with yttrium, Russian patent 2048590. CA 124:295443z.

Molinatti, G. M., Camanni, F., Massara, F., and Messina, M., 1967. Implantation of the sella turcica with ^{90}Y in the treatment of Cushing's syndrome of pituitary origin, *J. Clin. Endocr.* 27:861–876.

Morgan, A., Holmes, A., and Gold, C., 1971. Studies of the solubility of constituents of chrysotile asbestos *in vivo* using radioactive tracer techniques, *Environ. Res.* 4:558–570.

Mori, E., and Takanashi, R., 1992. Nonradioactive element-containing microspheres for determining blood flow in tissues or organs. Japanese patent 04 05244. CA 116:190663c.

Morrow, J., R., 1996. Hydrolytic cleavage of RNA catalyzed by metal ion complexes, in *Metal Ions in Biological Systems*, (A. Sigel and H. Sigel, eds.), Marcel Dekker, New York, Vol. 33, pp. 561–592.

Murray, J. R., 1987. Lasers for spectroscopy, in: *Laser Spectroscopy and Its Applications*, (L. J. Radziewski, R. W. Solarz, and J. A. Paisner, eds.), Marcel Dekker, New York, pp. 91–174.

Myers, S., Sparks, J. W., and Makovski, E. L., 1986. Factors affecting radioactive microsphere measurements of blood flow in pregnant guinea pigs, *Lab. Anim. Sci.* 36:522–526.

Nother, G., 1959. A technique for destruction of the hypophyses using ^{90}Y spheres, *Acta Radiol.* 184 (Suppl.):1–128.

Ogg, C. S., Pearson, J. D., and Veall, N., 1968. A method for measuring the gastrointestinal absorption of ^{47}Ca using ^{47}Sc as an inert marker, *Clin. Sci.* 34:327–332.

O'Neill, L., Murphy, M., and Gallagher, R. B., 1994. What are we? Where did we come from? Where are we going? *Science* 263:181–183.

Order, S. E., Klein, J. L., Leichner, P. K., Frincke, J. M., Lollo, C., and Carlo, D. J., 1986. ^{90}Y antiferritin. A new therapeutic radiolabelled antibody, *Int. J. Radiat. Oncol. Biol.* 12:277–281.

Oshikawa, S., Suzuki, M., Amihoshi, J., and Hirayama, G., 1995. Two wavelength antireflection film, Japanese patent 07 244204. CA 123:35427z.

Pandey, R. K., Sing, K. P., and Rathore, V. S., 1996. Measurement of thickness and water content of soybean leaves by beta gauging, *J. Nucl. Agric. Biol.* 25:1–8.

Pearson, J. D., 1966. Use of ^{51}Cr-labelled haemoglobin and ^{47}Sc as inert faecal markers, *Int. J. Appl. Radiat. Isot.* 17:13–16.

Perry, D. L., and Phillips, S. L., eds., 1995. *Handbook of Inorganic Compounds*, CRC Press, Boca Raton, Florida, pp. 341–342, 445–448.

Pond, K. R., Ellis, W. C., James, W. D., and Deswysen, A. G., 1985. Analysis of multiple markers in nutrition research, *J. Dairy Sci.* 68:745–750.

Quink, T. F., Abbott, H. G., and Mellen, W. J., 1970. Locating tree seed caches of small mammals with a radioisotope, *For. Sci.* 16:147–148.

Rakovic, M., Pouckova, P., and Silar, J., 1991. Can yttrium ions serve as a tumor localizer? *Isotopenpraxis* 27:332–334.

Rayman, R. R., Clavo, A. C., Crawford, S. C., Recker, B., and Wahl, R. L., 1997. Magnetically-enhanced radionuclide therapy (MERIT). *In vitro* evaluation. *Int. J. Radiat. Oncol. Biol. Phys.* 37:1201–1206.

Reardon, D. T., Meares, C. F., Goodwin, D. A., McTigue, M., David, G. S., Stone, M. R., Leung, J. P., Bartholomew, R. M., and Frincke, J. M., 1985. Antibodies against metal chelates, *Nature* 316:265–268.

Rehder, D., and Hink, K., 1989. The interaction of $Sc(OH)^{2+}$.aq with serine and small peptides investigated by ^{45}Sc NMR spectroscopy, *Inorg. Chim. Acta* 158:265–271.

Remmenga, S. W., Colcher, D., Gansow, O., Pippen, C. G., and Raubitschek, A., 1994. Continuous effusion chemotherapy as a radiation-enhancing agent for ^{90}Y-radiolabelled MAb therapy of a human tumor xenograft, *Gynecol. Oncol.* 55:115–122.

Roberson, P. L., Ten Haken, R. K., McShan, D. L., McKeever, P. E., and Ensminger, W. D., 1992. Three-dimensional tumor dosimetry for hepatic ^{90}Y-microsphere therapy, *J. Nucl. Med.* 33:735–738.

Rösch, F., Herzog, H., Plag, C., Neumaier, B., Braun, U., Mùller-Gartner, H.-W., and Stöcklin, G., 1996. Radiation doses of ^{90}Y citrate and ^{90}Y EDTMP as determined via analogues ^{86}Y complexes and positron emission tomography, *Eur. J. Nucl. Med.* 23:958–966.

Roselli, M., Schlom, J., Gansow, O. A., Raubitschek, A., Mirzadeh, S., Brechbiel, M. W., and Colcher, D., 1989. Comparative biodistributions of yttrium- and indium-labeled MAb B72.3 in athymic mice bearing human colon carcinoma xenografts, *J. Nucl. Med.* 30:672–682.

Roskill Information Services, 1985. *The Economics of Scandium*, Roskill Information Services, London.

Rowlinson-Busza, G., Snook, D., and Epenetos, A. A., 1994. ^{90}Y-labeled antibody uptake by human tumor xenografts and the effect of systemic administration of EDTA, *Int. J. Radiat. Oncol. Biol. Phys.* 28:1257–1265.

Rubin, R. H., and Strauss, W. H., 1987. Method for the diagnosis and treatment of inflammations using labeled immuno-globulins, PCT patent 87.04.351.

Salako, Q. A., and DeNardo, S. J., 1997. Radioassay of ^{90}Y radiation using the radionuclide dose calibrator. *J. Nucl. Med.* 38:723–726.

Salashchenko, N. N., and Shamov, E. A., 1997. Short-period x-ray multiplayers based on CrSc, *Opt. Commun.* 134:7–10. CA 126:110536g.

Sawada, Y., Patlak, C. S., and Blasberg, R. G., 1989. Kinetic analysis of cerebro-vascular transport based on indicator diffusion technique, *Am. J. Physiol.* 256:794–812.

Sax, S., and Lewis, R. J., 1992. *Dangerous Properties of Industrial Materials*, Van Nostrand Reinhold, New York.

Scheinberg, D. A., Strand, M., and Gansow, O. A., 1982. Tumor imaging with radioactive metal chelates conjugated to MAb, *Science* 215:1511–1513.

Schmidberger, H., Buchsbaum, D. J., Blazar, B. R., Everson, P., and Vallera, D. A., 1991. Radiotherapy in mice with ^{90}Y-labeled anti-Ly1 MAb, *Cancer Res.* 51:1883–1890.

Schubiger, P. A. Beer, H.-F., Geiger, L., Rosler, H., Zimmermann, A., Triller, J., Mettler, D., and Schilt, W., 1991. ^{90}Y-resin particles. Animal experiments on pigs with regard to the introduction of superselective embolization therapy, *Nucl. Med. Biol.* 18:305–311.

Shanberg, A, M., Lee, I. S., Tansey, L. A., and Sawyer, D. E., 1994. Depth of penetration of the Nd:YAG laser in the human prostate at various dosimetry, *Urology* 43:809–812.

Shapshay, S., Aretz, H. T., and Setzer, S. E., 1990. Soft tissue effects of the Ho:SGG laser in the canine trachea, *Otolaryngol. Head Neck Surg.* 102:251–256.

Shen, S., DeNardo, G. L., Yuan, A., DeNardo, D. A., and DeNardo, S. J., 1994. Planar gamma camera imaging and quantitation of ^{90}Y bremsstrahlung. *J. Nucl. Med.* 35:1381–1389.

Sheridan, M., and Tamarin, R. H., 1986. Kinships in a natural meadow vole population, *Behav. Ecol. Sociobiol.* 29:207–211.

Shimizu, Y., 1993. Antimicrobial antifouling thermoplastic articles and their manufacture, Japanese patent 05163382. CA 119:205168h.

Singleton, J. H., and Yannopoulos, L. N., 1975. The use of yttrium and scandium for fabrication of radioactive electron emitters, *J. Vac. Sci. Technol.* 12:414–418.

Skarstein, A., 1979. Effect of indomethain on blood flow distribution in the stomach of cats with acute gastric ulcer, *Scand. J. Gastroenterol.* 14:905–911.

Smith, T., Veall, N., and Wootton, R., 1982. Bladder wall dose from administered radio-pharmaceuticals. *Radiat. Protein Dosim.* 2:183–189.

Snook, D. E., Rowlinson-Busza, G., Meares, C., and Epenetos, A. A., 1991. ^{111}In and ^{90}Y-labeled macrocycle chelating agents, in: *Monoclonal Antibodies* (A. A. Epenetos, ed.), Chapman & Hall Medical, London, pp. 157–166.

Soloway, A. H., Balcius, J. F., and Ojemann, R. G., 1963. Separation of ^{49}Sc from a calcium target, *Int. J. Appl. Radiat. Isot.* 14:245–249.

Spivak, J. M., Grande, D. A., Ben-Yishay, A., Menche, D. S., and Pitman, M. I., 1992. The effect of low-level Nd:YAG laser energy on adult articular cartilage *in vitro*, *Arthroscopy* 8:36–43.

Stein, R., Blumenthal, R., Sharkey, R. M., and Goldenberg, D. M., 1994. Comparative biodistribution and radioimmunotherapy of MAb RS7 and its $F(ab')_2$ in nude mice bearing human tumor xenografts, *Cancer* 73:816–823.

Sterenborg, H. J. C. M., Van den Akker, H. P., van der Meulen, F., W., Van Swol, C. F. P., Van Leeuwen, A. G. J. M., and Van Genert, M. J. C., 1990. Laser lithotripsy of salivary stones. A comparison between the pulsed dye laser and the Ho:YSGG laser, *Lasers Med. Sci.* 5:357–363.

St. Raymond, P. A., 1990. ^{47}Sc chelated MAb, PhD thesis, University of Maryland. Diss. Abstr. 52:873b.

Su, Q. I., 1991. Yttrium, its applications, separation, in: *Rare Earth Minerals and Minerals for Electronic Uses* (B. Siribumrungskha *et al.*, eds.), Translations of Technical Publishers, Switzerland, pp. 421–431.

Syed, I. B., and Hosain, F., 1975. ^{44}Sc. A possible new radionuclide for bone scanning, *Appl. Radiol.* 3–4:82–83.

Talairach, I., and Szikla, G., 1965. *Neurochirurgy* 11:233, cited by V. F. Terekhova and E. M. Savitski, 1975, *Yttrium*, Israel Program Scientific Translations, Jerusalem.

Tetens, J., Tulleners, E. P., Ross, M. W., Orsini, P. G., and Martin, B. B., Jr., 1994. Transendoscopic contact Nd:YAG laser treatment of tympany of the auditory tube diverticulum in two foals, *J. Am. Vet. Med. Assoc.* 204:1927–1929.

Thakur, M. L., 1990. Novel use of a radiolabeled antibody against stage specific embryonic antigen for the detection of occult abscesses in mammals, U.S. patent 4917878. CA 113:128752r.

Tian, J.-H., Xu, B.-X., Zhang, J.-M., Dong, B.-W., Ping, L., and Wang, X.-D., 1996. Ultrasound-guided internal radiotherapy using ^{90}Y-glass microspheres for liver malignancies, *J. Nucl. Med.* 37:958–963.

Tilyou, S. M., 1991. ^{90}Y for cancer therapy studies secured through government and commercial suppliers, *J. Nucl. Med.* 32:16–20.

Tomenius, L., 1973. A method for determining the retention of inhaled particles in a bronchus, *Environ. Physiol. Biochem.* 3:1–12.

U.S. Food and Drug Administration (FDA), 1993. Indirect food additives: Polymers, *Fed. Register* 58:48598–48599. CA 120:7087p.

Vallera, D. A., Schmidberger, H., Buchsbaum, D. J., Everson, P., Snover, D. C., and Blazar, B. R., 1991. Radiotherapy in mice with ^{90}Y-labeled anti-Ly1 MAb. Therapy of established graft-versus-host disease induced across the major histocompatibility barrier, *Cancer Res.* 51:1891–1897.

Vander, W., and Stephen, B., 1995. Dynamics of yellow pine chipmunk (*Tamias amoenus*) seed caches. *Ecoscience* 2:261–266.

Vaughan, A. T. M., Yankuba, S. C. S., and Anderson, P., 1987. Antibodies labeled with metallic radionuclides, *Natl. Cancer Inst. Monogr.* 3:141–144.

Vögtle, F., 1991. *Supramolecular Chemistry. An Introduction*, John Wiley and Sons, Chichester.

Vriesendorp, H. M., Quadri, S. M., Andersdon, B. S., and Dicke, K. A., 1996. Hematologic side effects of radiolabeled immunoglobulin therapy, *Exp. Hematol.* 24:1183–1190.

Wehner, A. P., Wilkerson, C, L., Mahaffey, J. A., and Millman, E. M., 1980. Fate of inhaled fly ash in hamsters, *Environ. Res.* 22:485–498.

White, C. A., Halpern, S. E., Parker, B. A., Miller, R. A., Hupf, H. B., Shawler, D. L., Collins, H. A., and Royston, I., 1996. Radioimmunotherapy of relapsed B-cell lymphoma with ^{90}Y anti-idiotype MAb, *Blood* 87:3640–3649.

Wiener, C. M., Kirk, W., and Albert, R. K., 1990. Prone position reverse gravitational distribution in dog lungs with oleic acid-induced injury, *J. Appl. Physiol.* 468:1386–1392.

Wike, J. S., Guyer, C. E., Ramey, D. W., and Phillips, B. P., 1990. Chemistry for commercial scale production of ^{90}Y for medical research, *Appl. Radiat. Isot.* 41:861–863.

Wilson, E. D., 1998. *Consilience, The Unity of Knowledge*, Alfred A. Knopf, New York.

Wollner, I., Knutsen, C., Smith, P., Prieston, D., Chrisp, C., Andrews, J., Juni, J., Warber, S., Klevering, J., Crudup, J., and Ensminger, W., 1988. Effects of hepatic arterial ^{90}Y glass microspheres on dogs. *Cancer* 61:1336–1344.

Wu, C., Virzi, F., and Hnatowich, D. J., 1992. Investigations of N-linked macrocycles for ^{111}In and ^{90}Y labeling proteins, *Nucl. Med. Biol.* 19:239–244.

Wu, E., 1989. Plant virucides, insecticides and growth regulators containing triacontanol, Chinese patent 1030857. CA 113:54365t.

Xi, W., Tang, F., He, X., and Li, Z., 1996. Influence of sintering process and doping on properties of hydroxyapatite ceramics, *Gongneng Cailiao* 27:274–276 (Chinese). CA 126:175427s.

Xu, G and Xiao, J., eds, 1985. *New Frontiers in Rare Earth Science and Applications*, Science Press, Beijing, Vols. 1 and 2.

Xu, Y., 1992. Lanthanide fluorescence assay for bioaffinity compounds, PCT patent 9216840. CA 118:76599c.

Yamamoto, H., and Ishihara, K., 1996. Method for producing esters, Japanese patent 08157416. CA 125:194720c.

Yamamoto, H., and Ishihara, K., 1998. Preparation of acetals using scandium complexes, Japanese patent 10 45657. CA 128:204619e.

Yan, S.-C. P., Zhou, Z.-Z., and Wu, Y.-D., 1985. Application of scandium oxide in an electron emission material, in: *New Frontiers in Rare Earth Science and Application* (G. Xu and J. Xiao, eds.), Science Press, Beijing, Vol. 2, pp. 1476–1480.

Yan, Z. P., Lin, G., Zhao, H. I., and Dong, Y. H., 1993. ^{90}Y glass microspheres injected via the portal vein, *Acta Radiol.* 34:395–398.

Yang, X, 1991. Laser ablation of human arterial atheromas with the Nd:YAG laser, PhD thesis, Kuopio University, Finland, 79 pp, Diss. Abstr. Int. 53:322C.

Zamora, P., and Griffith, J. K., 1995. Fusion proteins containing metallothionein and targeting-protein structural components, PCT patent 9504753. CA 122:283859k.

Zharikov, E. V., 1991. Growth features of rare earth scandium garnets, *Proc. Ind. Natl. Sci. Acad.* 57:117–124.

Suggested Further Reading

Let us come now to references to authors, which other books contain and yours lacks. The remedy for this is very simple; for you have nothing else to do but look for a book that quotes them all from A to Z, as you say. Then you put same alphabet into yours . . . And if it serves no other purpose, at least that long catalogue of authors will be useful to lend authority to your book at the outset.

Miguel de Cervantes

Chapter 7

Appleyard, C. B., Williams, J. L., Hatway, C. A., Percy, W. W., Cunat, L., Membre, H., Marchal, L., Chaussidon, M., and Burnel, D., 1999. The effects of aluminium, iron, chromium, and yttrium on rat intestinal smooth muscle *in vitro*, *Biol. Trace Elem. Res.* 64:43–59

Borchardt, P. E., Quadri, S. M., Freedman, R. S., and Vriesendorp, H. M., 1997. Intralesional radio labeled human monoclonal IgM in human tumor xenografts, *Radiother. Oncol.* 44:283–293.

Borchardt, P. E., Quadri, S. M., Freedman, R. S., and Vriesendorp, H. M., 1998. ^{111}In- and ^{90}Y-labeled human monoclonal immunoglobulin M targeting of human ovarian cancer in mice, *J. Nucl. Med.* 39:476–484.

Chinol, M., Paganelli, G., Sudati, F., Meares, C., and Fazio, F., 1997. Biodistribution in tumor-bearing mice of two ^{90}Y-labeled biotin using three-step tumor targeting, *Nucl. Med. Commun.* 18:176–182.

De Jong, M., Bernard, B. F., De Bruin, E., *et al.*, 1998. Internalization of radio labeled [DTPA^0]octreotide and [DOTA^0, Tyr^3]octreotide: Peptides for somatostatin receptor-targeted scintigraphy and radionuclide therapy, *Nucl. Med. Commun.* 19:283–288.

DeNardo, G. L., Kukis, D. L., Kroger, L. A., O'Donnell, R. T., Lamborn, K. R., Miers, L. A., De Nardo, D. G., Mears, C. F., and DeNardo, G. L., 1997. Synergy of Taxol and radio-immunotherapy with ^{90}Y-labeled chimeric L6 antibody: Efficacy and toxicity in breast cancer xenography, *Proc. Natl. Acad. Sci. U.S.A.* 94:4000–4004.

DeNardo, S. J., Kukis, D. L., Miers, L. A., Lairds, A., Winthrop, M. D., Kroger, L. A., Salako, Q., Shen, S., Lamborn, K. R., Gumerlock, P. H., Meares, C. F., and DeNardo, G. L., 1998. ^{90}Y-DOTA-peptide-chimeric L6 radioimmunoconjugate efficacy and toxicity in mice bearing p53 mutant human breast cancer xenografts, *J. Nucl. Med.* 39:842–849.

Goddu, S. M., Howell, R. W., Giuliani, D. C., and Rao, D. V., 1998. Biological dosimetry of bone marrow for incorporated ^{90}Y, *J. Nucl. Med.* 39:547–552.

Hao, S., Wang, X., Wang, L., Dai, L., Li, Z., and Cheng, Y., 1997. Bioconcentration of rare earth elements lanthanum, gadolinium and yttrium in algae (*Chlorella vulgare* Beijering). Influence of chemical species, *Chemosphere* 34:1753–1760.

Ho, S., Lau, W. Y., Leung, T. W. T., Chan, M., Chan, K. W., Lee, W. Y., Johnson, P. J., and Li, A. K. C., 1997. Tumor-to-normal uptake ratio of ^{90}Y micro spheres in hepatic cancer assessed with ^{99m}Tc macro aggregated albumin, *Br. J. Radiol.* 70:823–828.

Kobayashi, H., Wu, C., Yoo, T. M., Sun, B., Drumm, D., Pastan, I., Paic, C., Gansow, O., Carrasquillo, J., and Brechbiel, M., 1998. Evaluation of the *in vivo* biodistribution of yttrium-labeled isomers of CHX-DTPA-conjugated MAb, *J. Nucl. Med.* 39:829–836.

Kolsky, K. L., Joshi, V., Meinken, G. E., Khandelwal, B., Mausner, L. F., Sweet, M. P., and Srivastava, S. C., 1998. Effect of trace metal contaminants on labeling antibodies with ^{47}Sc, *215th ACS National Meeting, Dallas*, March 29–April 2, 1998, Abstr. Nucl. 060.

Lewis, M. R., and Shively, J. E., 1998. Maleimidocysteine-DOTA derivatives. New reagents for radiometal chelate conjugation to antibody sulfhydryl groups undergo pH-dependent cleavage reactions, *Bioconjug. Chem.* 9:72–86.

Maruta, S., Homma, K., and Ohki, T., 1998. Conformational changes of the highly reactive cystine and lysine regions of skeletal muscle, *J. Biochem.* 124:578–584.

Quadri, S. M., Malik, A. B., Chu, H. B., Freedman, R. S., and Vriesendorp, H. M., 1996. Intraperitoneal ^{111}In and ^{90}Y human IgM (AC6C3-2B12) in nude mice bearing peritoneal carcinomatosis, *J. Nucl. Med.* 37:1545–1551.

Rizoiu, I. M., Eversole, L. R., and Kimmel, A. I., 1996. Effects of an erbium, chromium:yttrium, scandium, gallium, garnet laser on mucocutaneous soft tissues, *Oral Surg.* 82:386–395.

Salako, Q. A., O'donnell, R. T., and DeNardo, S. J., 1998. Effects of radiolysis on ^{90}Y-labeled Lym-1 antibody preparations, *J. Nucl. Med.* 39:667–670.

Smith-Jones, P. M., Stolz, B., Albert, R., Rainer, G., Briner, U., Macke, H., and Bruns, C., 1998. Synthesis and characterization of [^{90}Y]-Bz-DTPA-oct: A ^{90}Y-labeled octreotide analog for radiotherapy and somatostatin receptor-positive tumors. *Nucl. Med. Biol.* 25:181–188.

Sotogaucin, N., Endo, K., Hirunuma, R., Enomoto, S., Ambe, S., and Ambe, F., 1999. Biochemical reactions of various trace elements with blood components and transport proteins, *J. Radioanal. Nucl. Chem.* 239:429–432

Verstraeten, S. V., Nogueira, L. V., Schreier, S., and Oteiza, P. I., 1997. Effect of trivalent metal ions on phase separation and membrane lipid packing. Role in lipid peroxidation, *Arch. Biochem. Biophys.* 338:121–127.

Wester, H.-J., Brockmann, J., Roesch, F., Wutz, W., Herzog, H., Smith-Jones, P., Stolz, B., Bruns, C., and Stoecklin, G., 1997. PET-pharmacokinetics of ^{18}F-octreotide. A comparison with ^{67}Ga-DFO- and ^{86}Y-DTPA-octreotide, *Nucl. Med. Biol.* 24:275–286.

Winthrop, M., DeNardo, S. J., Muenzer, J. T., Chi, S.-G., and Gumerlock, P. H., 1997. p53-independent response of a human breast carcinoma xenograft to radio immunotherapy, *Cancer* 80(Suppl. 12):2529–2537.

Zijmans, H. J., Bonnet, J., Burton, J., Kardus, K., Vallit-Niedbals, R. S., and Tanke, H. J., 1999. Detection of cell and tissue surface antigens using up-converting phosphors, *Anal. Biochem.* 267:30–36

Chapter 8

Amano, R., Oishi, S., Enomoto, S., and Ambe, F., 1996. Biodistribution of trace elements in normal, aluminium-overload and cadmium-overload mice. An application of the radioactive multi tracer technique, *Ann. Clin. Lab. Sci.* 26:531–541.

Favaro, D. I. T., Hui, M. L. T., Cozzolino, S. M. F., Maihara, V. A., Armelin, M. J. A., Vasconcellos, M. B. A., Yuyama, L. K., Bonaventura, G. T., and Tramonte, V. L., 1997. Determination of various nutrients and toxic elements in different Brazilian regional diets by NAA, *J. Trace Elem. Med. Biol.* 11:129–136.

Hao, S., Wang, X., Wang, Q., Wang L., Chen, Y., Dai, I., Li, Z., and Mi, C., 1998. The species of spiked rare earth elements in sediment and potential bioavailability to algae (*Chlorella vulgaris Beijerinck*), *Chemosphere* 36:329–337.

Hirunuma, R., Endo, K., Enomoto, S., Ambe, S., and Ambe, P., 1999. Study on the distribution of radioactive trace elements in vitamin D-overloaded rats using the multitracer technique, *Appl. Radiat. Isot.* 50:543–549.

Kamijo, M., Suzuki, T., Kawai, K., and Murase, H., 1998. Accumulation of yttrium by Variovorax paradoxus, *J. Ferment. Bioeng.* 86:564–568.

Li, M., Luo, D., Weng, S., Chuang, T., and Yu, K. N., 1997. Transfer coefficients from soil to cabbage of Ba, Sr, Y and Zr, *Zhongshan Daxue Xuebao Ziran Kexueban* 36:109–112 (Chinese).CA 127:260166.

Li, Y., Hu, R., Rong, X., Yi, Z., and Zhao, Z., 1996. Effect of ionic rare earths on rhizosphere biological magnetics and growth of rice, *Hunan Nongye Daxue Xuebao* 22:341–345 (Chinese).CA 128:22311r.

Oish, S., Amano, R., Anoo, A., Enomoto, S., and Ambe, F., 1999. Simultaneous biobehavior of trace elements Sc, Mn, Fe, Co, Zn, Se, Rb, and Zr in the brain and other organs of C57BL/6N mice, *J. Radioanal. Nucl. Chem.* 239:411–416.

Peterson, J. J., and Meares, C. F., 1999. Enzymatic cleavage of peptide-linked radiolabels from immunoconjugates, *Bioconjugate Chem.* 10:553–557.

Qu, X., Chen, X., Li, Y., and Chen, K., 1996. Effects of different ion type rare earths and cerium-rich rare earths on the growth of broilers, *Hunan Naongye Daxue Xuebao* 22:177–181 (Chinese).CA 128:34113r.

Ren, Y., Zhang, Z., Ren, Y., Li, W., Wang, M., and Xu, G., 1997. Diagnosis of lung cancer based on metal contents in serum and hair using multivariate statistical methods, *Talanta* 44:1823–1831.

Shinonaga, T., Prohl, G., Muller, H., and Ambo, S., 1999. Experimental determination of mobility of trace elements in soybean plants, *Sci. Total Environ*, 225:241–248.

Sun, H., Wang, X., Wang, W., Wang, H., Wang, L., Chen, Y., Dai, L., and Mi, C., 1997. The effects of chemical species on bioaccumulation of rare earth elements in wheat grown in nutrient solution, *Chemosphere* 35:1699–1707.

Tsezos, M., Remoudaki, E., and Angelatou, V., 1996. A study of the effects of competing ions on the biosorption of metals, *Int. Biodeterior. Biodegrad.* 38:19–29.

Wang, O., Sun, H., Wang, H., Wang, X., Dai, L., Chen, Y., and Mi, C., 1997. Bioaccumulation of two speciations of rare earth elements in wheat seedlings, *Nanjing Daxue Xuebao Ziran Kexue* 33:544–548 (Chinese). CA 128:241837v.

Yin, D., Yang, X., Zhou, F., and Xue, Y., 1998. Efficiency of rare earth elements on algal growth in eutrophic water. *Huanjing Kexue* 19:56–59. CA 130:1145596g.

Zaichick, V., Tsyb, A., Matveenko, E., and Chernichenko, I., 1996. Instrumental neutron activation analysis of essential and toxic elements in child and adolescent diets in the Chernobyl disaster territories of the Kaluga Region, *Sci. Total Environ.* 192:269–274.

Chapter 9

Marubashi, K., Hirano, S., and Suzuki, K. T., 1996. Comparison of pulmonary toxicity of yttrium chloride between saline and YCl_3 pretreated rats, *Metal Ions in Biology and Medicine*, Proc. Int. Symp. 4th, pp. 493–495.

Marubashi, K., Hirano, S., and Suzuki, K. T., 1998. Effects of intratracheal pretreatment with yttrium chloride on inflammatory response of the rat lung following intratracheal instillation of yttrium chloride, *Toxiol. Let.* 99:43–51

Mugenburg, B. A., Boecker, B. B., Hubbs, A. F., Hahn, F. F., Snipes, M. B., Diel, J. H., Newton, G. J., and Griffith, W. C., 1998. Toxicity of inhaled $^{91}YCl_3$ in dogs. *Radiat. Res.* 150:212–226.

Nakamura, Y., Tsumura, Y., Tonogai, Y., Shibata, T., and Ito, Y., 1997. Differences in behavior among the chlorides of seven rare earth elements administered intravenously to rats, *Fundam. Appl. Toxicol.* 37:106–116.

Ramakrishna, V. V. S., Vivek Singh, and Garg, A. N., 1996. Occupational exposure among locomotive shed workers and welders using NAA of scalp hair. *Sci. Total Environ.* 192:259–267.

Reid, R. J., Rengel, Z., and Smith, F. A., 1996. Membrane fluxes and comparative toxicities of aluminium, scandium and gallium, *J. Exp. Bot.* 47:1881–1888.

Sigg, M., Crompton, N. G. A., and Burkart, W., 1997. Enhanced neoplastic transformation in an inhomogeneous radiation field. An effect of the presence of heavily damaged cells, *Radiat. Res.* 148:543–547.

Yang, Z., Yang, N., and Wang, S., 1998. Experimental study on the radioinjury action of ^{90}Y-labeled resin, *Hejishu* 21:704. CA 130:349099m.

Chapter 10

Amols, H. I., Reunstein, L. E., and Weinberger, J., 1996. Dosimetry of a radioactive balloon dilatation catheter for treatment of neointimal hyperplasia, *Med. Phys.* 23:1783–1796.

Asavatanabodee, P., Sholten, D., and Davis, P., 1997. ^{90}Y radiochemical synovectomy in chronic knee synovitis, *J. Rheumatol.* 24:639–642.

Cavouras, D., Kandarakis, I., Panayotakis, G. S., Evangelou, E. K., and Nomicos, C. D., 1996. An evaluation of the $Y_2O_3:Eu^{3+}$ scintillator for application in medical x-ray detectors and image receptors, *Med. Phys.* 23:1965–1975.

Chinol, M., Franceschini, R., Paganelli, G., Pecorale, A., and Paiano, A., 1997. Simple production of ^{90}Y in a chemical form suitable to clinical grade radio conjugates, *Radioact. Isot. Clin. Med. Res.*, Proc. Int. Badgastein Symp., 22nd, pp. 327–332.

Cremonesi, M., Ferrari, M., Chinol, M., Stabin, M. G., Grana, C., Prisco, G., Robertson, C., Tosi, G., and Paganeli, G., 1999. Three-step radioimmunotherapy with ^{90}Y biotin: Dosimetry and pharmakinetics in cancer patients, *Eur. J. Nucl. Med.* 26:110–120.

De Lacharriere, O., and Breton, L., 1997. Salts of lanthanides, tin, zinc, manganese, yttrium, cobalt, barium, strontium in a skin care composition, Canadian patent 2188688. CA 127:210205n.

Griffits, G. L., and Hansen, H. J., 1997. Targeted combination immunotherapy of cancer, PCT patent 9741898. CA 128:39530j.

Ishido, M., and Okada, S., 1997. Method for manufacture of scandium with low α radiation and low oxygen content, Japanese patent 09324227. CA 128:64311.

Jahangie, Z. N., Jacobs, J. W. G., Van Isselt, J. W., and Bijlsma, J. W. J., 1997. Persistent synovitis treated with radiation synovectomy using ^{90}Y. A retrospective evaluation of 83 procedures for 45 patients, *Br. J. Rheumatol.* 36:861–869.

Kobayashi, S., ed., 1999. Lanthanides: Chemistry and Use in Organic Synthesis, Springer Verlag, Berlin.

Lau, W. Y., Ho, S., Leung, T. W. T., Ho, R., Johnson, P. J., and Li, A. K., 1998. Selective internal radiation therapy for nonresectable hepatocellular carcinoma with intraarterial infusion of ^{90}Y microspheres, *Int. J. Radiat. Oncol. Biol. Phys.* 40:583–592.

Leavitt, R. D., and Avila, L. Z., 1997. Polymeric delivery of radio nuclides and radio pharmaceuticals, PCT patent 9733628. CA 127:283401h.

Leichner, P. K., Akabani, G., Colcher, D., Harrison, K., Hawkins, W., Eckblade, M., Baranowska-Kortylewicz, J., Augustine, S., Wisecarver, J., and Tempere, M., 1997. Patient-specific dosimetry of ^{111}In and ^{89}Y-labeled monoclonal antibody CC49, *J. Nucl. Med.* 38:512–516.

Mamers, H., and McCarthy, K. J., 1998. Preparation and testing of diffusable metal chelate fungicides, *Holzforschung* 52:18–26.

Mausner, L. F., Kolsky, K. L., Joshi, V., and Srivastava, S. C., 1998. Radionuclide development at BNL for nuclear medicine therapy, *Appl. Radiat. Isot.* 49:285–294.

Paganelli, G., Grana, C., Chinol, M., Cremonesi, M., De Cicco, C., Robertson, C., and Veronesi, U., 1999. Antibody-guided three-step therapy for high grade glioma with 90Y biotin, *Eur. J. Nucl. Med.* 26:348–357.

Platzek, J., Speck, U., and Niedballa, U., 1999. Lipophilic metal complexes for necrosis infarct imaging, PCT patent WO99-16.473. CA 130:27893s.

Qian, W., and Laurence, P. C., 1996. A restoration algorithm for ^{32}P and ^{90}Y bremsstrahlung emission nuclear imaging, *Med. Phys.* 23:1309–1323.

Ripert, V., Papiernik, R., Hubert-Pfalzgraf, L. G., and Belleville, P., 1998. The quest for nanosized crystalline colloidal suspensions of scandium and yttrium oxide for optical applications, *215th ACS National Meeting, Dallas*, March 29–April 2, 1998, Abstr. Inorg. 540.

Rösch, F., Herzog, H., Stolz, B., Brockmann, J., Köhle, M., Mühlensiepen, H., Marbach, P., and Müller-Gärtner, H. W., 1999. Stability, uptake kinetics and radiation doses of the somatostatin receptor ligand [^{86}Y]DOTA-DPhe^{1}-Tyr^{3}-octreotide using positron-emission-tomography in nonhuman primates, *Eur. J. Nucl. Med.* 26:358–366.

Semenov, S. A., Reznik, A. M., and Bukin, V. I., 1997. Technological aspects of scandium raw material resources, *Izv. Vyssh. Uchebn. Zaved. Tsvetn. Metall.* 1997(6):23–25 (Russian).

Shukla, S. K., Limouris, G. S., Cusumano, R., Acconcia, A., Cipriani, A., Atzei, G., Argiro, G., and Boemi, S., 1997. Advanced prostate cancer diagnosis and therapy with ^{67}Ga and ^{90}Y, respectively, *Anticancer Res.* 17:1731–1734.

Sills, J. M., and Nagle, C. C., 1998. Extrapolation of ^{90}Y and ^{90}Sr activity in older fuel fragments using a vented ion chamber, *Radiat. Prot. Manage.* 15:53–64.

Soto, J., and Patton, J. T., 1997. Yttrium containing metal complexes and olefin polymerization process, PCT patent 9743294. CA 128:35159x.

Teichman, J. M. H., Rogenes, V. J., Mciver, B. J., and Harris, J. M., 1997. Holmium–yttrium–aluminum garnet laser cystolithotripsy of large bladder calculi, *Urology* 50:44–48.

Todorova, G. K., and Apblett, A. W., 1997. Synthesis and properties of yttrium metaloorganic polymers, *213th ACS National Meeting, San Francisco*, April 13–17, 1997, Abstr. Inorg. 612.

Voges, J., Sturm, V., Treuer, H., Gauss, C., and Berthold, F., 1997. Cystic craniopharyngioma. Long-term results after intracavitary irradiation with stereotactically applied colloidal β-emitting radioactive sources, *Neurosurgery* 40:263–270.

Watanabe, N., Oriuchi, N., Igarashi, H., Higuchi, T., Yukihiro, M., Fukushima, Y., Tomivoshi, K., Hirano, T., Inoue, T., and Endo, K., 1997. Preparation of ^{90}Y-labeled human macro aggregated albumin for regional radiotherapy, *Nucl. Med. Biol.* 24:465–469.

Zhang, J., 1997. Manufacture of superhigh-purity yttrium oxide, Chinese patent 1140148. CA 128:259097p.

Index

MIX
Papier aus verantwortungsvollen Quellen
Paper from responsible sources
FSC® C105338

FSC
www.fsc.org

If you have any concerns about our products,
you can contact us on
ProductSafety@springernature.com

In case Publisher is established outside the EU,
the EU authorized representative is:
Springer Nature Customer Service Center GmbH
Europaplatz 3, 69115 Heidelberg, Germany

Printed by Libri Plureos GmbH
in Hamburg, Germany